MANUFACTURING SYSTEMS ANALYSIS WITH APPLICATION TO PRODUCTION SCHEDULING

Selected titles from the YOURDON PRESS COMPUTING SERIES
Ed Yourdon, *Advisor*

BAUDIN Manufacturing Systems Analysis with Application to Production Scheduling
BELLIN AND SUCHMAN Structured Systems Development Manual
BLOCK The Politics of Projects
BODDIE Crunch Mode: Building Effective Systems on a Tight Schedule
BOULDIN Agents of Change: Managing the Introduction of Automated Tools
BRILL Building Controls Into Structured Systems
BRILL Techniques of EDP Project Management: A Book of Readings
CHANG Principles of Visual Programming Systems
CONSTANTINE AND YOURDON Structured Design: Fundamentals of a Discipline of Computer Program and Systems Design
DEMARCO Concise Notes on Software Engineering
DEMARCO Controlling Software Projects: Management, Measurement, and Estimates
DEMARCO Structured Analysis and System Specification
DESALVO AND LIEBOWITZ Managing Artificial Intelligence and Expert Systems
DICKINSON Developing Structured Systems: A Methodology Using Structured Techniques
FLAVIN Fundamental Concepts in Information Modeling
FOLLMAN Business Applications with Microcomputers: A Guidebook for Building Your Own System
FRANTZEN AND MCEVOY A Game Plan for Systems Development: Strategy and Steps for Designing Your Own System
INMON Information Engineering for the Practitioner: Putting Theory into Practice
KELLER Expert Systems Technology: Development and Application
KELLER The Practice of Structured Analysis: Exploding Myths
KING Creating Effective Software: Computer Program Design Using the Jackson Method
KING Current Practices in Software Development: A Guide to Successful Systems
LIEBOWITZ AND DESALVO Structuring Expert Systems: Domain, Design, and Development
MAC DONALD Intuition to Implementation: Communicating About Systems Towards a Language of Structure in Data Processing System Development
MC MENAMIN AND PALMER Essential System Analysis
ORR Structured Systems Development
PAGE-JONES Practical Guide to Structured Systems Design, 2/E
PETERS Software Design: Methods and Techniques
RODGERS UNIX Database Management Systems
RUHL The Programmer's Survival Guide: Career Strategies for Computer Professionals
SCHMITT The OS/2 Programming Environment
SHLAER AND MELLOR Object-Oriented Systems Analysis: Modeling the World in Data
THOMSETT People and Project Management
TOIGO Disaster Recovery Planning: Managing Risk and Catastrophe in Information Systems
VESELEY Strategic Data Management: The Key to Corporate Competitiveness
WARD Systems Development Without Pain: A User's Guide to Modeling Organizational Patterns
WARD AND MELLOR Structured Development for Real-Time Systems, Volumes I, II, and III
WEAVER Using the Structured Techniques: A Case Study
WEINBERG Structured Analysis
YOURDON Classics in Software Engineering
YOURDON Managing Structured Techniques, 4/E
YOURDON Managing the System Life Cycle, 2/E
YOURDON Modern Structured Analysis
YOURDON Structured Walkthroughs, 4/E
YOURDON Techniques of Program Structure and Design
YOURDON Writing of the Revolution: Selected Readings on Software Engineering
ZAHN C Notes: A Guide to the C Programming

Michel Baudin

MANUFACTURING SYSTEMS ANALYSIS WITH APPLICATION TO PRODUCTION SCHEDULING

YOURDON PRESS
Prentice Hall Building
Englewood Cliffs, New Jersey 07632

Library of Congress Cataloging-in-Publication Data

Baudin, Michel.
Manufacturing systems analysis : with application to production scheduling / Michel Baudin.
p. cm.
"Yourdon Press."
Bibliography: p.
Includes index.
ISBN 0-13-556382-8
1. Scheduling (Management)--Data processing. 2. Production planning--Data processing. 3. System analysis--Data processing.
I. Title.
TS157.5.B38 1990
658.5'3--dc19 89-3488
CIP

Editorial/production supervision: Karen Bernhaut
Cover design: Wanda Lubelska Design
Manufacturing buyer: Mary Ann Gloriande

The publisher offers discounts on this book when ordered in bulk quantities. For more information, write:

Special Sales/College Marketing
Prentice-Hall, Inc.
College Technical and Reference Division
Englewood Cliffs, NJ 07632

Printed in the United States of America

10 9 8 7 6 5 4 3 2 1

ISBN 0-13-556382-8

Prentice-Hall International (UK) Limited, *London*
Prentice-Hall of Australia Pty. Limited, *Sydney*
Prentice-Hall Canada Inc., *Toronto*
Prentice-Hall Hispanoamericana, S.A., *Mexico*
Prentice-Hall of India Private Limited, *New Delhi*
Prentice-Hall of Japan, Inc., *Tokyo*
Simon & Schuster Asia Pte. Ltd., *Singapore*
Editora Prentice-Hall do Brasil, Ltda., *Rio de Janeiro*

CONTENTS

PREFACE **XV**

Part One — MANUFACTURING SYSTEMS ANALYSIS **1**

Chapter 1 — COMPUTERS IN MANUFACTURING **1**

1.1. THE STAKES IN MANUFACTURING SYSTEMS ANALYSIS 2
1.2. THE SYSTEMS NIGHTMARE 2
1.2.1. Computer Integrated Manufacturing 2
1.2.2. How Incompatible Systems Proliferate 4
1.3. WHY DO PRODUCTION SCHEDULING SYSTEMS FAIL? 8
1. 3.1. Conflict with Performance Measures 8
1.3.2. Management Commitment — A Double-Edged Sword 10
1.3.3. Inadequate Factory Model 11
1.3.4. Scheduling Is not the Problem 12
1.3.5. The System Cannot Model the Factory 13
1.3.6. Poor Software Quality 15
1.4. CONCLUSION AND DIRECTION 15

Chapter 2 — MANAGING THE NEEDS ASSESSMENT PROCESS 17

2.1. ORGANIZATION VERSUS TECHNOLOGY 18
2.2. ORGANIZING THE COMMUNICATION 18
2.2.1. The Analysts 18
The Sales Representative 18
The Consultant 19
The Systems Engineer 20
2.2.2. The Manufacturing Organization 21
What It Must Do to Support the Analysts 21
Analysis Paralysis 21
Corporate Culture 22
2.2.3. The Analysis Process 23
2.3. PRAGMATICS 25
2.3.1. Syntax, Semantics, and Pragmatics 25
2.3.2. Comparing a System with a Requirements Document 26
2.3.3. Current Factory versus Future Factory 26
2.4. THE ANALYST'S TOOLS 27
2.4.1. Manufacturing Engineering 27
2.4.2. Software Engineering 28
2.5. CONCLUSION 29

Chapter 3 — STRUCTURED ANALYSIS IN MANUFACTURING 30

3.1. INTRODUCTION 31
3.1.1. A Hardware Diagram 31
3.1.2. A Concepts Diagram 32
3.1.3. An External Context Diagram 33
3.2. STRUCTURED ANALYSIS 36
3.3. OBSOLETE ALTERNATIVES TO STRUCTURED ANALYSIS 36
3.3.1. Boilerplates 36
3.3.2. Flow Charts 37
3.4. VARIANTS OF STRUCTURED ANALYSIS 38
3.4.1. Petri Nets 38
3.4.2. Gane and Sarson's SADT 39
3.5. CONCLUSION 39

Chapter 4 — DATA FLOW MODELING 41

4.1. HIERARCHICAL EXPLOSION OF PROCESSORS 42
4.2. DE MARCO'S CONVENTIONS 42
4.3. WHEN IS A DATA FLOW DIAGRAM USEFUL? 44
4.3.1. External Context Specification 44
4.3.2. Reports 44
4.3.3. Diagnostic Systems 45
4.3.4. Transactions 45
4.3.5. Administration 46
4.4. DATA FLOW VERSUS CONTROL FLOW 47

4.4.1. Why the Distinction Matters 47
4.4.2. Telling Control Flow from Data Flow 48
4.4.3. Relativity of the Meaning of Control Flow 49
4.5. DIAGRAM AESTHETICS AND CROSSED ARROWS 50
Appendix A — Three Houses and Three Utilities 51

Chapter 5 — STATE TRANSITION MODELS IN MANUFACTURING ***54***

5.1. OBJECT MODELING 54
5.1.1. What an Object Is and Isn't 54
5.1.2. State Descriptions 55
5.1.3. Events 58
External Events 58
Temporal Events 59
Internal Events 59
Events Induced by History 59
5.1.4. State Transition Models and the Markov Property 59
5.2. REVIEWING A STATE TRANSITION DIAGRAM 61
5.2.1. Sources and Sinks 61
5.2.2. Reducibility 62
5.3. APPLICATION EXAMPLES 63
5.3.1. A Vehicle 63
Modeling Location by Coordinates 63
Modeling Location by Station 64
Modeling Activity Types 64
5.3.2. Engineering Changes 65

Chapter 6 — MODELING THE FACTORY'S MEMORY ***68***

6.1. MODELING THE SYSTEM MEMORY 69
6.1.1 What the System Remembers 69
6.1.2. Modeling for Retrieval 70
6.2. SETS AND MEMORY MODELS 70
6.2.1. From Sets to Databases 70
Foundations 70
Pragmatic Neutrality 71
Database Atoms 72
Properties 72
Tuples 72
Unions and Intersections 73
Subsets and Enumeration 73
Infinite Sets, Factory Models, and Database Size 73
Choice 74
6.2.2. Relations 74
6.2.3. Mappings 76
Definition 76
Rule versus Lookup 76
Many-to-Many Relationships 76
6.2.4. Relational Algebra 77

6.2.5. Need for Graphic Notations 78
6.3. GRAPHICS 78
6.3.1. Entity-Relationship Diagrams 78
6.3.2. Diagrams for Relations of Arity >2 79
6.3.3. Representing Set Inclusion 80
6.4. SECURITY 81
Appendix A — RELATIONAL ALGEBRA OPERATIONS 81
6.A.1. Selection 81
6.A.2. Projection 82
6.A.3. Join 82
6.A.4. Division 82

Chapter 7 — DATA DICTIONARIES IN MANUFACTURING ***83***

7.1. DEFINING ATTRIBUTES IN A DATA DICTIONARY 84
7.1.1. Purpose of Data Dictionaries 84
7.1.2. De Marco's Conventions 84
7.1.3. Optional Attributes 86
7.1.4. Relations as Attributes 87
7.1.5. Textual Explanations 87
7.1.6. Redundancies 88
7.2. DEFINITIONS 88
7.2.1. The Reader's Point of View 88
7.2.2. Defining Primitives 89
7.2.3. Defining Inferred Results 89
7.2.4. Established Usage 90
7.3. LIMITATIONS OF THE DE MARCO CONVENTIONS 92
7.3.1. Variant Structures 92
7.3.2. Bills of Materials and Recursion 93
Appendix A — AN EXCURSION INTO SOFTWARE DESIGN 95
7.A.1. Variant Types in PASCAL 95
7.A.2. First "Relational" Structure 96
7.A.3. Variants in PROLOG or LISP 97
7.A.4. Variants with No Redundancy 97
7.A.5. Second Relational Structure 98
7.A.6. What Is Actually Done 99

Chapter 8 — MODELING PROCEDURES AND DECISIONS ***100***

8.1. MODELING PROCEDURES 100
8.2. STRUCTURED NATURAL LANGUAGE 101
8.3. DECISION TREES 103
8.5. ALGORITHMS 106

Chapter 9 — ESSENTIAL SYSTEMS ANALYSIS 108

9.1. THE ESSENTIAL MODEL OF A WORK CELL 108
9.1.1. Event Partitioning 109
9.1.2. Object Partitioning 110
9.1.3. Integration 111
9.1.4. Conclusions on Work Cell Modeling 111
Work Cells as Attributes of Other Objects 112
Work Cells as Independent Data Objects 112
Work Cells in Object-Oriented Programming 112
Dedicated Hardware 112
9.2. EXTRACTING THE ESSENCE OF A SYSTEM 112
9.2.1. The Perfect Technology Assumption 112
9.2.2. Fundamental and Custodial Activities 113
9.2.3. Administration 114
9.2.4. Logical Core and Interface Ring 114

Chapter 10 — MATERIAL FLOW MODELING 117

10.1. MATERIAL FLOWS IN FACTORIES 117
10.2. THE ASME SYMBOLS AND THEIR VARIANTS 121
10.2.1. Vocabulary 122
10.2.2. Convergence and Divergence 124
10.2.3. JIS Symbols 125
10.2.4. Representation of Deepening Commitments 126
List of States 126
List of Operations 127
Addition of Inspections, Scrap, and Rework 127
Assignment of Equipment and Transport Steps 128
10.3. TRADITIONAL USES OF MATERIAL FLOW DIAGRAMS 129
10.3.1. Graphic Coding of Documents for Identification 129
10.3.2. Process Analysis 129
10.3.3. Physical Flow Analysis 130
10.4. HIERARCHICAL MATERIAL FLOW MODELING 132

Chapter 11 — MATERIAL FLOW VERSUS DATA FLOW 134

11.1. DIFFERENCES BETWEEN MATERIAL AND DATA FLOWS 135
11.2. FROM MATERIAL FLOW TO DATA FLOW 137
11.3. TOWARDS A LAYERED MODEL OF MATERIALS HANDLING 143
11.3.1. The ISO-OSI Model of Data Communication 143
Layer 7 — Application 144
Layer 6 — Presentation 146
Layer 5 — Session 146
Layer 4 — Transport 146
Layer 3 — Network 146
Layer 2 — Link 147
Layer 1 — Physical 147

11.3.2. Ambiguities in the ISO-OSI Model 148
11.3.3. Similarities and Differences between the Models 148
11.3.4. A Layered Model for Materials Handling 149
11.4. CONCLUSION 149

Chapter 12 — MANUFACTURING FUNDS FLOW **150**

12.1. A MISSING PERSPECTIVE 151
12.1.1. The Triple Flow Model 151
12.1.2. Manufacturing Funds Flow Analysis Issues 152
12.2. THE BUSINESS OF MANUFACTURING 153
12.2.1. Business Performance in General 153
12.2.2. The Paradox of the Pilot Line 156
12.3. THE FACTORY AND ITS OPERATING FUNDS FLOWS 156
12.3.1. Effect of Throughput Time Reductions 159
12.3.2. Price Erosion and Experience Curves 161
12.3.3. The Example of a Memory Factory 164
12.4. FINANCIAL ANALYSIS METHODS IN MANUFACTURING 170
12.4.1. Cost Accounting and Manufacturing 170
12.4.2. Payback Period 172
12.4.3. Discounted Cash Flows 173
12.5. RELATION TO DATA FLOWS AND MATERIAL FLOWS 175

Part Two — APPLICATION TO PRODUCTION SCHEDULING **177**

Chapter 13 —COMPONENTS OF PRODUCTION SCHEDULING **177**

13.1. TIME 178
13.2. THE PRODUCTION NETWORK 181
13.3. PART AGGREGATES 181
13.3.1. Lot 181
13.3.2. Load 182
13.3.3. Process Batch 182
13.3.4. Transfer Batch 182
13.3.5. Kit 182
13.4. ACTIVITY TYPES 182
13.4.1. Fabrication 183
13.4.2. Assembly 186
13.4.3. Disassembly 186
13.4.4. Inspection, Sorting, and Binning 187
13.4.5. Transport 188
13.4.6. Storage and Retrieval 188
13.5. RESOURCES 189
13.5.1. Hierarchical Model 189
13.5.2. Work Centers versus Work Cells 189

13.5.3. Resource Characterization 189
13.5.4. Little's Law 192
13.5.5. Infinitely Divisible Resources 194
13.5.6. Composite Resources 196
13.5.7. Setup Times and Setup Matrices 197
13.5.8. Interruptions of Service 198
13.6. INVENTORY 199
13.6.1. Schedule-Based Buffer Analysis 200
13.6.2. Age Distribution Analysis 200
13.6.3. Batch Size Adjustment Buffers 202
13.6.4. Safety Stocks 204
13.6.5. Queue Organization 205
13.7. PUTTING IT ALL TOGETHER 207

Chapter 14 —THE BACKWARD SCHEDULING LOGIC OF MRP-II **209**

14.1. MRP'S IMPLEMENTATION RECORD 209
14.2. THE PHILOSOPHY OF MRP-II 210
14.3. WIGHT'S DISCUSSION OF MRP VERSUS REORDER POINT 211
14.4. THE EXTERNAL CONTEXT OF MRP-II 212
14.5. MODULAR BREAKDOWN 213
14.5.1. The Master Production Schedule (MPS) 214
14.5.2. Backward Explosion of Demand 217
14.5.3. Net Inventories 218
14.5.4. The Time-Zero Spike 219
14.6. CAPACITY CALCULATIONS 221
14.6.1. Capacity Modeling in MRP-II 221
14.6.2. Capacity Requirements Planning 221

Chapter 15 —DISPATCHING IN MRP-II **224**

15.1. FROM PLANNING TO DISPATCHING 225
15.2. REAL-TIME, ON-LINE, AND OFF-LINE DISPATCH 226
15.3. DISPATCHING ALGORITHMS 227
15.3.1. First-In-First-Out (FIFO) 227
15.3.2. Earliest Due Date (EDD) 228
15.3.3. Shortest Processing Time (SPT) 228
15.3.4. Slack 229
15.3.5. Critical Ratio 230
Appendix A. EDD, SPT, AND SLACK FOR N JOBS ON 1 MACHINE 232
15.A.1. EDD Minimizes the Maximum Lateness 232
15.A.2. SPT Minimizes the Average Throughput Time 233
15.A.3. Slack Maximizes the Minimum Lateness 234
Appendix B. WAITING TIMES IN QUEUES WITH POISSON ARRIVALS 234

Chapter 16 —APPRAISING MRP-II 239

16.1. WHY CRP IS RARELY IMPLEMENTED 239
16.2. THE INTERPRETATION OF STANDARD THROUGHPUT TIMES 240
16.3. INADEQUACY OF SINGLE PARAMETER MODELS 240
16.3.1. Throughput Time Modeling 240
16.3.2. Yield Modeling 242
16.4. THE TIME PERIOD DILEMMA 243
16.5. THE 10,000-KNOB CONTROL PANEL 245
16.6. RESPONSE TO ACTIVITY LEVEL CHANGES 245
16.7. WHEN DOES MRP WORK? 246

Chapter 17 —THE THEORY OF OPT 248

17.1. OPT IN THE UNITED STATES 248
17.2. FROM BUSINESS ANALYSIS TO SCHEDULES 250
17.3. THE OPT CONCEPT OF BOTTLENECK 250
17.4. FROM BOTTLENECKS TO CAPACITY CONSTRAINT RESOURCES 252
17.5. THE UNBALANCED PLANT 253
17.5.1. Eli Goldratt's argument 253
17.5.2. External versus Internal Variability 253
17.5.3. The Cost of Capacity 254
17.5.4. Flow Balance and Time 255
17.5.5. Process Batches and Transfer Batches 256
17.5.6. Considering All Constraints Simultaneously 256
17.6. THE SECRET ALGORITHM 256
Appendix A TIME LOST DUE TO BREAKS 256

Chapter 18 —THE OPT PRODUCTION SCHEDULING SOFTWARE 259

18.1. FUNCTIONS OF THE OPT SOFTWARE PRODUCT 259
18.1.1. External Context 259
18.1.2. The OPT Factory Model 260
18.1.3. The Essential Modules of OPT 263
18.1.4. Backward Scheduling with SERVE 265
18.2. GENERATING SCHEDULES WITH OPT 269
18.2.1. Pure SERVE Run 269
18.2.2. Focused Data Cleaning 269
18.2.3. Split-OPT-Serve-Reports 269
18.3. COMMENTS ON THE OPT PRODUCT DESIGN 270
18.3.1. The OPT Software — The Price of Portability 270
18.3.2. OPT's Nonstandard Terminology 272
18.4. CONCLUSION 272

Chapter 19 — FINITE FORWARD SCHEDULING AND DISCRETE EVENT SIMULATION 274

19.1. FINITE FORWARD SCHEDULING 275
19.1.1. What is "Finite Forward Scheduling"? 275
19.1.2. Job-Shop Scheduling versus Production Scheduling 275
19.1.3. Nondelay Schedules 276
19.2. TASK AND RESOURCE MODEL 276
19.3. Producing a Schedule 280
19.4. DISCRETE EVENT SIMULATION 282

Chapter 20 — MATERIAL FLOWS IN JUST-IN-TIME 283

20.1. JUST-IN-TIME AS A HOLISTIC APPROACH 284
20.1.1. Just-In-Time — What's in a Name? 284
20.1.2. Seek Repetitiveness 285
20.1.3. Manufacturing, Marketing, and JIT 287
20.1.4. Job-Shops, MRP, and JIT 288
Pros and Cons of Job-Shop Layouts 288
JIT and Factory Response to Change 289
20.1.5. Holistic Manufacturing 289
20.2. FUNDAMENTALS OF FLOW LINES 290
20.2.1. Demand Structure Analysis 290
20.2.2. Flow Line Organization 291
Comparison of JIT with Group Technology 292
JIT Cell Structure 292
20.2.3. Setup Time Reduction 294
Dedication by Families 294
Standardization of Procedures and Training 294
Machine Modifications 295
20.3. AUTOMATION 295
20.4. INTEGRATION WITH EQUIPMENT MAINTENANCE 296
20.4.1. Motivation 296
20.4.2. Training and Operation Standards 297
20.4.3. Housekeeping 297
20.5. INTEGRATION WITH QUALITY CONTROL 299
20.5.1. Effects of Flow Lines on Quality 299
20.5.2. Just-In-Time and SQC 299
20.5.3. Foolproofing the Process with Pokayoke 300
20.6. PEOPLE AS THE MOST CRITICAL RESOURCE 300
20.7. BACK TO PRODUCTION SCHEDULING 302

Chapter 21 —JUST-IN-TIME PRODUCTION SCHEDULING 303

21.1. DAMPING EXTERNAL VARIABILITY 303
21.1.1. The Overbooking Model 304
21.1.2. Demand and Production Forecasting 305
21.2. GENERATION OF LEVEL SCHEDULES 307

21.2.1. First Blender Algorithm 308
21.2.2. The Monden Algorithm 309
21.3. FLOW CONTROL ON THE SHOP FLOOR 311
21.3.1. The Kanban system 311
Kanban Flows between Two Cells 311
Exception Handling 313
Kanban State Transitions 313
Data Flow Model 313
21.3.2. The Timetable Planning System 316
Forward Scheduling of Resources 316
Monthly Production Calendars 317
Weekly Production Calendars 317
Scope of Applicability 319
21.4. TO WHAT EXTENT SHOULD JIT BE COMPUTERIZED? 320

Chapter 22 —PERFORMANCE MODELING AND JUST-IN-TIME 322

22.1. OBJECTIVES AND METHODS 323
22.1.1. Inferring Steady-State Throughput Parameters 323
22.1.2. Three Approaches 323
Back-of-the-Envelope Formulas 323
Queueing Network Algorithms 324
Discrete Event Simulation 328
22.2. EFFECT OF LEVELING ON PROCESS BATCHES 329
22.2.1. Run Length in a Multiproduct FIFO Queue 329
22.2.2. Effect of the Service Rate 331
22.2.3. Maximum Run Length in a Level Schedule 332
22.2.4. The Zealot's Paradox 333
22.3. PERFORMANCE ANALYSIS OF THE KANBAN SYSTEM 335
22.3.1. What the Literature Says 335
22.3.2. Towards a Queueing Network Model 336
Simplifying Assumptions 336
The Cyclic Queue Model 338
Interpretation 340
Appendix A. RUN LENGTH WITH MULTICLASS SINGLE SERVER 341
Appendix B. EQUILIBRIUM OF A TRUNCATED CYCLIC QUEUE 345

REFERENCES 349

INDEX 353

Preface

A GUIDED TOUR

This book is intended to help operations managers, industrial engineers, and systems engineers faced with the challenge of effectively using computers in manufacturing. To meet this goal, what I write must be thorough and general, but to be worth sharing, it must also be drawn from my personal experience. These requirements are in conflict, and the result is a tradeoff.

After studying this book, the reader should be able to do the following:

- Formulate a factory problem so as to make its solution as easy to find as possible.
- Present proposed solutions in a clear and concise manner to those who may adopt or implement them.
- Detect specification errors which would impair implementation.

Although systems analysis is generally viewed as the art of specifying *software*, its techniques are presented as a means to describe any combination of resources and methods that can solve a problem, whether or not software is involved.

Computers are viewed as tools for manufacturing. The tools of systems analysis can be used to formulate the real problems of manufacturing, identify where computer systems can help, and solve the problems, even if the final decision is not to use computers. Particular computer applications are viewed as worth pursuing only insofar as they actually help manufacturing performance.

CHAPTER BY CHAPTER OVERVIEW

The book has two parts. Chapters 1 to 12 describe analysis tools and concepts, and are illustrated with examples drawn from various aspects of manufacturing. In Chapters 13 to 22, these tools are all used in a coordinated attack on the problem of production scheduling. In particular, several approaches in use are examined by the same methods.

Part I — Manufacturing Systems Analysis

The examination of a number of analysis tools leads up to a global view of factories as a three-level system of funds, data, and material flows which is formulated in Chapter 12. The reverse sequence could have been chosen, with the most global and abstract models introduced first and the details of each level afterwards.

The top-down approach was not adopted for two reasons. First, readers might have been skeptical of broad, general statements made without building up to them. Second, it might have given a misleading impression of the ambition of this book. It presents not a general theory of manufacturing but only a consistent set of tools that I have found useful in the description, communication, and solution of manufacturing problems. It does not claim to be comprehensive and is open to further additions.

Chapters 1 and 2 cover topics of interest to all readers. Through two extensive examples, Chapter 1 shows the type and severity of the problems the techniques described in the rest of the book are intended to avoid. Chapter 2 discusses the organizational environment of manufacturing systems analysis.

Chapters 3 to 9 are technical and describe the tools of structured analysis as they apply to manufacturing. They are meant for readers who *write* specifications. Chapters 10 and 11 examine the modeling of *material* flows, as opposed to data flows. While structured analysis works in building a model of the data manipulations associated with manufacturing, the foundation of this model is a formal representation of how parts move through machines, and extensions to structured analysis that are needed for this purpose.

In Chapters 3 to 11, those who only *read* specifications will find more details than they need but may find the discussion nonetheless enlightening as a behind-the-scenes look at how an analyst thinks.

Chapter 12 discusses the *financial* side of manufacturing and describes factories as systems in which financial goals are pursued by directing and monitoring a material flows through data.

Part II — Application to Production Scheduling

Chapter 13 lays the groundwork for Part II by defining concepts that are used in the rest of the book to describe and critique production scheduling systems. The components of a production scheduling system are the same routes, operations, and resources introduced in Chapters 10 and 11 as building blocks of material flow models. Only this time they are viewed exclusively from the point of view of their throughput, throughput time, and inventory behavior. Other characteristics of these objects would be emphasized in, for example, a book on quality control.

When viewed in isolation, a resource or an operation can be modeled in detail and with realism. When it becomes a node within a large network, as happens in Chapters 14 to 22, the details blur and only a selection of summary characteristics remains.

The decisive reason why the methods reviewed in Part II are MRP-II, OPT, and Just-in-Time is that I have written production scheduling software based on MRP logic, implemented OPT, and consulted on Just-in-Time. This itinerary, however, was not a personal choice, but was dictated by the needs of manufacturing organizations. That I was led to them is, by itself, a statement of the relevance of these approaches.

The order in which they are discussed is obligatory. The backward scheduling and capacity requirements planning logic of MRP-II is used in OPT and Just-in-Time and must therefore be explained first, along with the shortcomings that are the reasons for seeking alternatives. This is done in Chapters 14 to 16.

Many companies may have proprietary production scheduling systems based on optimization concepts from Operations Research. Eli Goldratt's OPT, however, is the only one I know of to have been marketed as a software product. As such, it had a strong start in the mid-1980s in the United States, but was not able to gain broad acceptance and have a lasting impact. Today, it retains value as an enlightening case study. Chapters 17 and 18 review its theory and functional architecture.

Part of OPT is a proprietary algorithm for finite forward scheduling, but the "secret algorithm" referred to in Chapter 18 leaves a frustrating gap in the description of the system's logic. Chapter 19 is an attempt to fill this gap by describing a method for finite forward scheduling. The algorithm of Chapter 19 is not part of OPT, and is in a separate chapter to make this clear to the reader.

Now that we are sensitized to the oversimplifications of MRP-II and OPT, we turn our attention to Just-In-Time as an approach that really works. Unlike MRP-II or OPT, Just-In-Time includes an industrial engineering overhaul of the shop floor that cannot be ignored and is described in Chapter 20.

In Chapter 21, we assume the shop floor to have been streamlined, and describe the production scheduling side of Just-in-Time. The Just-in-Time philosophy is implemented by *several* scheduling methods, applicable to different types of factories. The chapter describes variants of MRP used for high level planning in a supplier network, the leveling of master schedules, and shop floor scheduling using cards, as in the Kanban system, or timetables.

Since we owe Just-in-Time to factory practitioners and not to Operations Research, there are few theoretical results about its performance. At the beginning of Chapter 22, we review some concepts of performance modeling that had been used piecemeal in earlier chapters, and explore what we can make them tell us about Just-in-Time.

BOOK STYLE

Focus on Concepts

The concepts behind manufacturing decisions are discussed, but not their implementation details. In the case of MRP, for example, we discuss what it *means* to schedule based on standard throughput times but we do not tell the reader how to enter these parameters in a specific software product.

Unlike most discussions of MRP, this book does not contain a sample printout of a critical ratio dispatch report. On the other hand, it discusses the meaning of critical ratio and compares its effect on a factory with other sequencing methods, which even Orlicky or Wight [Wig74] do not do.

The reader will likewise look in vain for the listing of a program to control a robot arm. On the other hand, what an automated guided vehicle and a host system may have to say to each other is discussed as an application of state transition modeling in Chapter 5.

Locating Information

Aside from the table of contents and index, this book has an italicized abstract in front of every chapter to help the reader not only find information, but also decide whether to read the whole chapter. The abstract contains a statement of every major point made in the chapter, stripped of any supporting arguments and examples. It is written to minimize the information loss to a reader who proceeds directly to the next chapter.

Perspective on Methodology

Stumbling blocks inevitably appear as the result of applying a formal method to real problems, because some ideas simply do not fit the mold. In Chapter 7, for example, we see that the history of an object cannot be properly represented with the relational model. In describing the Kanban system in Chapter 21, we encounter a case where, even though cards are data elements, their movements cannot be effectively depicted by a data flow model.

In such situations, to get the job done, some practicing analysts will bend the rules without a second thought and without realizing the significance of doing so. Others may be blinded by adherence to formal rules, and fail to notice what does not fit.

Whatever tools an analyst uses, they are thinking aids. I believe that they should be used with zeal and vigor, but also with lucidity. A specification is difficult to write for several reasons. First, if the tools are not applied rigorously, then the specification is ambiguous and its translation into software is impaired. Second, if the analyst does not understand the limitations of the tools, the need to develop new ones may be overlooked. It is in this spirit that I chose examples that show both where tools work and where they do not.

This book hopes to depart from the mainstay of manufacturing literature in its use of systems analysis techniques. Busy drawings of bubbles, hexagons, boxes, and arrows are found in practically every published paper, and most of these drawings are free of any discernible meaning. Every diagram in this book is drawn in a specific graphic language for the purpose of getting a point across faster than words would.

Proving Assertions about Manufacturing

In manufacturing, it is not possible to *prove* that a particular technique for quality control or production scheduling is optimal or even better than another one.

When we have mathematical models, we also have theorems describing their properties. Any reader who is willing and able to follow their proofs will be convinced of their truth. Other readers may take them on faith and trust the author's mathematics. The whole academic discipline of Operations Research is built on that approach, but, unfortunately, the solidity of a mathematical proof is an illusion. It establishes that a result is deducible from a model, but not that the model has anything to do with reality.

A more common attitude in the manufacturing world is to take a purely empirical view and accept only the competitive superiority of a technique's users as proof of its worth. In sports, the soundness of a team's strategy is measured by its ability to make the team win.

This approach, however, also has its limitations. First, it cannot be used to appraise anything that has not been done before. Second, in manufacturing, there is more than one way to win or lose. *By definition*, the winner of the World Series is the best baseball team, but there is no analogous, unambiguous measure of success in manufacturing. Finally, the credit for success and the blame for failure in manufacturing are assigned in a subjective manner and by participants.

When factories are successful, the cause is rarely obvious. Everyone will agree that the Japanese automobile industry is successful and a case can be made that it is due to Just-in-Time. Some, however, will argue that other factors were more important, such as low labor costs, the absence of a defense industry, or the national culture.

The empirical view is clearly correct in the sense that the value of a manufacturing technique lies in its ability to improve performance, whatever measure is used. But we need all the analytical and mathematical skills we can muster to infer this from history, and to appraise the potential of new techniques.

This book uses explicit mathematical models to a degree not usually found in the manufacturing literature. When advanced mathematics are used, only the *results* are presented in the body of the text. The reader who is willing to accept them on faith is welcome to do so; proofs are given in appendices for the skeptic.

Hardware Independence

A system using an 80386 microprocessor to control a wafer stepper becomes obsolete faster than one using a relational database to perform statistical quality control. The relational model and the concepts of quality control are timeless, but their incarnation in a particular generation of hardware is dated; to maintain the relevance of this book through several generations of hardware, such discussions have been avoided.

The speed of evolution of computer hardware is evidenced by such empirical rules as "Moore's Law," which says that the density of memory circuits quadruples every two years. By comparison, software has been progressing slowly: the database industry is still implementing the relational model defined two decades ago.

The common features of such activities as patterning silicon wafers and casting iron are hidden by differences in process technology so great that a metallurgist and a semiconductor engineer do not understand each other. However, the identical concerns of production managers on both sides for meeting due dates or reducing inventory point to deep similarities.

The quality control issues in semiconductors may be keeping the width of a line within 0.1μ of a standard. On an aluminum casting, it will be within 3 mils of the standard. In both cases there are standards, tolerances and allowances, and the same analysis tools apply. While examples from any specific industry necessarily use its vocabulary and are difficult to follow for a reader working in another, they are needed to keep a discussion of manufacturing issues from being too abstract.

Choice of Examples

Most books on manufacturing only use examples from metal working; this one draws them mostly from semiconductors and the "planar process" of integrated circuit fabrication. These examples are used to illustrate ideas that are applicable elsewhere. The use of photolithography and printed-circuit board stuffing should not restrict the scope anymore than the almost exclusive references to metal working have in the manufacturing literature published to date.

Before describing a particular method, we try to make all the assumptions explicit. Hall's Zero Inventories [Hal83] contains a wealth of information about the Just-In-Time system, with examples exclusively from mechanical fabrication and assembly. Hall, however, never states his assumptions; for example, he refers to "final assembly" as if there were always such an operation, and without saying whether its existence is needed for his concepts to apply.

ACKNOWLEDGMENTS

My gratitude goes first to those who gave me the opportunity to work on manufacturing problems. From my years at Fairchild Semiconductor, I wish to thank Murray McLachlan, who guided my first steps, and Randall A. Hughes, who invited me to join the INCYTE project. In the South-Portland plant, thanks are due Colin Buxton and Paul Edmonds. By relying on my software in their daily activity, they gave me the engineer's ultimate satisfaction. Later, at General Motors, Robert Herr expanded my horizons beyond semiconductors.

For many late night talks about production scheduling, help in researching the literature, and reviews of my early drafts, I must thank Prof. J. Michael Harrison of Stanford Business School. His support and his friendship have been vital to the completion of this project. For chapters 20 and 21, I owe a great deal to Kei Abe, of MTJ, my partner and mentor in Just-in-Time implementation consulting.

This book has been four years in the making, during which my associates have shown a patience I hereby acknowledge. Thanks are due in particular to Anne Oja and Dave Mandel, of Consilium, and to You Kohata, of Fuyo Data Processing and Systems Development in Tokyo.

Thanks also to the Prentice Hall editorial team, and in particular to Ed Yourdon, Pat Henry, and Ed Moura for their support. Houston Carr, of Chicago Pneumatic Tool, was kind enough to review three drafts of this book, and his contribution is gratefully acknowledged.

Michel Baudin

Part One: Manufacturing Systems Analysis

1

COMPUTERS IN MANUFACTURING

The problems of CIM and the reasons why most production scheduling systems fail are reviewed to show what is at stake in manufacturing systems analysis. CIM is the activity of a factory in which all data manipulations are done by computer. Factory data are comprised of specifications, schedules, operational status, and history. Teamwork between the departments of a manufacturing organization is a prerequisite to CIM, and, if it is not present, computer systems only automate the lack of coordination.

Scheduling production is telling a factory what to make, and a major target for computer systems. Yet success stories are rare. A scheduling system may fail because it prescribes behavior that is not favored by the performance measures in use. Management commitment may be lacking or, if present, it may discourage the open discussion of problems. Some key features of the factory may be omitted when it is modeled in the scheduling system. Finally, other issues, such as equipment reliability, may require attention ahead of scheduling.

Manufacturing problems have no easy, universal solution — be it called "CIM" or automation. The methods used by industry leaders should be studied and "leapfrogging the Japanese" from a position of ignorance is wishful thinking. Finally, analysts should not be entangled in discussions of issues such as distributed systems before establishing what computers are needed for.

1.1. THE STAKES IN MANUFACTURING SYSTEMS ANALYSIS

In the following sections we discuss (1) the consequences of the disorganized introduction of computers in a factory, and (2) the major reasons why so many production scheduling systems fail. Then we will draw some conclusions as to the direction we should go in to improve manufacturing.

1.2. THE SYSTEMS NIGHTMARE

There is no need to argue that computers should be used in manufacturing. Programmable controllers on the shop floor are commonplace; engineers use workstations; production control departments are equipped with computerized planning systems. A situation in which all these systems are not only present but cooperating in some sense has come to be called *Computer Integrated Manufacturing* (CIM) and to be perceived as sufficiently desirable to spawn a new industry.

In reality, unless great attention has been devoted to the goal of integration, this cooperation is not achieved. Before analyzing CIM implementation, let us first review what the concept covers.

1.2.1. Computer Integrated Manufacturing

Computer Integrated Manufacturing is the activity of a factory in which all data manipulations are done by computer. "Data manipulation" in this definition is intended to cover all input, processing, and output activities. The phrase "computer integrated manufacturing" was coined in 1973 by J. Harrington [Har73]* .

CIM is not a goal in itself. Although it is likely that at some time in the future it will be commonplace, CIM should not be pursued for its own sake. The economics of manufacturing, and no other consideration, should drive the assumption of more and more functions by computers in factories.

Whether manual or computerized, the data manipulations associated with manufacturing include

- Product design and manufacturing engineering.
- Materials management, and production and maintenance scheduling.
- Issuance of instructions to production personnel and automated equipment.
- Material, equipment, and tool status tracking.
- Quality control.

* See reference section for an explanation of all abbreviated sources

The CIM literature is long on the basics of data collection and communication but short on interpretation and use in decision making. If we focus for a moment on quality control, we can see the following challenges:

1. Deciding what data to collect.
2. Automating instruments for data collection.
3. Communicating measurements to computer systems.
4. Using the measurements to make decisions.
5. Implementing the decisions.

1. and 4. are the types of functional issues *we* shall be concerned with throughout this book. At stake in 1. and 4. is the reduction of observation requirements by optimized sampling, early warning of process problems, and rapid diagnosis. 2., 3. and 5. have to do with execution and have received the most attention so far in the CIM industry.

3. and 5. are *communication* problems. The CIM industry has been bogged down in the development of standards to solve them in a generic manner, and its greatest achievement may have been the specification of the Manufacturing Automation Protocol (MAP) for the shop floor, and the Technical Office Protocol (TOP) for the engineering side [Jon88].

2. is an *automation* problem which must be solved in different ways in each industry, depending on whether the parameters to measure are film thicknesses and sheet resistivities, as in semiconductors, or part dimensions, as in mechanical assembly.

CIM and automation are not synonymous. A "lights-out" factory is inconceivable without CIM, but the converse is not true. In a CIM factory, a shop floor may be found on which work instructions are issued to people by a computer system, where the work is executed manually, and where the results are reported back to the computer system through video terminals.

Some companies give the title of "automation managers" to employees whose assigned task is the implementation of CIM. The title reflects a general intention to eventually build lights-out factories. The real content of the job shows that CIM is in fact a higher priority.

The level of automation varies between industries. For semiconductor processing, automated equipment is available for most steps, and direct labor goes primarily into materials handling and data collection. In conferences on semiconductor manufacturing automation, discussions of CIM are mixed with talk of robot arms and guided vehicles.

To the extent possible, we shall avoid getting entangled in automation and communication issues, and stay focused on the essential questions of functionality, such as

- What instructions should the computer system issue?
- What messages should it accept?
- What actions should it take in response to these messages?

- What services should it provide to manufacturing managers, engineers, maintenance personnel, and other support groups?

1.2.2. How Incompatible Systems Proliferate

The MAP/TOP protocols are among the more visible achievements of the CIM industry. V.C. Jones [Jon88] opens a book on this subject with an example in which he contrasts product redesign before and after CIM.

In the "before CIM" version, the sales department is unable to make the customers' frustrations known to engineering, which therefore mistakenly assumes the product to be perfect. In the meantime, a new CAD system has made old designs inaccessible for engineers, and they are forced to start from scratch. In the "after CIM" version, the customers' concerns have been passed on to engineering by electronic mail and the product design, updated in the CAD database, is validated against the factory's process capability.

Thus we are told that more automated and better integrated computer systems will take a manufacturing organization from an uncoordinated, chaotic mode of operation to one characterized by intimate cooperation between departments, simply by making it technically feasible.

But if sales does not tell engineering what customers find wrong with a product, is it truly for lack of electronic mail? The telephone, facsimile, or even ordinary mail could carry the information. Could it not be instead that engineers are not receptive to feedback from the field, or that sales representatives feel they have more urgent things to do than to provide it?

The belief that computers can fundamentally alter the way a business organization operates is wishful thinking, because organizations only accept computer systems built in their own image. CIM cannot be that which unites a collection of warring fiefdoms; it can only be that by which an already strong team communicates and supports its decisions. Let us now examine specifically how, in the absence of a CIM strategy, incompatible computer systems proliferate.

Engineering document control is done on one system, material requirements planning on another, and quality control on yet another. Meanwhile, many employees spend their time programming their own personal computers. This situation has been referred to as a "systems nightmare," because it is characterized by disappointing results, obtained at the following high costs:

1. Bills of materials are kept in every system, and employees can never be quite sure which ones are correct. Ninety-five percent agreement between the different systems becomes an ambitious goal. Even if it is reached, the remaining 5% of inconsistent data causes measurable losses in the form of

   - Stocks of unneeded parts.
   - Shortages of needed parts.
   - Assemblies with wrong parts.

2. The synchronization of the different systems (also known as *data reconciliation*) becomes a regular, labor consuming activity. It is common to transfer part move data from a shop floor tracking system to a cost accounting system every week or every month. In the intervening period, manufacturing processes usually have undergone changes which have not been forwarded to the accounting system.

3. The credibility of the computer systems is hurt by this situation, and manufacturing personnel are reluctant to depend on them. The old manual system is maintained along with the new computer systems, and still is considered *the* source of reliable data.

4. Different systems have different user interfaces, and training needs are multiplied.

5. The poor performance of the systems causes deterioration of employee morale as an additional hidden cost.

The mechanisms by which the "systems nightmare" is brought about are summarized in Figure 1.1. In the 60s and 70s, Management Information Systems (MIS) emerged as the first attempt to bring business under computer control. The state of technology at the time dictated an approach in which all the data processing needs of a company would be met by a central mainframe computer. It was possible to have the production reports of a New England plant generated in California, and vice versa.

The small number of rigidly defined outputs was supplied to end users in hardcopy. When used, these reports were treated as inputs for further manual data processing. Managers would be seen doing arithmetic on figures read from listings, and did not think it possible to get the software modified to produce the desired results directly. The tensions created by this unsatisfactory situation were disincentives to any kind of coordination when the technological justification for the central mainframe disappeared.

Personal computers gave individual employees power over machines that could produce results comparable to those of central MIS. As personal computers proliferated, it became common for essential operational data to be stored on private floppy disks. This allowed a proliferation of inconsistencies and jeopardized data security.

Another major contributor to this "islands of information" phenomenon is the bias towards incremental projects due to the capital acquisition process and noted by Kaplan [Kap85]. The ceiling placed on the size of the expenditures which may be approved at each level of management encourages subordinates to submit requests just below their direct supervisors' limits.

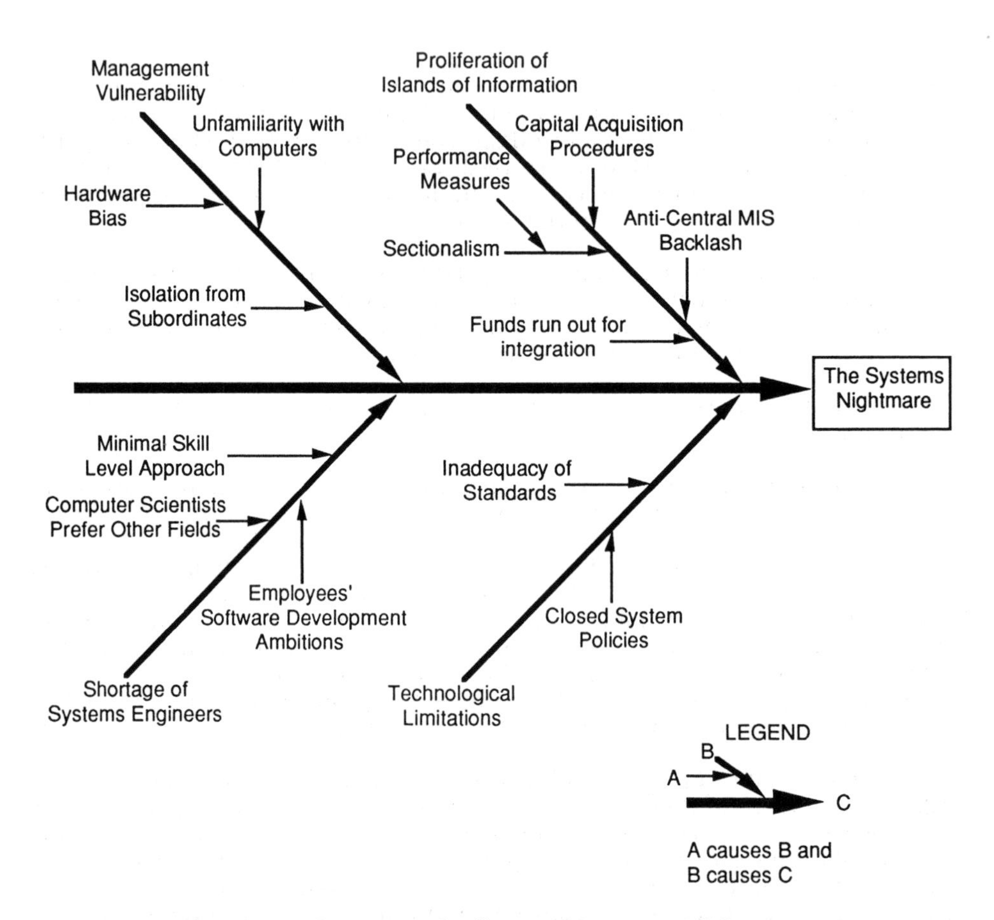

Figure 1.1. Causes and Effects of Systems Nightmare

A large budget can be spent on personal computers, graphics workstations, and incompatible minicomputers. If the organization is lucky, networking products may exist, but there may be no funds left to buy them. Sectionalism is the term used by Ishikawa [Ishi85] to designate the tendency of departments within an organization to pursue goals that are in contradiction with those of other departments or of the organization as a whole.

When such attitudes are present, the expressed system needs of each department are unlikely to be compatible. Furthermore, each department is determined to keep control over its own system. Performance measures are the most easily identifiable sources of sectionalism. A production department rated by its throughput and a quality control group rewarded for discovering rejects have conflict built into their charters.

If the same production department, while praised for efficiency, has no penalty associated with high levels of work in process, it is unlikely to initiate or support the implementation of Just-In-Time, whether manually or with the help of a computer.

The senior managers of manufacturing organizations today generally have had minimal exposure to computers. They tend to agree to the goals of CIM, but are confused on the approach to follow. The difficulty of integrating islands of automation is generally underestimated. Many managers do not feel compelled to be connected with whatever systems are put in place, and do not have terminals in their offices.

By remaining cut off, they deprive themselves of access to the data and their subordinates of a motivation to make the systems work. Meanwhile, electronic mail rapidly upstages the telephone and the written memorandum as the primary means of communication.

A new culture develops within the organization in which progress reports, invitations to meetings, agendas, and comments on document drafts are all done through electronic mail, except at the top level where the slower, traditional methods remain in force. The top managers still give hand-written drafts to their secretaries for "word processing."

When asked about their activities, the CIM managers of many factories describe their hardware. The equivalent way to present a manual system would be in terms of staff size, file cabinets, typewriters, and stationery. Hardware appears more tangible than software, and its acquisition is easier to justify to controllers with the same bias. Unfortunately, a hardware description says little about the function of a system. Rather than knowing what brand of computer is used, it is more important to know what data the system produces for whom, and from what inputs.

In spite of a recent renewal of interest, manufacturing still is not a glamorous field for computer scientists to enter. Employees transferred to systems engineering from other fields tend to acquire the minimum level of skills required to get by and go no further. Some employees with no prior experience of software development suddenly stop doing their regular job and attempt to write software instead. If this is tolerated, the amateur software thus developed becomes a contributor to system fragmentation.

Data communication between computer systems and manufacturing equipment requires standard formats to be adopted by the industry and by the equipment makers. The lack of universally accepted protocols ranks has been a tenacious obstacle in the path of CIM in the semiconductor industry.

Also standing in the way of CIM are the policies of some computer manufacturers who do not facilitate communications between their equipment and that of others. Whether openly stated or not, this "closed system policy" is aimed either at imposing their own practices as a de facto industry standard or at encouraging customers to work with them as a single supplier for all their needs.

1.3. WHY DO PRODUCTION SCHEDULING SYSTEMS FAIL?

Let us assume from now on that every effort has been made to avert the systems nightmare, and that a manufacturing organization decides to implement a computer-based production scheduling system. In such endeavors, success stories are disappointingly rare. In this section, we review some of the pitfalls in general.

Once a demand is placed on a factory in the form of customer orders or a marketing forecast, decisions have to be made in order to respond to it. These decisions include (1) timing releases of raw materials to the floor, (2) allocating machines, tools, and personnel to tasks, and (3) setting the work calendar. The execution of these decisions must then be monitored to detect deviations. This process is what we will mean by "production scheduling."

Success or failure, in production scheduling as in other areas of manufacturing, is relative to a competitive environment. A manufacturing organization can prosper without being effective if its competitors aren't either, or if other strengths compensate for this weakness. The implementation of a production scheduling system is a success if it provides a competitive advantage sufficient to justify its purchase and cover its operation costs. The main causes of failure are shown in Figure 1.2.

1. 3.1. Conflict with Performance Measures

A production scheduling system has no impact on events on a shop floor until workers and first-line supervisors accept its output as a guide to their activities. The most frequent apparent cause of failure is their refusal to do so. The following example illustrates a common conflict between a production scheduling system and a measure of worker performance.

Example 1: We consider a single machine, operated by one worker, and loaded with jobs of various durations arriving at random times in the form of customer orders. Before every job, a setup of duration S must be performed. Upon job completion, the output is immediately shipped to customers. This factory is shown in the Figure 1.3. The worker is rated on efficiency, calculated as the fraction of the time the machine is kept busy producing. On the other hand, the factory itself is rated by its average lead time. Let us examine what happens.

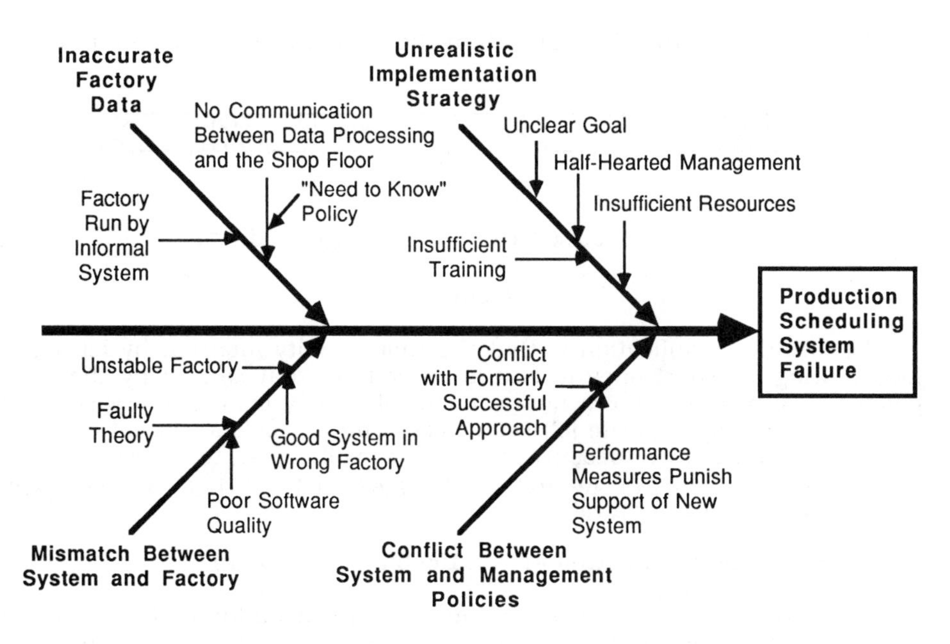

Figure 1.2. Causes of Production Scheduling System Failure

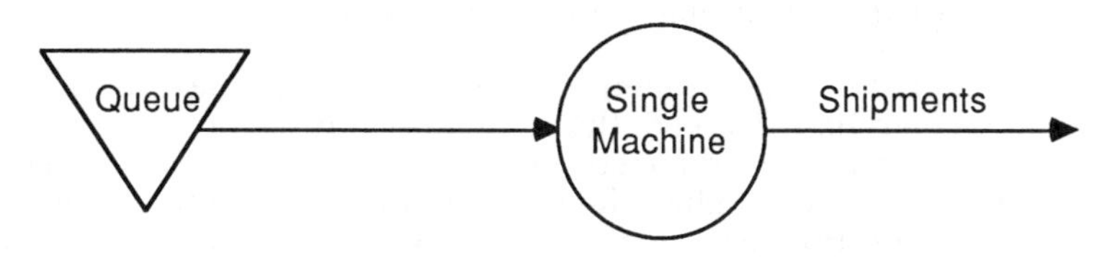

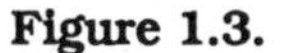

Figure 1.3.

The ratio of production time to setup time is highest for long jobs, and therefore the worker will tend to give them priority over short jobs, resulting in the timeline of Figure 1.4. If another long job arrives before the current one is completed it will be processed next.

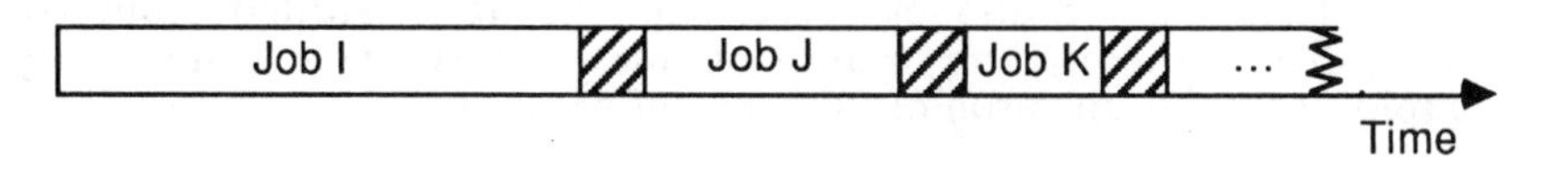

Loading for High Efficiency

Figure 1.4.

On the other hand, in this trivially simple factory, it can be shown mathematically that the average lead time of the factory would be minimized by always selecting the shortest job in the queue, resulting in the timeline of Figure 1.5.

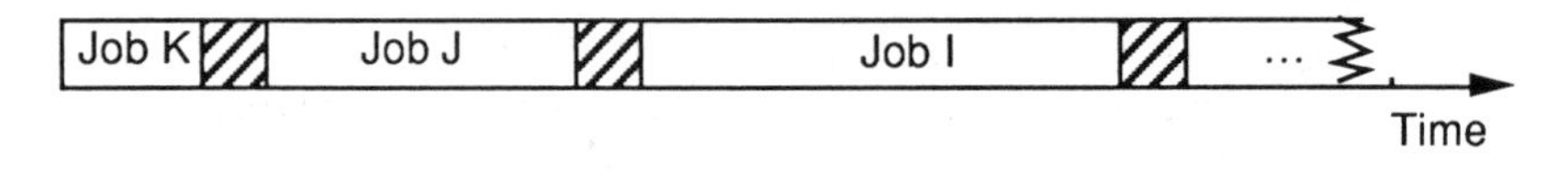

Loading for Minimum Average Lead Time

Figure 1.5.

The intuitive justification is that the time penalty incurred by letting long jobs waiting for short ones to end is lower than that caused by the reverse policy. This example shows how two not obviously contradictory goals may lead to prescribing opposite job sequences. Unless efficiency is not done away with as a performance measure, the worker will refuse to follow a schedule which, though it may be for the greater good of the company, may deprive him or her of a bonus or a raise.

1.3.2. Management Commitment — A Double-Edged Sword

The need for management commitment to a scheduling system is obvious. The plant manager must supply adequate resources and make ritual appearances at training sessions. Pep talks from the top convey the message that everyone is expected to make the system work.

Needed though this commitment may be, it is a double-edged sword: once the plant manager has decreed that implementation of system X will be a success, then it will be reported as such no matter what. The lack of management backing for a scheduling system may be due to the lack of a clear connection with the business goals of the factory. The implementation needs to be perceived as part of the execution of a strategy, and not as driven by the desire of a group of engineers to use the latest technology.

Example 2: The organization structure is as in Figure 1.6. The vice-president of manufacturing has been sold on production scheduling system X and informs the plant managers that its implementation will be a success. The company has a policy of rotating its managers between facilities for three-year tours of duty to broaden their experience and prepare them for promotion. Below the manager level, each plant is staffed from the local community. The company has been in business for 100 years, and many of the local employees have upwards of 20 years of seniority.

The plant managers are under pressure from above to make the system work but cannot afford conflict with their local staffs, because a reputation for making waves would be more harmful to their careers than the failure of system X. The local employees know that the managers have no strong ties to the community. They are passing through on their way to other assignments. This perception restricts the managers' authority. In the modus vivendi that evolves, everyone goes through the motions of using system X, but, beneath the surface, the factory operates as it did before.

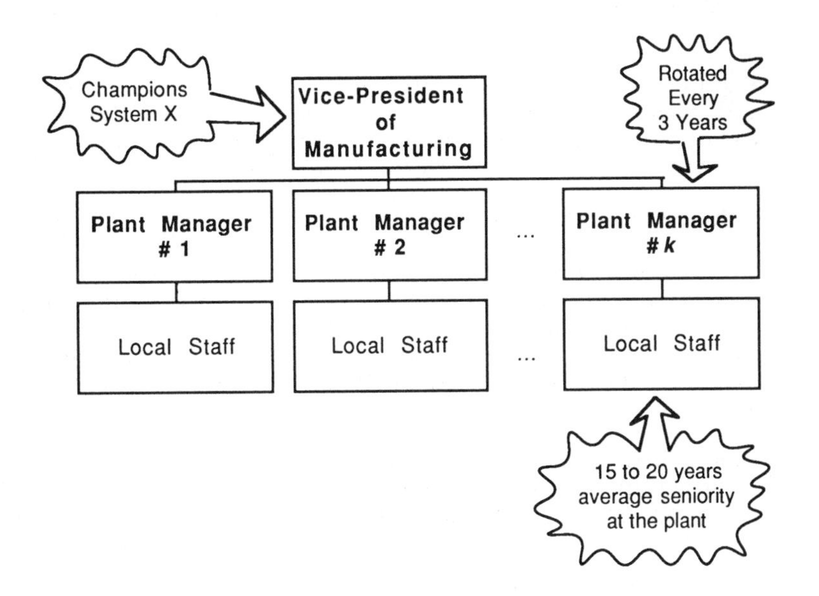

Figure 1.6.

1.3.3. Inadequate Factory Model

The schedules published by the system may be impossible to follow because relevant elements of the factory structure have been omitted from the model. Every day, each manufacturing supervisor has a different and valid reason why the schedule is impossible: an attachment, assumed available in large quantities, really must be shared between several machines, equipment capacities have been overestimated, or the effect of batching has been overlooked.

If an implementation is carried out with sufficient tact for employees not to reject it out of hand and the system is theoretically sound, but nonetheless produces impossible schedules, then the reason must be that its factory model is inaccurate.

Frequently, such a model already exists when the new system arrives. For cost accounting, or for a now obsolete scheduling system, bills of materials and routings have already been modeled for the products of the factory. Yet this data is often so inaccurate as to be unusable. How can such a situation arise? A stroll through the clean, air-conditioned data processing department and the noisy, grimy shop floor of a metal working factory reveals worlds each populated by quite different human beings and each frequently having no contact with the other.

Programmers who are told only what they "need to know" write software manipulating factory data without knowing what it means. The workers on the shop floor have never seen an output from the data processing department and do not depend on it. In this situation, the factory model undergoes no reality check.

On the other hand, in those rare cases when an integrated document control and parts tracking system is used, errors in the database of manufacturing specifications are immediately brought to management attention because they hinder the flow of production.

1.3.4. Scheduling is not the Problem

The level of data accuracy required for production scheduling should, however, not be overestimated. Not every detail of every routing needs to be known. Detailed information is needed only on the capacities and requirements of critical resources. If this data is available, and the system still fails to produce usable schedules, then the question of the system's fitness to the factory is raised.

Production scheduling may not be the factory's problem. If process quality is too low or equipment failures too frequent, then the factory's human resources are better spent on the solution of these problems than on attempts to schedule around scraps, reworks, and broken machines.

Example 3: Work center A is comprised of six identical machines, each of which is down 20% of the time, disrupting production even though it is loaded only to 80% of its nominal capacity. Can this problem be solved by clever scheduling?

A probability calculation shows that the capacity of the work center oscillates between that of two machines and that of six machines as shown in Table 1.1:

Table 1.1. Availability of Unreliable Machines

Number of Machines Up	Fraction of the Time
6	26.2%
5	39.3%
4	24.6%
3	8.2%
2	1.5%
1	0.15%

Since the full average capacity of the work center is required, a buffer in front of the work center is needed to make sure that all available machines have work to do. Another buffer is needed after the work center to protect the downstream work centers from starvation when many machines are down. This situation is illustrated in the following diagram. At this stage, we still do not know how big the buffers should be.

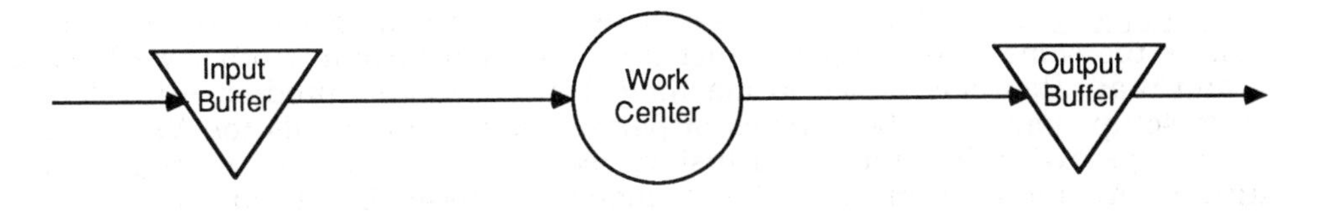

Figure 1.7.

Since 1.2 machines are down on average, there should be a repair technician dedicated to the work center. However, the maintenance requirement will vary over time with the number of malfunctioning machines, so that the maintenance department must have personnel in reserve.

The percentage of downtime is not sufficient information to set the buffer sizes. If the equipment goes up and down quickly, then small buffers will be able to smooth the flow through the center. With a given downtime rate, the longer the times between failures and the times to repair, the larger the size of the required buffers. This information is usually not available.

This example shows not only the extent to which equipment downtime complicates scheduling, but also the limited benefits to be expected from scheduling, as opposed to engineering the problems out of the equipment. The machines are down 20% of the time for reasons that engineers can identify and remove, provided such an effort is given the requisite management support. With equipment uptime brought up to 95%, many of the preceding difficulties vanish.

1.3.5. The System Cannot Model the Factory

Every scheduling system is built on assumptions about the factories it serves. Assumptions that are valid in a high volume, repetitive manufacturing environment will not be adequate for a job-shop with low equipment utilization.

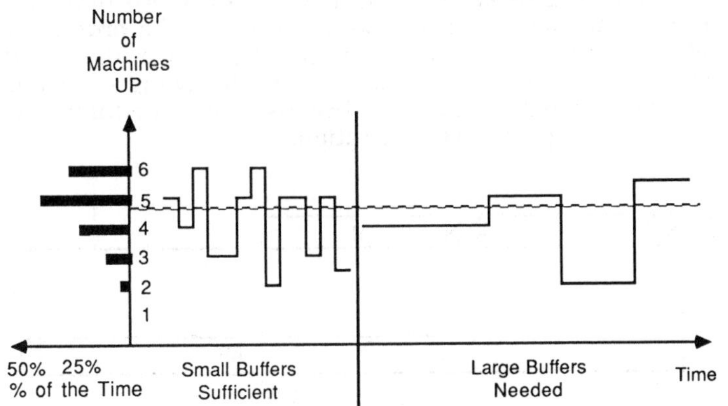

Figure 1.8.

Example 4: Parts A and B are fabricated through the same machine F and require the same time per part. Machine F does not have any other work load. A and B are then assembled in machine G into a salable product AB. F has a large setup time S for both types of parts. The raw materials for A's and B's are always available. Production history records show that, on average, F can turn out A's and B's sufficiently fast to match G's assembly speed.

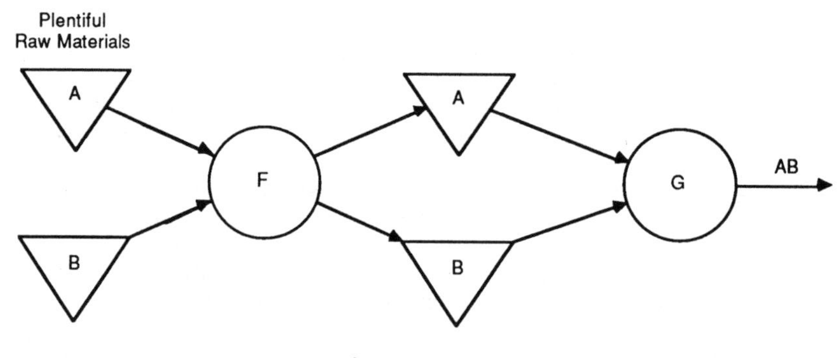

Figure 1.9.

The line is currently empty, and an order for 1,000 units of AB's arrives. Historical records show that the factory has been able to produce 12,000 units in a year. Considering this, the planner promises delivery of the order within one month.

In using average rates, the planner has implicitly assumed that F was equivalent to two machines, each with half the capacity of F, working concurrently on A's and B's. However, F is really one single processor: it makes A's or it makes B's, but not both at the same time, and the changeover time S from one to the other is not negligible.

Let us assume that, to keep up with G, F can afford only two setups per month. For the first half-month following receipt of the order, F will produce only A's and G will be idle. G will only begin to produce AB's after F has been changed over to produce B's, as shown in the following diagram. Delivery of the order will be one half month late because the scheduling model did not take batching and setups into consideration.

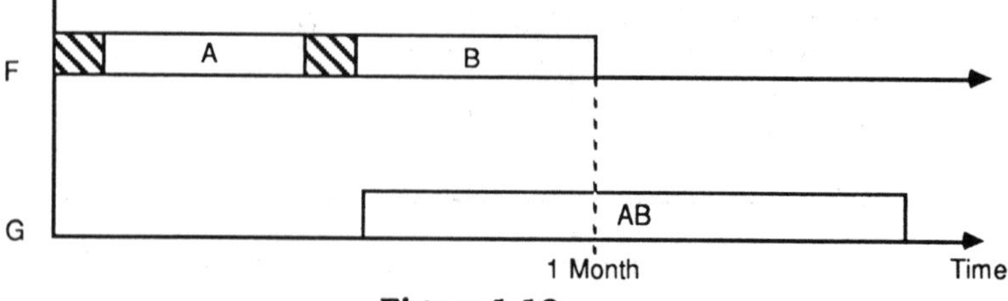

Figure 1.10.

1.3.6. Poor Software Quality

Finally, a scheduling system may be based on a sound theory and fit a particular factory but be incarnated in software of such poor quality as to be unusable. Its outputs might be able to improve the factory's performance if only they were produced in a timely manner, formatted in understandable reports, and if bugs did not occasionally cause runs to abort.

1.4. CONCLUSION AND DIRECTION

We have seen in the preceding some of the difficulties of CIM in general, and of getting one's money's worth out of production scheduling system in particular. If we now step back and, starting from *manufacturing* needs, take a global view, we are led to challenge the conventional wisdom of the CIM and automation industry.

Much of the manufacturing literature — and of the corporate strategy it influences — is based on assumptions which do not withstand a critical examination. In particular, contrary to popular opinion, (1) a headlong rush into CIM and automation is not the best way to raise productivity, and (2) the manufacturing techniques of the *leaders* of Japanese industry should be implemented in other countries.

One widely held belief in the United States is that automation is the way to restore manufacturing competitiveness. Some experts actually find solace in the fact that the U.S. is "more advanced that Japan" in such areas as automated material handling or artificial intelligence even though it is lagging in Just-In-Time and Total Quality Control.

In the United States, solving problems by automation is a conditioned reflex based on a long history of success. In reality, premature, ill-conceived automation projects result in low productivity and high costs. Far from being the answer to foreign competition today, the policy of automating obsolete practices is part of the problem.

According to another frequently heard theory, it is not necessary to study and implement the manufacturing techniques of successful Japanese competitors. It would only be playing catchup. Instead, new techniques should be invented to leapfrog them. In reality, the feasibility of a technological leapfrog based on ignorance is dubious at best.

Furthermore, manufacturing technology is not all there is to the business of manufacturing. Proximity to market, cheap real estate, and readily available natural resources also count. If, by implementing Just-In-Time, a struggling U.S. manufacturer simply approaches the productivity of a Japanese competitor, nonmanufacturing advantages may still give him or her an edge.

Finally, we buck yet another trend in the CIM literature by ignoring the issue of distributed systems. As systems analysts, we worry about what systems do, and leave to others the decision on whether they should be implemented in software running on one processor or on a network of processors.

Since this was not touched on in the body of this chapter, and treating distributed systems as a red herring is a controversial position, we will explain it briefly here. A centralized database — that is, with all queries and updates funnelled through a single processor — is the most straightforward implementation of a database. There are two reasons why this may not be satisfactory:

- Reliability : The processor in charge of the database is a single point of failure for the system. When it is down, the whole system is down, and so is a factory that depends on it.
- Performance : A centralized database frequently is a performance bottleneck for a CIM system. It coordinates all retrieval and update requests, and when it is overloaded, so is the CIM system.

Reliability and performance are both perceived as problems with the current generation of CIM systems in the semiconductor industry, and *distributed systems* are widely believed to be a solution. There are many ways in which a system may be distributed, none of which is in the field of expertise of a manufacturing manager or engineer.

Distributed processing is compatible with a centralized database. A possible architecture is one in which a dedicated database machine exchanges data with other computers. Caching schemes may be used where, for example, processing recipes are communicated to production equipment only at startup time or when a new revision is implemented. Measurement data may be stored in the database in the form of summaries generated by local data acquisition systems. In a distributed database, the database itself is partitioned between different sites; queries and updates are routed to the proper site in a manner that is invisible to the user.

Are distributed databases a solution to the reliability and performance restrictions of current CIM systems? As of this writing, special purpose distributed database systems are in operation at banks and airlines, but general purpose, distributed database management systems (DBMS) are still in research. The absence of commercially available systems is a sign that the development of a distributed DBMS is very difficult.

A CIM development organization would have to build one from scratch, targeted at its application. Assuming such an effort to be successful, its usefulness is predicated on the assumption that hardware technology will not in the meantime improve to the point that reliability and performance are no longer an issue. The limitations of centralized DBMSs have to be lived with in the short term. In the long run, the risk of undertaking the development of a special purpose DBMS seems higher than that of waiting for hardware technology to make the problems go away.

In other words, the reliability and performance of computers may be unsatisfactory for many manufacturing applications, but their improvement by distributed architectures or otherwise is the job of computer suppliers, not of manufacturing systems analysts. As the following chapters will show, we have our hands full with our own business.

2

MANAGING THE NEEDS ASSESSMENT PROCESS

The process of CIM needs assessment in a factory is a dialog which must be managed between, on one side, managers and engineers and, on the other, analysts. The analyst's role may be played by sales representatives, consultants, or internal systems engineers, all with biases induced by their relationships to the factory. The plant manager must support the analysis and refrain from presuming on its outcome, while guarding against analysis paralysis, keeping information flowing, defusing conflicts, and resisting special interest groups.

The analyst must dig below users' sometimes contradictory statements to identify the factory's needs and present them in documents users approve. This requires attention to the physical quality of the drafts, revision control, and security. To make the process work, each iteration should involve presentations and review deadlines. The documents should be concise and have enough substance to sustain the users' interest. CIM systems differ from financial systems in that manufacturing involves the flow of materials, not simply data. Manufacturing analysts need to be well versed in industrial engineering, production scheduling, or quality control, as well as the methods of structured analysis.

2.1. ORGANIZATION VERSUS TECHNOLOGY

A plant manager appoints a selection committee for a CIM system. The committee consults future users and writes an acceptance specification. This document is sent to suppliers in a Request for Proposals (RFP). Sales representatives from suppliers come to explain how their systems solve the plant's problems. After implementation, systems engineers listen to users' feedback and describe required enhancements to suppliers.

In all these circumstances, someone on the computer side of the fence must listen, interpret, filter, and model requirements expressed by the people of a manufacturing organization. Listening to users is a goal espoused by every systems engineer and analyst. To do it effectively requires not only mastery of the appropriate technical tools discussed in the next chapters, but also an understanding of the relative position of analysts and their counterparts in manufacturing. With this in mind, this chapter is dedicated to the organization of manufacturing analysis.

In manufacturing, the term *systems analysis* refers both to the activity resulting in the definition of software requirements and to the construction of a factory model with a given software system. In the first case the analyst describes what the system should be; in the second case, how to set it up once it has been acquired. Deciding that an office needs a desk with a particular configuration of drawers is similar to analysis in the first sense; choosing what folders with what labels to put into the drawers is akin to analysis in the second sense.

Most of the discussions of this chapter apply to both activities, but the following chapters are dedicated to the first. Although the techniques of this book were developed for the purpose of defining software systems, they are not rigidly coupled with the way computers work. They are applicable to the design of manual factory management systems, should that choice be made.

2.2. ORGANIZING THE COMMUNICATION

2.2.1. The Analysts

The Sales Representative

The cheapest way for a manufacturing organization to get help analyzing its computer needs is to call in suppliers' sales representatives. As long as they feel that the customers are not just shopping for information, they will, at their company's expense, spend time demonstrating their product and explaining how it solves the factory's problem.

The recurring themes in the sales presentations of three or four suppliers should provide a good feel for the capabilities of current, commercially available systems. This, however, does not necessarily match the needs of the factory. Sales representative likes to present themselves as analysts, but their conclusion is predetermined: the factory must purchase the system they are selling.

Unless the CIM needs of a factory are assessed without such constraints, management will not realize what compromises are made in the choice of a currently available system. The options of integration of separately purchased subsystems or of custom development, although generally undesirable, should not be excluded before starting.

The depth of analysis that can be expected from sales representatives is further curtailed by limited access to proprietary information. His or her relationship with the factory is too superficial to justify sharing information that might be of use to competitors.

The Consultant

On the other hand, outside consultants provide unbiased help, at a cost. Having nothing to sell other than their expertise, they can, theoretically, do a no-holds-barred analysis of the factory. As outsiders with no stake in the manufacturing organization itself, they are free of the influence of corporate politics.

But their contributions are not universally appreciated. "A consultant is someone who borrows your watch to tell you the time and walks away with it." said Robert Townsend [Tow64] to underscore his recommendation never to use any. Consultants' outsider status has its advantages, but, on the flip side, their credentials are not as well understood by management as those of employees.

The prosperity of this profession, however, attests that it must have something to offer. Indeed, engineers with a deep knowledge of one or more aspects of manufacturing technology are unlikely to find professional fulfillment within a single organization. Such people do bring, on a temporary basis, expertise that cannot be developed in-house.

When the employees of the manufacturing organization find out that the "industrial engineering expert" dispatched by a consulting firm was a detective in a big city police force until a year before, the credibility of the whole effort is destroyed, especially among engineers who are themselves specialists in the field.

While on assignment in a factory, consultants are under a microscope, with their every move watched by the client's employees, some of whom are looking for faults. The consultant's nightmare is not knowing something he or she should, and the person may be tempted to try and cover up shameful gaps.

The client should not have unrealistic expectations: a consultant can be useful without being omniscient. A systems analyst does not need a deep understanding of the plant's process technology. The willingness to confess ignorance on a particular point shows in fact the confidence of a true expert.

When using consultants, a factory is faced with the decision of how much access should be granted to confidential information. The consultants may sign a nondisclosure agreement, but it is certain that they will use, if not actual figures, at least the conclusions they arrive at in work for other clients.

Someone who stays in residence within an organization for three months is not a consultant but a contractor. Instead of sowing seeds in a two-day visit, to grow through the efforts of employees, he or she becomes a temporary addition to the staff, directly doing the work, and not transferring technology as a consultant should. Such a "consultant" may be cheaper by the day but not for a whole project. The true consultant remains an outsider, coming every few months for brief follow-up visits until the project is complete.

The size of a consulting organization is also a factor. The larger it is, the higher the risk that the needs of its own payroll will influence its recommendations and that the knowledge of its employees will be an edulcorated version of the founder's.

The Systems Engineer

A plant manager can avoid resorting to outsiders by drawing the analysis team from the ranks of the manufacturing engineering group or the computer systems staff if there is one already. We shall call such inside analysts *systems engineers*, even though the labels may vary. In various respects, insider status is both a blessing and a curse.

As part of plant personnel, a systems engineer theoretically has access to all he or she needs to know. In practice it will be true if the company offers lifetime employment, the factory is geographically isolated from competitors, and the systems engineer is a native of the area. In Silicon Valley, such conditions do not hold and the insider advantage is limited.

Unlike the outside consultant, the systems engineer has a career stake in the factory, and can be promoted or passed over in future personnel movements. This situation cannot be without an influence over results. In particular, there is no simple, quantitative way to measure the contribution of the computer systems staff to the factory. Production managers build parts and meet schedules. Engineers design products or processes. Systems engineers have nothing similar to claim credit for, unless they develop software. They do tend to find all commercially available systems inadequate.

To the systems engineer, as well as to its other managers and engineers, each factory is unique in so many respects that solutions developed for other factories could not possibly work for it. With a deep knowledge of one factory, and usually none of any other, the systems engineer is biased. The outside consultant is more likely to see the factory's problems as generic.

How much does the internal analyst cost? Accountants measure this by an hourly rate, with fringe benefits and an allocation for overhead. This would be the economic cost only if he or she were hired specifically for this purpose or if the alternative were dismissal. If other engineers can pick up the slack in the new analyst's old position, then no cost can be directly allocated to the assignment.

2.2.2. The Manufacturing Organization

What It Must Do to Support the Analysts

The best analyst can only help a willing factory. The plant manager can show support by letting subordinates know beforehand that this is their opportunity to influence the systems policy. While analysis is in progress, impromptu appearances at meetings can show how much they matter. By being accessible to the analysts, a manager sends the message that others are expected to be accessible, too.

Until the analysis is completed, the plant manager should avoid making decisions which presume its outcome. It may be tempting to hire five programmers in July because there is a budget and they are available just after graduation but will not be four months later. However, the factory's needs will not be known until the analysis is complete. In particular, it is not established that five programmers will be needed at all. Hiring them disrupts the analysis by changing its goal to finding a way to keep the programmers busy.

Analysis Paralysis

Rash acts, such as those just mentioned, are frequently triggered by the fear of *analysis paralysis*, that is, a never-ending analysis process. Instead of leading to a decision, the requirements document leads to another study by another committee, which in turn leads to another one, and so on without ever reaching a conclusion.

Analysis paralysis is a real danger; managers need a feel for how much is too much and how long is too long. A rough guideline could be that the a priori analysis of functional requirements should require roughly the same resources as a hardware investment of the same amount. Poring over data flow diagrams is like reviewing an architect's blueprints. The latter is done as a matter of course, and so should the former, with the same degree of thoroughness.

When not due to personal indecisiveness, analysis paralysis may be due to the lack of adequate technical tools. Without tools such as those of structured analysis, discussed in the following chapters, analysts tend to produce overly lengthy documents which nonetheless fail to give a clear picture.

A recent business best-seller [Pet83] makes a surprising argument in favor of taking action without worrying about the consequences. "Ready, shoot, aim" is the proper sequence of actions, say the authors. Applied to CIM, such a "bias for action" would lead to buying hardware and software without knowing what it is for. The experience to be gained is allegedly valuable enough to make it worthwhile even if implementation fails. To the objection that this is irrational, the authors actually respond that rationality is not the only appropriate basis for business decisions. This is not the point of view of this book.

Corporate Culture

In factory X, each draft of the requirements document is distributed to a group of reviewers. A week later, they meet to discuss their questions, comments, and objections, and a collated list is given to the analyst who then revises the document. This process is continued until the reviewers agree that the document faithfully models their requirements.

This is how it is supposed to work, but in factory Y, which may be within the same company, the same document fails to bring any echo. "We don't have time for this," say the people of factory Y when pressured to respond. Many organizations only pay lip service to the analysis process. There is not much an analyst can do but wait for the organization to learn from experience.

Another cultural hurdle may be an excessive secretiveness. Some Japanese companies consider their production scheduling methods to be a competitive advantage and therefore a secret, but it is impossible to automate a method that the user will not reveal. In practice, several competitors may use the same production scheduling software and yet perform that function differently, just as several players may use the same piano and yet create different sounds.

Competitors frequently have the same processing equipment which they use each in a special way. Every CIM system in operation contains a unique factory model that is not communicated to any outsider. Generic approaches can be revealed to outsiders without jeopardizing a factory's competitive position.

Some organizations thrive on confrontation. Meetings start with ultimatums: "Your software must to this for me or else." But the analyst has no authority to agree. His role is not to make promises but to listen, which cannot be done in a shouting match. The analyst builds a consistent model of the perfect software solution; he or she does not make the compromises, but establishes a reference against which others may negotiate in an informed way.

A manufacturing organization is a political microcosm, complete with special interest groups. Some of those may push the systems strategy in a questionable direction, while others struggle to make their views heard. An analyst's conclusions, no matter how rational, have little chance of being accepted without the backing of a sufficiently powerful faction.

The constituency for quality control in a U.S. factory usually is a "quality control department" with no hope for its members to become top management; there, quality control software is a hard sell. By contrast, many large manufacturing companies in Japan are run by engineers who started their careers implementing quality control and do not have to be sold on it.

2.2.3. The Analysis Process

"We hired consultants. They came to our plant and asked people what they did. Two weeks later they came back and offered to write software to automate exactly what we had been doing. I refused: for us, there was no value added." Thus spoke the president of a small semiconductor manufacturing company. If he had been satisfied with the current practice of his factory, he would not have called in the consultants. He expected them to know more about manufacturing than his own staff.

To find out the true needs of a factory, the analyst must dig below the explicitly stated requests because those are often initially formulated in a language so vague as to provide next to no information. The analyst who takes users' wishes verbatim to a software supplier has failed to do his or her part.

"I want a simulation capability," says the production control manager. In her mind, this may mean anything from changing parameters in a spreadsheet to reproducing every part movement and equipment state change on the shop floor. The analyst must find out what problem she is trying to solve.

As discussed in Chapter 1, users frequently demand that a CIM system be distributed, which is irrelevant to its function. Even the most flexible manufacturing hardware has a narrower purpose than a computer. A request for a machining center is an expression of a need to drill, grind, or mill metals in various ways. On the other hand, the description of a hardware configuration says next to nothing about what a computer system is for.

Production, engineering, maintenance, and other departments have the common goal of making the factory prosper, but, in day-to-day operations, they are frequently in conflict: Production wants to run parts through a machine that is due for preventive maintenance, or engineering wants to run an experiment on a new product and preempt resources needed to ship the old ones.

In this atmosphere of permanent negotiation, the analyst is faced with contradictory requirements from different quarters. The plant manager, for example, wants CIM software to give direct, unfiltered access to work in process status, but the production supervisors do not want to relinquish control over the upward flow of information.

The analyst must interpret the functional requirements of "the factory," but the factory speaks with many different voices. Theoretically, the highest manager involved has the final say, but in practice subordinates have the power to block the implementation of software they do not like.

Systems analysis results in confidential documents undergoing multiple revisions, and these documents need to be distributed to a sufficiently large number of reviewers to get all the relevant feedback. This process must be administered to supply reviewable documents, in the same version to all, and to minimize the risks of security leaks.

The substance of a draft or a strawman specification is open to debate; its form, however, must be that of a professional business document: it must have a title, author names, a date, a table of contents, a version number, and page numbers. Some analysts dispense with such niceties on the grounds that the documents are drafts. But in the absence of page numbers, the reviewers could not say "On page 6, paragraph 2, you write that...," and seven managers in a meeting would be seen trying to agree on what part of the document they were discussing instead of what it said.

If a specification is handed out with one word changed in response to a reviewer's comments, it must have a new version number. It is an easy discipline to forget, but otherwise reviewers may have different documents in hand with identical labels. The frustration and confusion this creates may be lethal to the analysis process.

Word processing with integrated text and graphics is needed, because it simplifies revision control. The separate production of text and diagrams multiplies opportunities for error and version mismatch. The physical quality of the documents also needs attention. Beautiful printing cannot salvage poor concepts, but low-resolution dot-matrix characters, further blurred by a copy machine that is short on toner, can render a good document illegible.

The prevention of security leaks is almost impossible. Partitioning the documents so that each reviewer sees only what pertains to his or her area of responsibility minimizes opportunities for inadvertent leaks. Numbering all copies, stamping them "confidential," and asking for their return after reading may raise the reviewer's consciousness, and make the documents appear more interesting, but also motivate deliberate leaks.

Some software suppliers have the philosophy that "no one reads these specs anyway," and proceed to write them in a way that is guaranteed to fulfill this prophecy. Their purpose in writing at all is to impress customers with the bulk of their work rather than to carry on a dialogue. Busy pictures are hastily assembled with what the authors themselves refer to as "verbiage."

If the first draft of a functional requirements document elicits no objection, it means that it has not been read, not that it has been read and agreed to. For documents to be read and responded to, they must be written with this goal in mind, and that means concisely, with a specific reader in mind, and with each word or graphic symbol meaningful and relevant to the target reader's work.

Documents should not be just handed out, but introduced to reviewers by an analyst's presentation of its content, with a date set for a wrap-up meeting. To keep the dialog going, the analysts must answer each written comment, either by incorporating it in the document or by providing a written explanation of why it cannot be.

To avoid controversy, the temptation is great for the analyst to stick to statements no one can disagree with. Consensus on truisms is easy to attain, but useless. Everybody will agree that the factory should have "the right parts, at the right place, in the right quantity," or that "costs should be reduced," but, once such things have been said, what more do we know about the needs of the factory?

At the other extreme, a statement that "the production control manager should be replaced" is guaranteed to generate the wrong kind of controversy. It is irrelevant to systems analysis and can cause severe distress to the targeted person.

An adequate balance could be found in a statement such as: "Our CAD system should download product definitions to the shop floor control system, which in return should upload process capabilities to CAD." It is imprecise, with many details to be filled in, but it says enough for a devil's advocate to make a case that such an interface is not needed at all, prompting a discussion of benefits to be derived from it.

2.3. PRAGMATICS

2.3.1. Syntax, Semantics, and Pragmatics

The process of manufacturing analysis results in a functional requirements document which, more than a wish list, should be a consistent model of a system. This means, using terms we are about to define, is that it needs to have a syntax, semantics, and pragmatics.

The pieces of a computer model of a factory are defined according to a *syntax* — for example, "a lot-ID is a sequence of 12 characters," or "a process is a sequence of operations." *Semantics* refers to rules by which these objects interact within the system, as if in a game — for example, a lot moves along a process one operation at a time.

But playing this game is of value to a manufacturing organization only to the extent that there are *pragmatics* — that is, rules mapping those logical objects to physical objects on the shop floor: there must be boxes of parts whose motions are emulated by the lots defined in the computer model.

Poker has a syntax: a hand is made of five cards drawn from a pack of four suits of 13 cards and two jokers. It has semantics: four of a kind beats a full house, which beats two pairs. Poker has no pragmatics: it exists only in cards. There is no such thing as a "full house" in the poker sense outside of the game. On the other hand, the pragmatics of a CIM system are the only reason why anyone would want one, and they are established in the analysis process.

De Marco [DeM78] presents a case study of a software system to predict the performance of racehorses from their horoscopes. This system is syntactically and semantically well constructed, based on inputs from users selling the predictions. It failed, however, at the pragmatic level, because horoscopes did not predict winners. The manufacturing analyst is faced with the danger of being so caught up in the elegant semantics of a model as to lose touch with the not-so-elegant reality of the shop floor.

Banking systems are an example of clear pragmatics: each account maps to a person or a legal entity owning it, and each dollar in an account is convertible into a bill. Those systems are so successful that routine transactions between a customer and a bank today are almost invariably performed on a computer, be it by a human teller at a window or by the customer at a teller machine. By contrast, there are few factories where similarly simple parts tracking transactions are computerized.

A technical explanation for this discrepancy is that the routine work of a bank is almost entirely data processing, making it straightforward to map the abstractions it is already dealing with to software concepts. On the other hand, manufacturing has to contend with the flow and transformation of *materials*, physical entities with complex structures for which the work of abstraction has a much greater distance to cover. In a bank, the computer system does the work; in manufacturing, it keeps track of it.

2.3.2. Comparing a System with a Requirements Document

Before introducing the tools of manufacturing systems analysis, we can only give rough examples of the use of a requirements document. A commercially available system that cannot model a bill of materials is clearly a poor fit for an automobile assembly factory in a pragmatic sense, because it offers no way to model components.

An example of a less serious problem would be a system offering no way to assign a due date to a set of parts in a factory where this is needed. A creative sales representative could say: "You can put it in the cycle time field." But this would be a violation of the system's semantics, somewhat like playing checkers with a chess set. It is unlikely to work satisfactorily, because some reports would display whatever happens to be in the cycle time field in a column thus labeled, creating user confusion when it really is a due date.

On the other hand, if the factory is used to 20-character product names and the system limits this length to 12, this is a syntactic discrepancy which can probably be resolved with a dose of creativity. It does not disqualify the system. The first five characters may in fact be a product family prefix, and the second five be a technology code. If the system offers such fields in its product definition records, then it can model all the data contained in the 20-character names.

2.3.3. Current Factory versus Future Factory

An issue that is frequently brought up is that a perfect understanding only of today's factory needs is all an analyst can hope to get from its managers and engineers. If a CIM system is provided which meets all their needs, will it still do so five years from now? The first automobiles looked like buggies without horses, because their designs were retrofits of a new technology onto an old one. Likewise, retrofitting a CIM system to an existing factory is unlikely to yield the same result as designing a factory for CIM from scratch.

One method to involve the personnel of existing factories in the analysis of future needs is extended off-site brainstorming sessions, where the managers and engineers are encouraged to forget their day-to-day problems and construct their dream factories. However, before packing off the whole staff of a factory to a week-long off-site meeting, it is worth pondering whether manufacturing problems will truly undergo such a change within the next five years as to make a system obsolete. In an industry like semiconductors, many new equipment types will appear within five years, requiring the development of new interfaces with the CIM system.

On the other hand, manufacturing concepts do not change with processing equipment. A new machine may be faster and better, but its introduction does not invalidate an approach to production scheduling. Breakthroughs in manufacturing, as opposed to process technology, are decades in the making, as can be seen from examples such as assembly lines, total quality control, or the Toyota production system. Such time lags far exceed the life cycle of a CIM system.

2.4. THE ANALYST'S TOOLS

2.4.1. Manufacturing Engineering

A manufacturing analyst should have expertise both in manufacturing technology and in systems analysis. He or she should understand the theory of manufacturing at a level that its practitioners, tied up in day-to-day problems, do not have the leisure to reach. This, combined with the most modern approaches to systems analysis, is not a luxury when confronted with factories.

A common misconception, particularly in the artificial intelligence (AI) community, is that a deep theoretical knowledge of the field of application is not needed to specify or build software to serve it. Interviews with "domain experts" are supposed to be sufficient. This attitude does not reflect much respect for the experts.

A factory's managers and engineers are not in the business of training analysts in the basics of their trades. They are worried about equipment breaking down, parts coming out with wrong dimensions, and employees not coming to work. Taking the manufacturing organization as the source of theoretical knowledge in production scheduling or quality control is the mistake the company president quoted earlier was castigating. It is truly borrowing the user's watch to tell the time.

An analyst must know industrial engineering, production scheduling, or quality control, but this knowledge is reserve ammunition and to be spent sparingly. Firing off too much too soon may antagonize a manager unsure of his or her own proficiency and be perceived as forcing a canned solution onto the factory without learning the facts.

In the mind of analysts who are well versed in manufacturing, keywords from a user conjure up ideas which enable them to take shortcuts in their questions without jeopardizing the quality of their understanding. When engineers say "We use $\overline{X}$-R Charts," the analysts immediately know that (1) somewhere, process variables are measured, (2) someone plots statistics of the results, (3) each new value is tested against control limits, and (4) alarms are somehow communicated back to the shop floor. Subsequent questions can serve to fill in the remaining blanks.

By contrast, the next question of analysts who rely on the domain expert is "What's an $\overline{X}$-R Chart?" unless they choose to pretend to know in the hope of figuring it out from the user's words. The analysts' credibility is unlikely to survive many such occurrences. The engineers will conclude that their time is being wasted and that these analysts are incapable of understanding the factory's needs.

Analysts without independent knowledge of the field also have no means of gauging the quality of the user inputs. Engineers who discuss the $\overline{X}$-R Chart may have a deep understanding of that tool. On the other hand, they may produce them only because of a government requirement and have no idea what they are for. Analysts must be able to tell.

Although a factory is a highly controlled environment, its dynamics obey laws that management does not have the power to change. Analysts are sometimes faced with a demand for software with a logic that couldn't possibly work. Some software companies have the following philosophy: "We do what the customer wants, even if it's wrong. We tell him we think it's wrong, but we do it anyway."

Sooner or later, if the analysts' assessment was correct, reality will show it to the users. It is, however, a no-win situation for analysts, because the correct prediction of a failure is not a quotable achievement. It is even worse for the users who have borne the cost of implementing software that didn't work in a pragmatic sense.

2.4.2. Software Engineering

The analyst is a bridge between a manufacturing organization and software suppliers, and must be able to communicate with both sides. We have discussed so far the need for manufacturing expertise, and emphasized it because it tends to be overlooked. Turning now to the other side, let us ask ourselves what the manufacturing analyst needs to know about software engineering.

The manufacturing analyst is not responsible for the design of efficient algorithms. There is no point in worrying whether a program will run in 20 minutes on machine X before knowing (1) whether its logic pragmatically solves the manufacturing problem it is aimed at, and (2) whether computer X will be used. Finding the answer to (1) is part of the analyst's job. The year that may elapse between analysis and implementation is long enough to make both computer X and premature performance estimates obsolete.

2.5. CONCLUSION

This chapter is a long sequence of warnings. I have personally fallen into all those pitfalls of manufacturing analysis. Such problems are real; yet the analysis process can recover from any of them, if the parties involved are sufficiently motivated.

3

STRUCTURED ANALYSIS IN MANUFACTURING

Structured analysis is a concise specification technique based on graphic tools expressing software functions as perceived by users, and which is applicable to manufacturing. Hardware diagrams carry no functional information, and loosely drawn "concepts" diagrams fail to say more than lists of bullets with keywords. The example of an external context diagram shows the effectiveness of a rigorous graphic language, coupled with a focus on results for users.

Structured analysis is preferred to boilerplates and flow charts. The traditional technique of following a boilerplate table of contents set in a standards manual emphasizes frozen form over meaningful content and stifles engineers. Flow charts should be avoided, as they are obsolete and ineffective ways to describe procedures.

Petri nets are compatible with structured analysis and provide a unifying theory for many of its tools. However, they are more applicable to software design than to communication with users. McMenamin and Palmer's enhancements to structured analysis are used in this book. Finally, De Marco conventions for data and the ASME symbols for material flows are used in preference to Gane and Sarson's SADT and Harrington's I-DEF symbols.

3.1. INTRODUCTION

This chapter introduces tools described later in greater detail and outlines the reasons for preferring them to a number of alternatives. Authors such as De Marco [DeM78], McMenamin and Palmer [McM84], or Date [Dat81] do not specifically target manufacturing applications, and use examples from many other fields.

When using tools exactly as specified by a consistent theory, no matter how good that theory is, one inevitably encounters areas it fails to cover. The following discussion would be misleading if it did not give a feel for the "envelope" of applicability of the concepts. The examples are chosen in the spirit of highlighting difficulties as well as cases where they work well.

Most of the manufacturing software developed to date was designed on the fly by programmers. Once something remotely acceptable to users was coded, a "functional specification" was written for form's sake. The use of primitive systems analysis techniques resulted in what De Marco [DeM78] calls "Victorian novels"— that is, thick and obscure documents which would have brought analysis paralysis to any organization trying to use them for their intended purpose. Let us examine, in an example, the issues at stake.

3.1.1. A Hardware Diagram

Figure 3.1 is based on the first page of an advertising brochure for a CIM system. Due to its prominent position in the document, it immediately draws the reader's attention, and we can safely assume it to highlight those aspects of the system the supplier is most proud of. The reader has most likely seen such diagrams before, and is invited to take a look at each box and each connection on it from the point of view of a manager wondering how that system would contribute to a factory's profits.

If the diagram were taken out of context, the reader might not even guess that it has to do with manufacturing. It highlights names of computers and technical terms such as "terminal server," "local area network," or "Ethernet." The only clues that the system actually serves factories are equipment names in fine print.

A factory manager normally doesn't know and doesn't want to know the meanings of technical terms from the computer field. Therefore he or she cannot fathom what good those contraptions can do. With bars of various thicknesses, the diagram shows different kinds of connections between boxes of various shapes. The manager will probably assume them to represent cabling between pieces of electronic gear. From that, one can get an idea of the system's hardware costs, but not of the functions it would perform.

"Reports," "terminals" and process equipment names do indicate places where useful outputs might come out of the system. But what do the reports contain? What do the terminals show? What does the system say to process equipment? All we know from the diagram is that interfaces exist, but it gives no idea of what flows through them.

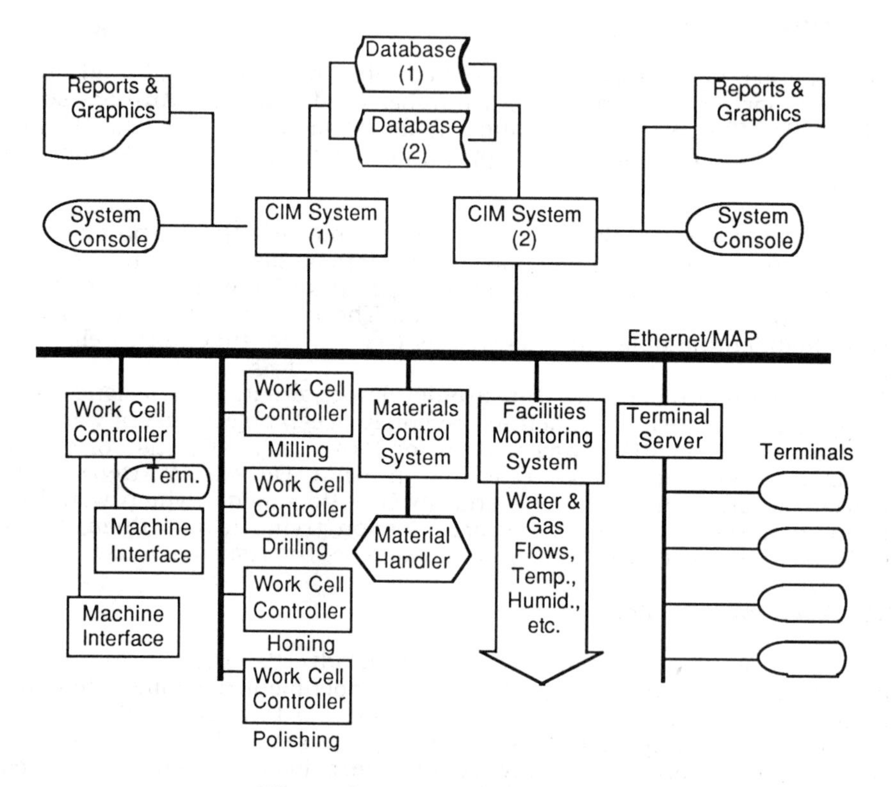

Figure 3.1. A Hardware Diagram

3.1.2. A Concepts Diagram

If the manager does not give up and reads further, he or she will find Figure 3.2 further in the same brochure. The boxes in it are labeled with manufacturing keywords: "Production Planning," "Floor Scheduling," "Engineering Analysis," or "Facilities Monitoring." The only computer term left is "Database."

This should get a manager's attention. But the odds are that he or she will retain no more from Figure 3.2 than these words, and that all the accompanying artwork will have been done in vain. This will certainly be the case if the manager gives the diagram a casual look and moves on. Let us assume, on the contrary, that he or she is determined to milk every possible bit of information out of that diagram. We shall see that it does not withstand scrutiny.

In practice, the manager will conclude that the diagram is too difficult to understand and move on, thus confirming the diagram's author in the belief that more rigor would be a waste of time. "Miscellaneous Reports," for example, is hardly an apt description of what the manager is likely to be most interested in.

There is no doubt that the boxes in the center represent data, because their labels say so. Presumably, the boxes in the periphery represent software, except for the "Miscellaneous Reports" box, which has to represent both reports and the software to generate them. The "Production Control" box is also puzzling, because it is also labeled "Process Instructions." Besides the fact that production control is an action while process instructions are data, those instructions are generally not issued by the production control department.

Figure 3.2 also contains arrows. Those could mean that the function in the origin box is performed before that in the destination box, or that some data flow between the boxes. In the absence of labels, the manager would have to guess the meaning. If we guess that the Production Planning module produces a Master Production Schedule, we can wonder where that schedule is stored: there is no arrow showing Production Planning writing to the database.

Figure 3.2 is an improvement over Figure 3.1 in that it shows at least what the system is about, if not what it does. Inconsistent graphics, and vague, incomplete captions prevent it from describing functions. Those seemingly minor flaws confuse the reader and prevent Figure 3.2 from saying more than a list of bullets and keywords.

3.1.3. An External Context Diagram

At this point, the reader may think this is nitpicking and that it is unrealistic to expect a single, uncluttered diagram to tell a better story. Let us now take a look at Figure 3.3 and consider what it tells our factory manager. Like Figures 3.1 and 3.2, it aims to give a first impression of a CIM system to a reader with no prior knowledge of it.

The first striking feature is that the system itself is in the background. Most of the entities shown in the diagram are its users: each rectangular box designates a category, identified by the name of a position or a department that exists in every factory. There is not a single computer term in the diagram, and none of the system's features are shown individually.

On the other hand, the data exchanged between the system and each category of users are described in as much detail as space allows. Not all vagueness is eliminated, but expressions like "Performance Indicators" leave less to the reader's imagination than "Miscellaneous Reports." The manager may also note that Figure 3.3 shows a Master Production Schedule as an input to the system, and that the system therefore does not work directly from customer orders.

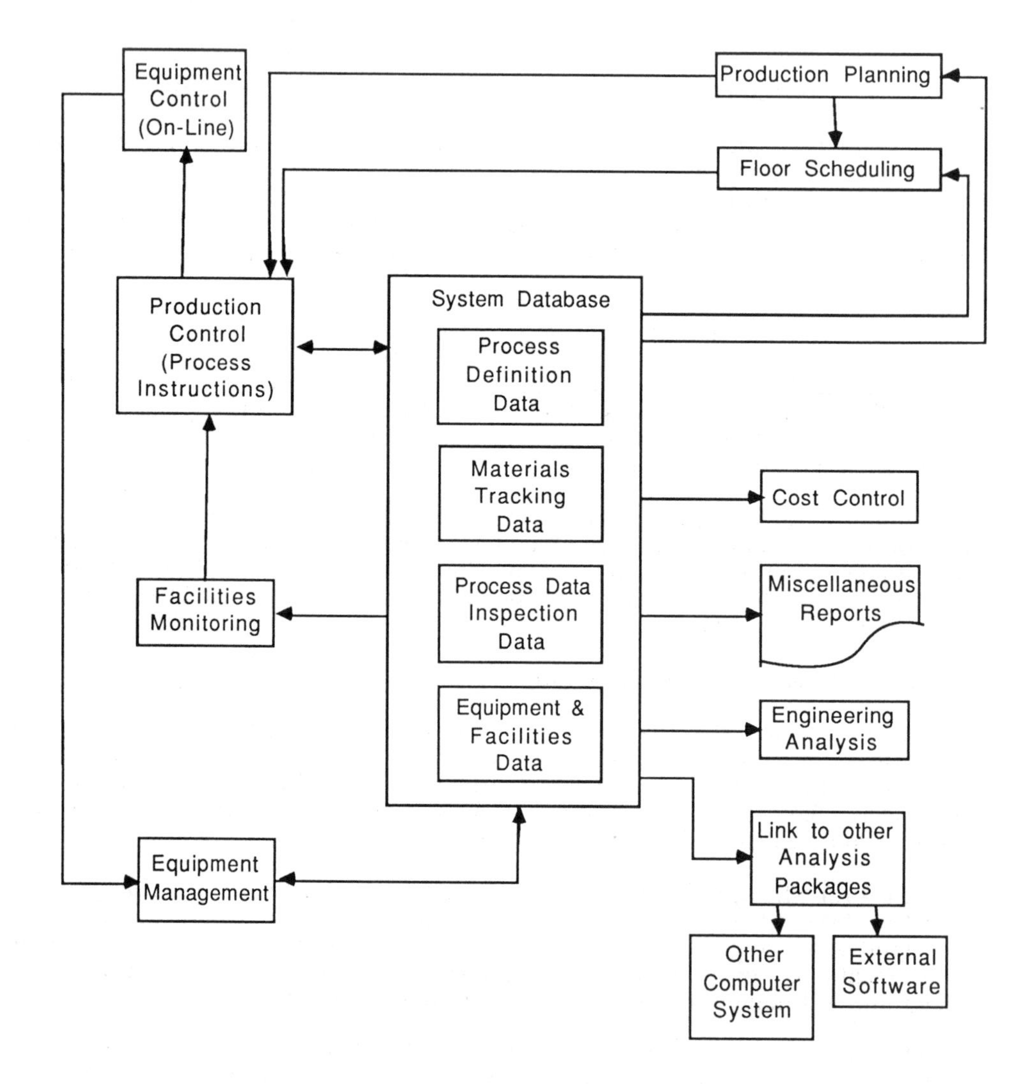

Figure 3.2. A Concepts Diagram

If misprocessing of parts and delays in the implementation of engineering changes have been causing the factory to lose customers, the manager will perceive the benefits of automated document control. In structured analysis, Figure 3.3 is called an *external context*. The philosophy of such a diagram is that, if you draw a boundary between a computer system and its environment, and describe all the data crossing this boundary, you know what the system does and for whom it does it.

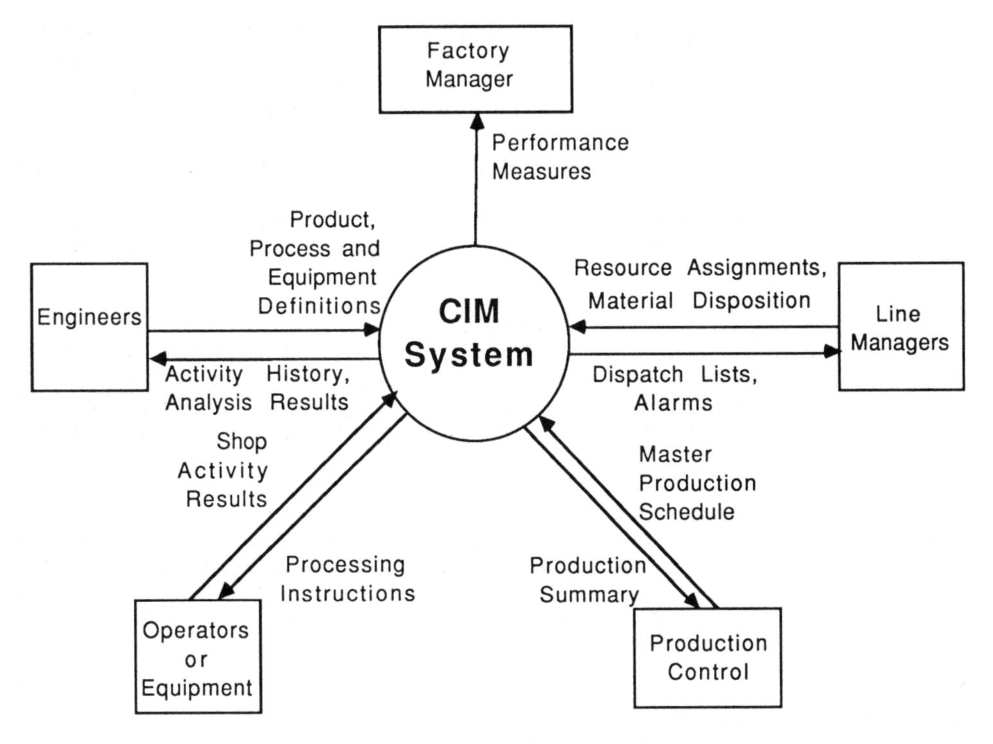

Figure 3.3. External Context Diagram

Software is to hardware as the content of a book is to the paper it is printed on. Within a computer system, software should not be viewed as an add-on to hardware. Rather, it is hardware that should be viewed as the physical support of software. Without printing there would be no books as we know them, and without hardware there would be no software, but even if a book is bound in leather and typeset in beautiful fonts, its usefulness still resides in the meaning of the words in it.

The preceding examples have illustrated the philosophy of structured analysis and introduced one of its tools. An external context diagram can be read almost without any training by a targeted user. Drawing one correctly, however, requires excruciating care, as does the use of the other tools we are about to describe. Much of this care goes into making itself invisible; the writer's challenge is to make difficult concepts appear clear and simple, without being misleading.

It is not always obvious which of the analysis tools fits a particular problem. The appropriate ones are found by trial and error. If the analyst encounters an essential feature that just can't be represented in a data flow model, then maybe a state transition diagram will throw light on it.

Sometimes, it is impossible to guess whether that will be the case without drawing it. The tools should be thought of not only as communication vehicles with users and software developers, but also as thinking aids for the analyst. They serve not only to describe a solution, but also to find it.

Discussions of what it is that we call "data" are usually avoided in the literature. The notion of data is viewed as so primitive that it needs no definition, and usually this causes no confusion. Don Knuth describes data as "the stuff that's input or output," and De Marco as "a wordsy-numbersy thing."

In software, the idea of the output of module A being the input of module B is naturally conceived as "data flowing from A to B." This pattern resembles a flow of parts from a machine A to a machine B, and it is tempting to conclude that, at some level of abstraction, both phenomena can be modeled the same way. Being "stuff that's input or output," materials satisfy Knuth's definition.

Yet it is disquieting to think of pieces of metal on a conveyor belt and of subrouting parameters as behaving the same way. One senses that this identification of materials and data doesn't quite work, but to clarify this point, we will need to think more deeply on what we mean by "data." We shall postpone this discussion to Chapter 10 and ignore these nuances for the time being.

3.2. STRUCTURED ANALYSIS

Structured analysis is a technique developed in the past 15 years and largely successful in remedying the shortcomings of the "Victorian novels." Structured specifications express functional models of software systems in a concise, complete, and consistent format, understandable to users, amenable to formal review, and providing clear guidance to software development.

The tools of structured analysis are (1) data flow diagrams, (2) state transition diagrams, (3) entity-relationship diagrams, (4) data dictionaries, (5) structured natural language, (6) decision trees, and (7) decision tables. The reader interested only in learning these tools may skip to the next chapters. The following sections discuss the reasons for not recommending older and better known methods. They are aimed at the reader who is curious about why this particular selection was made.

3.3. OBSOLETE ALTERNATIVES TO STRUCTURED ANALYSIS

3.3.1. Boilerplates

In their first attempts to get functional specifications out of software engineering, most organizations issue standards stating which parts such documents should have and who should approve them. The result is a boilerplate table of contents analysts are expected to follow. The stated intent of the standards is (1) to serve as a checklist and ensure that all relevant issues are discussed, and (2) to provide reviewers and software designers with documents in a consistent format.

Such standard attempt to substitute form for content, which, in practice, is not feasible. They say, for instance, that a specification should have sections describing inputs, outputs, processing, control, and security. For a small module, this may be adequate, but what should the analyst do about a large system?

A textual description of the inputs to all modules, followed by a description of outputs, is likely to be unreadable. On the other, if the prescribed five-part structure is followed module by module, then the system appears as a collection of unrelated modules, because the coordination structure is invisible.

Analysts feel stifled by such standards and balk at following them. Systems analysis cannot be reduced to filling blanks in a form. In each problem, some entries in the boilerplate are irrelevant, and coming up with a paragraph about them is pointless. Conversely, the boilerplate will always be missing one or two key points. The analyst needs a kit of tools, but also the freedom to pick and choose between them, like any other professional.

When the boilerplate approach is used, compliance with the standards becomes the only evaluation criterion. Functional specifications are reviewed not for their fitness as a road map for software development, but for the presence of all the numbered paragraphs required.

3.3.2. Flow Charts

From their name, flow charts are easily confused with data flow diagrams. The differences are highlighted in Figure 3.4. In the flow chart, the arrow from A to B indicates that A and B are carried out in succession by a single processor. B is initiated once A is completed. In the data flow diagram, the arrow indicates that x is both an output of A and an input to B. A and B are two simultaneously active processors, with A passing a stream of x values to B.

The most damning restriction of flow charts is their inability to represent concurrent activities. The target computer on which software is to run may have a single processor, but the user system being modeled certainly does not. Furthermore, except for personal computers, operating systems make even computers behave like a network of concurrent processors.

Flow charts are the most ancient graphic method for describing software and are still taught to students even though they have been obsolete for two decades. They do show sequencing, conditional branching, and iteration, but even that can be shown more effectively in other ways.

Flow charts almost never play any role in the specification or design of software systems. Conservative software organizations often make flow charts a formal documentation requirement, which is met by drawing them *after* the software has been built.

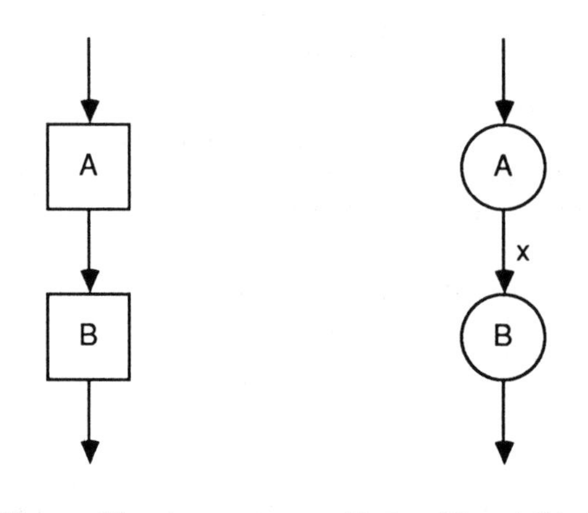

Figure 3.4.

3.4. VARIANTS OF STRUCTURED ANALYSIS

3.4.1. Petri Nets

As system modeling tools, the Petri nets described by Reisig [Rei83] are consistent with the principles of structured analysis in the sense that data flow and state transition models can both be expressed as special cases of Petri nets. However, this ability to describe many types of models in a single formalism is more useful for software design than for communication with users.

In Petri nets, tokens travel from place to place as in Figure 3.5. Petri nets with at most one token allowed in one place can model discrete state transitions. Nets with multiple tokens per place can model data and material flow systems. This theory is attractive for its many useful mathematical results.

Unity of form is, however, not desirable in an analysis situation with users. There, the more distinctive the graphic conventions are, the more effectively can confusion be avoided. The underlying unity, the ability to show data flow and state transition systems as two facets of the same thing, is irrelevant to users.

Furthermore, graphic representations of Petri nets can only be produced for small systems for the purpose of illustration. One of the great advantages of structured analysis is that its graphic tools are usable on large systems through hierarchical modeling.

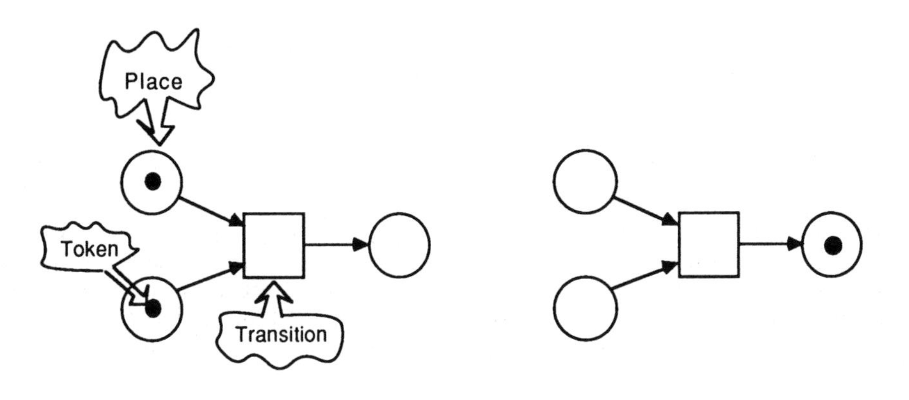

Figure 3.5.

3.4.2. Gane and Sarson's SADT

The advantages of structured analysis make it an easy choice over older methods. It is much more difficult when there is no overwhelming advantage one way or the other. De Marco, Gane and Sarson, and Weinberg are authors who have proposed different graphic conventions for structured analysis, and indecisive arguments could be made in favor of each.

3.5. CONCLUSION

When it is properly introduced, users take easily to structured analysis. However, this applies mostly to one of the tools of structured analysis: the data flow diagrams, which represent the modules of a system as bubbles connected by labeled input and output data flows. The other tools of structured analysis, namely data dictionaries, structured natural language, decision trees, and decision tables are also needed, but require more efforts from users and do not elicit as strong a response.

Structured analysis is not as well accepted yet on the software development side. The reasons why are open to debate, but do not seem related to technology. By bringing the software development process close to the level of organization of civil engineering, structured analysis makes programmers accountable to a level they are not used to. The diffusion of the technique is paced by the maturation of the software industry.

If one criticism of De Marco's exposition of structured analysis is to be made, it is that the clarity of his explanations make it look easy. Indeed, structured specifications may be easy for users to read, but they are not for analysts to write.

De Marco's tool kit is missing a few items that are needed in manufacturing. In particular, his book contains no treatment of state transition diagrams, which are indispensable to describe systems monitoring — for example, the availability of a machine (see Chapter 5).

McMenamin and Palmer's *Essential Systems Analysis* [McM84] makes several additions to De Marco's theory that are used freely in this book. To them, a requirement is truly functional if it would still exist for users with perfectly reliable, infinitely fast computers with unlimited memory. Requirements failing this test represent not what users want done, but someone's perception of the capabilities of current computer hardware. Likewise, their event and object partitioning approach, and their distinction between fundamental and custodial activities is illuminating in a manufacturing context (see Chapter 9).

Material flow modeling resembles data flow modeling, but with subtle differences. Harrington [Har84] uses a variant of Gane and Sarson's conventions, to which the older formalism developed by the American Society of Manufacturing Engineers (ASME) is preferred here. The ASME symbols for material flow modeling have the advantage of being already known to industrial engineers, not only stateside, but in Japan and Europe.

4

DATA FLOW MODELING

Data flow models represent a system as a tower of networks of processors exchanging messages of known structure along determined paths, without any coordination mechanism. The explosion of each processor in a high level network into a network of lower level processors is used to represent large systems graphically. A data flow model is presented top down, starting from data exchanges between the system and the factory, and ending with interactions between such small processors that further breakdown would be meaningless.

Data flow diagrams are built exclusively from symbols for (1) processors, (2) data stores, (3) entities external to the system, and (4) data flows. Elements of control, specifying when each function is activated, are omitted from a data flow model of an application. They can be shown separately, as output data flows of the control system under which the application is run.

We have already seen, in Chapter 3, an example of a data flow diagram and noticed how much information it can convey to a manufacturing specialist with no training in systems analysis. Now we need to see how to draw one.

What makes such a diagram informative is not abundance of information but selection. Data flow models represent a system as networks of processors exchanging messages of known structure along determined paths, without any coordination mechanism.

The choice of emphasizing this aspect of a system at the expense of others is the main difference between structured analysis and earlier methods. If structured analysis is successful, it is because data flow models most closely match users' perceptions of their data manipulations, whether manual or automatic.

The main tool for the expression of data flow models is the data flow diagram. Constraining though the rules for drawing them may seem, they are not more so than those used for figures in Euclidean geometry. The analogy goes further: data flow diagrams speak to the geometric intuition of the reader. To describe a system only in words is like describing a spiral staircase without moving one's hands.

4.1. HIERARCHICAL EXPLOSION OF PROCESSORS

The external context diagram appears first in a presentation of a finished specification to users. This does not mean that it is the first to be generated in the course of analysis. It is only after thrashing through numerous revisions that a model can be presented in a logical, top-down fashion.

The production of all the outputs shown in an external context diagram usually appears as a problem to be solved by "divide and conquer." The analyst breaks down the tasks of the system into meaningful smaller pieces and specifies the data exchanges between these pieces.

Figure 4.1 illustrates this process. The key consistency requirement is that the incoming and outgoing data flows of a bubble should match those of the diagram into which it is exploded. The revision of a diagram at one level frequently has consequences at higher and lower levels, at which revisions must also occur to maintain consistency.

4.2. DE MARCO'S CONVENTIONS

Before discussing data flow diagrams any further, let us look in Figure 4.2 at De Marco's definitions for the four components of data flow diagrams, three of which we have already encountered in Chapter 3. For a complete treatment, the reader can read De Marco [DeM78].

The data flow diagram symbol for files, or data stores may stand for a file cabinet with cardboard folders, a magnetic tape, a disk file, or a database. The symbol is meant to deliberately ignore such physical considerations and focus on a single aspect: there is a place where a certain type of information is kept. Anyone who would try to "improve" on the convention by introducing different symbols for each type of data store would in fact be missing the point.

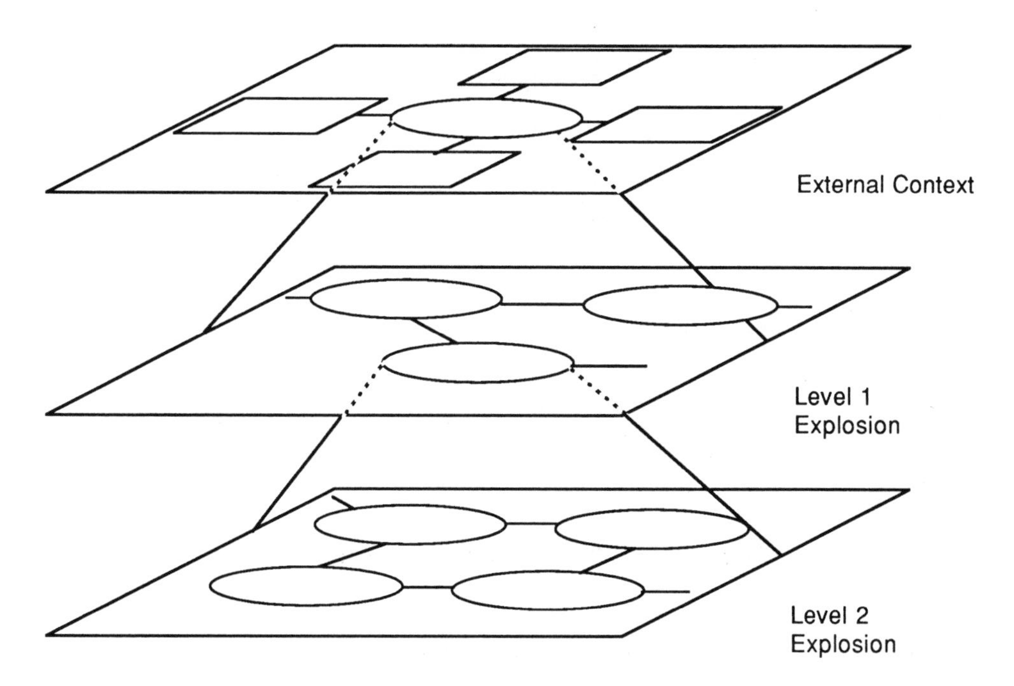

Figure 4.1. Hierarchical Explosion of Data Flow Diagrams

In a factory, the data sources and sinks are people, machines, or preexisting software systems. As seen in the external context diagram, those boxes point out who the system takes direction from or produces something for. The objects thus identified have in common that they are outside the system but interact with it.

A data flow going from one element of a diagram to another is similar to a road on a map, with the arrowheads indicating whether traffic goes one or two ways. The data flow symbol says that there is data flowing from one place to another. The label attached to it describes that data, and the usefulness of the symbol without a label is almost nil. Space limitations and overall clarity constraints cause labels to be terse, but complete definitions of the data flows are found in the data dictionary, described in Chapter 7.

A "bubble" in a data flow diagram can model a person, a department, a machine, or a software module. Objects modeled as bubbles, or processors, transform data received along input data flows into other data they ship on output data flows. The processors are internal to the system being modeled.

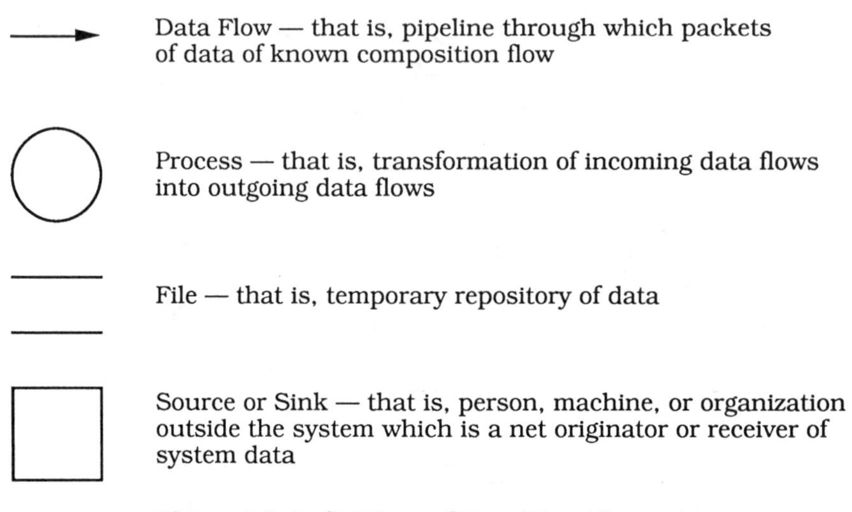

Data Flow — that is, pipeline through which packets of data of known composition flow
Process — that is, transformation of incoming data flows into outgoing data flows
File — that is, temporary repository of data
Source or Sink — that is, person, machine, or organization outside the system which is a net originator or receiver of system data

Figure 4.2. Definitions of Data Flow Elements

4.3. WHEN IS A DATA FLOW DIAGRAM USEFUL?

4.3.1. External Context Specification

The example in the introduction shows the use of a data flow diagram to model a whole CIM system. External context diagrams can also be used to describe one of its subsystems, with other subsystems taking on the role of sources or sinks.

4.3.2. Reports

Descriptions of reporting software frequently confuse the reports produced from a database with the software used to generate them, as in the following example:

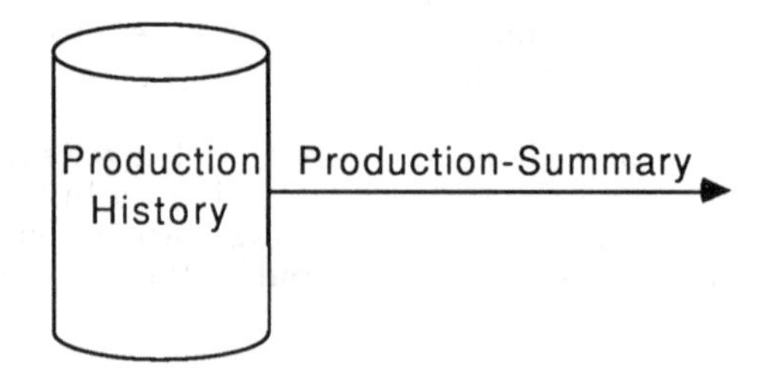

Figure 4.3. Confusion of Software with Its Output

The first problem with this approach is one of logic: a database, which drum icons frequently represent, is passive. It will not volunteer a report any more than a file cabinet will pull out a folder by itself. To get a production summary, an agent is needed to retrieve the relevant production results, summarize them, format the results, and put them out. Let us now look at a data flow diagram showing it.

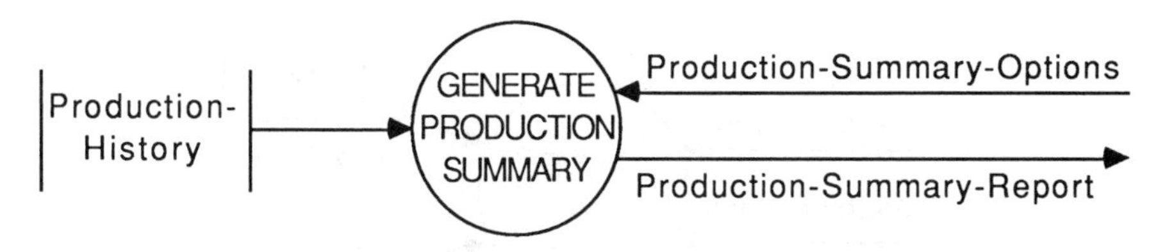

Figure 4.4. Data Flow Diagram of a Report Generator

The GENERATE-PRODUCTION-SUMMARY bubble refers to a piece of software which may represent many engineer-months of development. If it exists, it must be bought; if it doesn't, it must be built. In either case, it is a system component that cannot be ignored.

The bubble is also the target of options a user may enter to affect the content and the format of the report. Only the existence of these options is shown in the diagram, but the "Production-Summary-Options" label points to a data dictionary definition of all the possible choices. The amount of customizing a user can do to a report frequently determines how useful it is.

4.3.3. Diagnostic Systems

The following diagram shows a simple case of a diagnostic system represented by a data flow diagram. MONITOR PARAMETER receives Measurements, reads the applicable Limits from a file, and issues Alarms whenever the Measurements are not within Limits.

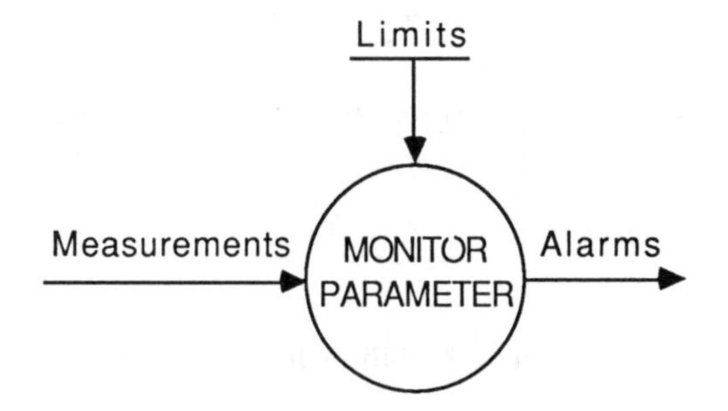

Figure 4.5. Diagnostic System

4.3.4. Transactions

The MOVE IN bubble in the following diagram coordinates two state changes associated with starting processing of a batch in a machine. MOVE IN reads the Work-In-Process-Status and the Machine-Status, and issues Error

Messages if, for example, the batch is on hold or the machine is down. Otherwise, it updates both files to record that the batch is in process and the machine in use.

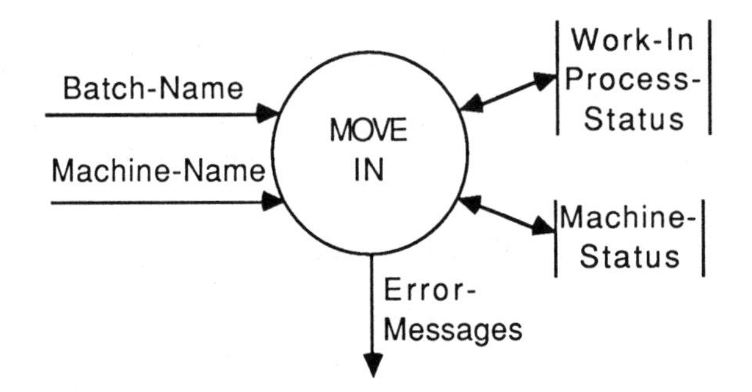

Figure 4.6. A Parts Tracking Transaction

4.3.5. Administration

VALIDATE NAME, in Figure 4.7, is a type of module that is included in a system whenever there is a possibility of input errors. It issues an Error Message when no match is found for the input name in the table of Valid Names. If a match is found, the Name is passed on along the Validated-Name data flow.

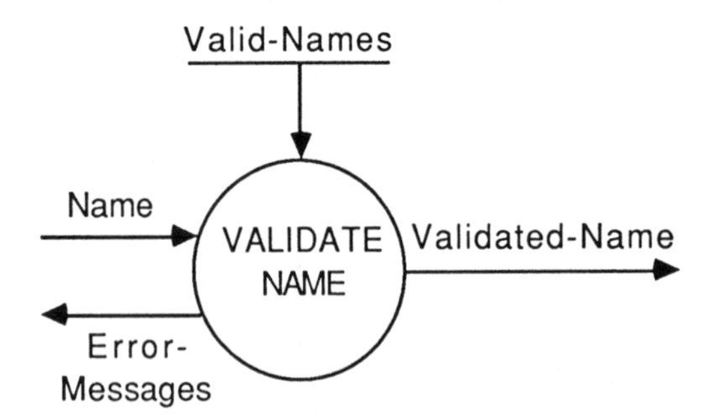

Figure 4.7. Validation of a Name

In the early stages of model building, administrative activities clutter diagrams and obscure the essence of the system [McM84]. It is preferable to assume inputs to be mistake-free until all greater issues are resolved, and then add back the administrative activities to the model.

4.4. DATA FLOW VERSUS CONTROL FLOW

4.4.1. Why the Distinction Matters

Most analysts have, at some point in their careers, worked as application programmers and acquired from that experience a view of problem solving that is biased by the tools they had to use, and not shared by users with other backgrounds, particularly manufacturing.

All commonly used application programming languages, including FORTRAN, COBOL, PL/1, and PASCAL, constrain solutions to sequences of operations performed one at a time. Depending on inputs, the sequence may vary, but every run of a program written in one of those languages produces a single thread of execution.

A machining center operates in a similar fashion: it uses one tool at a time drawn from a set of attachments. Processing a part through it takes on the form of a sequence similar to the execution of program steps by a computer. On the other hand, the factory as a whole usually has many machines in concurrent operation, each of which has its own thread of execution.

The programming of database management systems, operating systems, or distributed systems requires an understanding of concurrency, but specialists from those fields rarely become analysts. New languages, such as ADA and MODULA-2 can model the concurrent execution of operations, but their use is not widespread yet. Languages like LISP and PROLOG also permit more flexible programming, but they are currently not commonly used in commercial application development.

Programmers are conditioned to dedicate their attention to "control flow" — that is, the order in which the steps of a program are executed. Data flow modeling relegates control flow to a secondary consideration, on the following grounds:

1. There is no point in sequencing operations before knowing what they are.
2. Operations are characterized by their inputs and outputs.
3. Data transfers between operations must occur regardless of the control flow pattern.

Many engineers describe the details of how something is done to audiences who only care about the final result, and this tendency may be traced back to academic training. The answers to textbook exercises are already known, and the sole purpose of working them out is to force the minds of the students to go through the process of finding the solutions. That is what they must turn in and are graded upon. Later, in professional life, they do not always realize that the process is taken for granted and only the answer matters.

4.4.2. Telling Control Flow from Data Flow

The fact that y is an input of B that is produced by A is a property of the data flow model that constrains operation sequencing but does not determine it completely. The analyst should not try to force a sequential structure where it is extraneous. To ensure that data flow diagrams are free of control flow information, we must learn to recognize it when it is present in a diagram. Figures 4.8 and 4.9 show common examples.

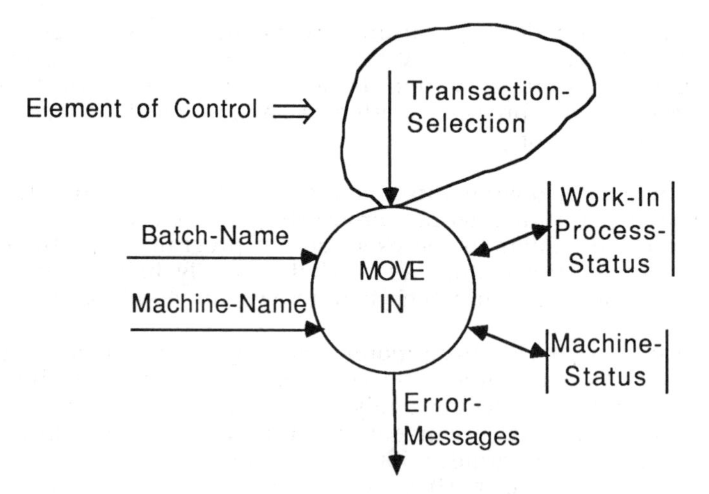

Figure 4.8. Start and End Signals

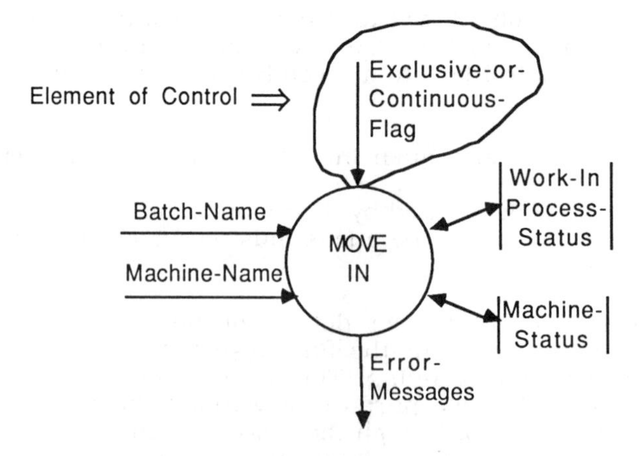

Figure 4.9. Switches and Flags

All the pieces of all input data flows somehow find their ways into output data flows. Any deviation from this principle indicates inputs that can and should be removed from the model. Elements of control are a case in point:

they are consumed by the bubble they point to. In the preceding examples, what flows along Batch-Name and Machine-Name goes into updates to the Work-In-Process and Machine-Status, or into Error-Messages. On the other hand, Transaction-Selection and Exclusive-or-Continuous-Flag disappear inside of MOVE-IN.

4.4.3. Relativity of the Meaning of Control Flow — A Menu System

Depending on the context, the same flow may be an element of control or an element of data. In a menu-driven system, the menu itself is a means of sequencing actions and as such is an element of control. It is omitted from a data flow model of the system as in Chapter 3.

There are many conceivable ways of telling the system which function to activate. Whichever method is chosen has no bearing on what the functions are. However, once functions have been specified, the menu system itself emerges as software that must be specified. In a simple case, its external context diagram could be as in Figure 4.10.

A menu is displayed to the user. Somehow, by clicking a mouse, typing a number, or pressing a function key, the user selects an entry from the menu. The menu system then either displays a submenu or issues a command to a transaction center telling it what program to execute.

This example shows how what appears as a control structure at one level lends itself to data flow modeling at another. Within the factory, the CIM system is a control structure. Within the CIM system, the menu is a control structure, and within the menu, the operating system of the host computer takes on that role.

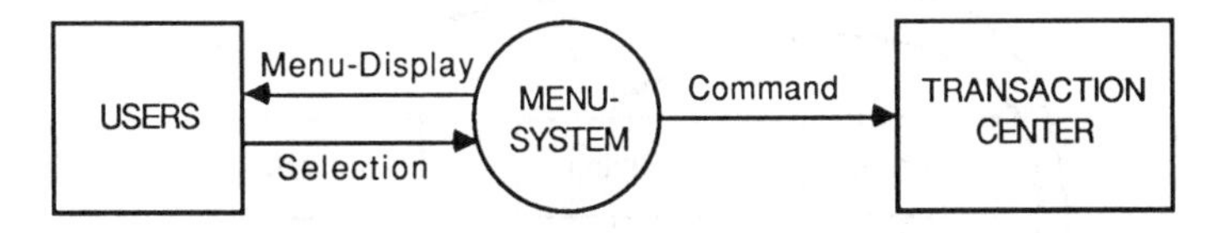

Figure 4.10. Simple Menu Data Flows

Ward and Mellor [War85] choose to represent the control structure of a system on the same diagram as its data flows, using dotted bubbles and arrows to avoid confusion, as in Figure 4.11.

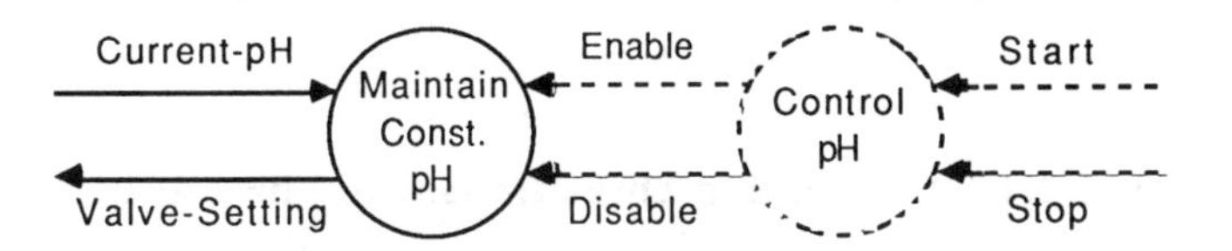

Figure 4.11. Ward and Mellor's Convention in Chemical Process Control

In some circumstances, there may be no other option. In general, this method is contrary to the principle of showing one thing per picture: two simple diagrams convey two simple messages more effectively than one busy diagram.

If the control structure is represented by the data flows of the control system, separately from the application functions, then the control system and the application can be updated independently. In other words, changes in the methods by which an application is activated can be specified without touching the description of the function itself. When it is possible, this decoupling is clearly desirable.

4.5. DIAGRAM AESTHETICS AND CROSSED ARROWS

The clarity of diagrams is jeopardized when arrows are allowed to cross, because the reader then has to disentangle them. De Marco goes so far as to make the absence of crossing arrows a requirement. However, this "requirement" is not intrinsic to data flow modeling, but to the representation of data flow models by means of two-dimensional diagrams.

In practice, the avoidance of crossed arrows is a draftsman's nightmare. To see why, the reader is invited to take a look at the problem illustrated in Figure 4.12, of connecting water, gas and electricity supply centers to three houses without crossing lines. What at first looks like a simple puzzle is in fact impossible, and proven so in topology. The interested reader will find a proof in Appendix A.

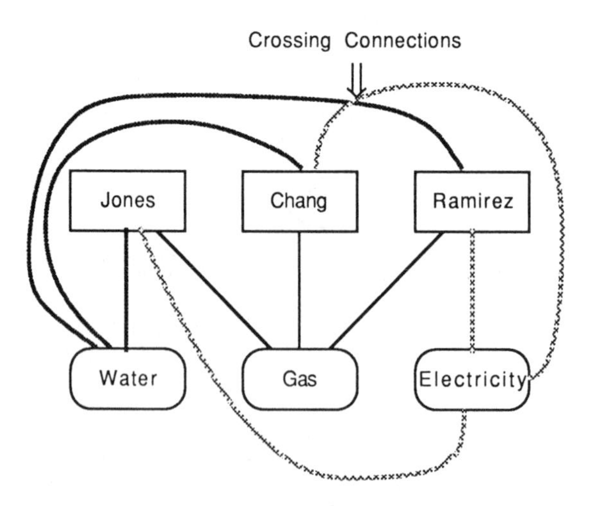

Figure 4.12. Connecting Three Houses to Three Utilities

An analyst can spend hours in vain bending data flows around obstacles. Figure 4.13 shows the type of solution that is both most meaningful and quickly implemented. If files A and B can be merged into an aggregate AB with some degree of cohesion, we have solution b. The internal structure of AB can then be shown in an entity relationship diagram, as explained further in Chapter 6. Another solution is available if the two processors X and Y can be viewed as a single processor XY, as in c. The distinction between X and Y is then made in the explosion of XY — that is, one level lower in the hierarchy of diagrams.

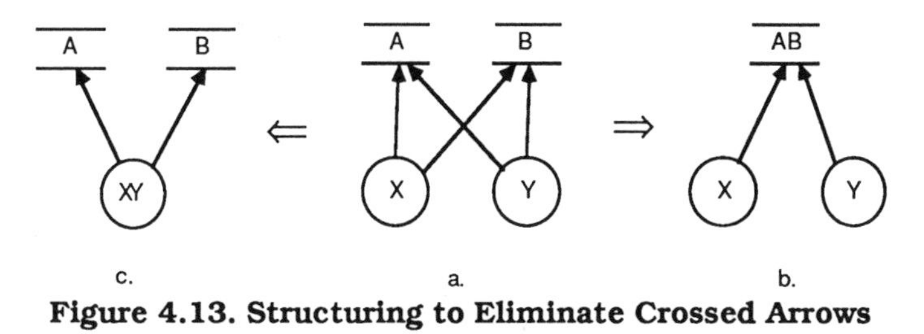

Figure 4.13. Structuring to Eliminate Crossed Arrows

Appendix A — Three Houses and Three Utilities

Let us consider a network of nodes with noncrossing arcs in the plane, and examine the number of regions it divides the plane into. In the example of three houses and three utilities, we have six nodes and nine arcs. Figure 4.14 shows how a planar network with this number of nodes and arcs divides the plane into *five* connected regions, one of which includes the outside of the network.

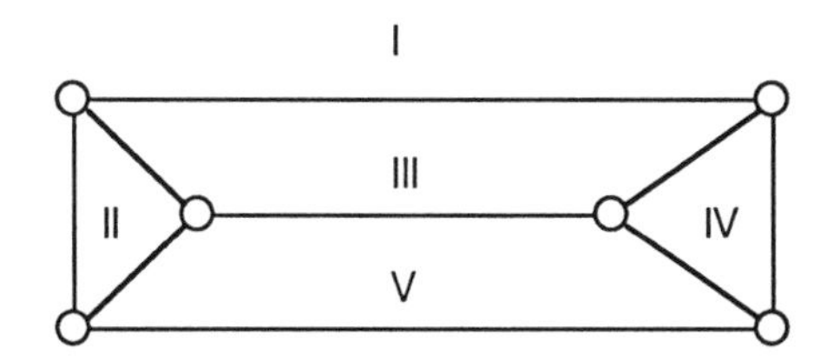

Figure 4.14. Regions Defined by a Network

The effect of deleting arcs one by one is shown in Figure 14.15. Whenever an arc is part of a *loop*, the network itself remains connected when it is deleted, but the number of regions it divides the plane into drops by one. In Figure 4.15.d., we see that, after four arcs have been thus deleted, there is only one connected region left, and deleting one more arc would split the network into unconnected subnetworks.

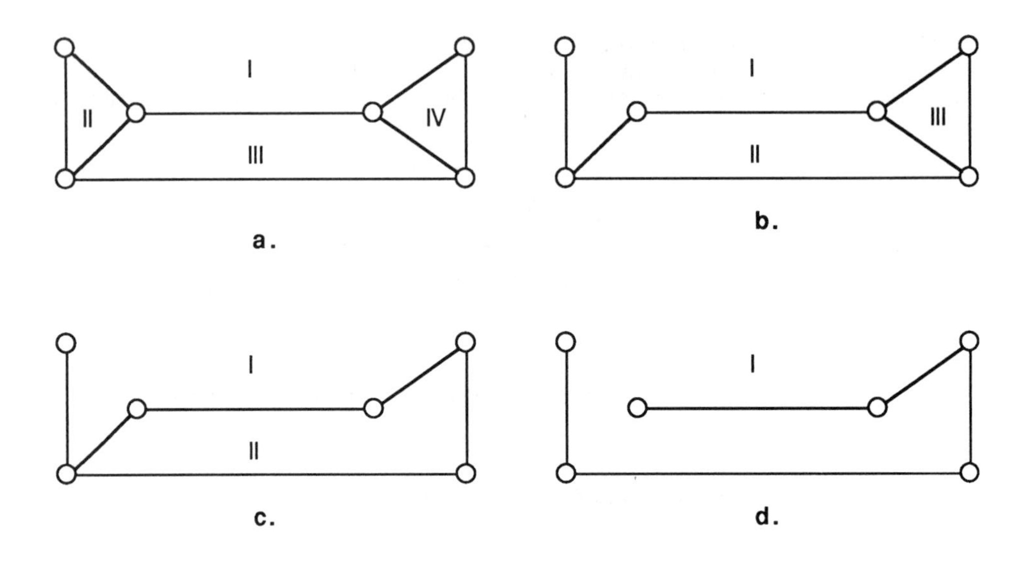

Figure 4.15. Effect of Deleting Arcs on Connectedness

Let us define the "connectedness number" c of a network to be the maximum number of arcs we can delete without splitting the network into two unconnected subnetworks. Since the arcs that can be deleted without splitting the network are those which are part of loops, the number r of regions into which the network divides the plane is $1 + c$.

For a network of 6 nodes and 9 arcs, $c \leq 4$ because 6 nodes cannot be connected by less than 5 arcs. Conversely, with loops disallowed, 5 arcs cannot help but connect the 6 nodes, and $c = 4$. Therefore, $r = 5$.

Let us assume that it is possible to connect three houses L, M, and N to three utilities A, B, and C without crossing paths. Then L, M, N, A, B and C are the 6 nodes of a planar network with 9 arcs. If we take only two utilities, Figure 4.16 shows one way to connect them to the three houses. An argument similar to the preceding shows that, for a planar network of 5 nodes and 6 arcs, $c=2$ and $r=3$, regardless of how the nodes and arcs are laid out.

Furthermore, there are no arcs connecting two houses together or two utilities together. Therefore, none of the loops can involve only three nodes. They must be "quadrangles" of the form House—Utility—House—Utility, with the third House outside the connected region enclosed by the loop.

Within the complete network, the third utility C cannot be inside one of the loops, because the arc linking to the house outside the loop would have to cross one of the boundary arcs. By symmetry, none of the utilities can be inside the loops defined by the other two.

Our full {L, M, N, A, B, C} network contains three such subnetworks, namely {L, M, N, A, B}, {L, M, N, A, C}, and {L, M, N, B, C}. We have just seen that the bounded regions in these subnetworks, equivalent to **II** and **III** in Figure 4.16, cannot be included in one another. They cannot intersect either without their boundary arcs crossing. The full network therefore divides the plane into at least *6* disjoint, bounded regions.

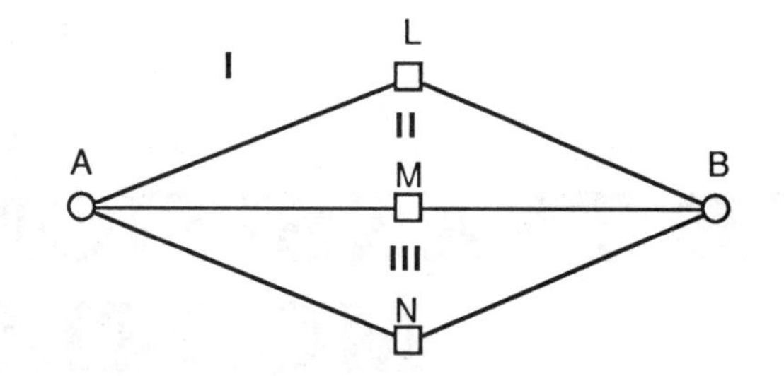

Figure 4.16. Regions of a Network of 3 Houses and 2 Utilities

Since, for the full network, $r = 5$, this is absurd, and therefore it is impossible to connect the three houses to the three utilities without crossing paths.

5

STATE TRANSITION MODELS IN MANUFACTURING

State transition models show the effect of factory events on individual components such as machines or processes. All states of an object may not require the same data. For clarity, states should be labeled with adjectives, and events with action sentences. Although some objects have a continuous state space, they can frequently be reduced to discrete state transition models by examining what the model is to be used for.

Events causing state transitions can be (1) of external origin, (2) due to the passing of time, (3) of internal origin, or (4) the product of history. In models with only internal and external events, the transitions an object may undergo depend on its present state but not on its past. It is often by error that states are sources or sinks in the state transition network. When legitimate, the presence of sources or sinks make the history of an object nonstationary and affect the analysis of its history.

5.1. OBJECT MODELING

5.1.1. What an Object Is and Isn't

The operation of a factory involves constant changes in the state of its inventories, machines, processes, and people. The simultaneous effects of those changes on several of these objects are represented in data flow models:

when processing starts in a machine, parts leave a queue, the machine becomes busy, and an operator may become available to do another task, all at the same time.

But if we switch to the point of view of the operator or the machine, we also need to model how starts, completions, shift changes, failures, and other events occurring in the factory affect it, regardless of the effect on other objects. It is clearly a functional requirement that each object in the factory be modeled in such a way that the changes affecting it are mirrored in changes to its model.

Carried away by enthusiasm, some analysts tell users how objects "can be anything." They are then surprised to find that this conjures up not the intended image of flexibility, but one of discomfort. It is in fact not true that everything can be modeled as an object, but, even if it were, the vision of an unrestricted universe of possibilities would be frightening. The implication would be that all the analysis work has been left undone.

The word "object" was used loosely in the preceding paragraphs. Replacing it with "thing," "component," or "entity" would have done no good because the usage of those words would have been equally loose. Rather than a rigorous technical definition, what the analyst needs is to recognize objects when they come up, and they do in the form of recurring noun phrases in user statements.

If engineers, operators, and production supervisors all refer to "jigs" when describing their work, then there is an object called "jig" that the analyst needs to model. As McMenamin and Palmer [McM84] point out, there are objects one can touch, such as jigs, and others that are ideas, such as fabrication processes. They are objects of the analyst's attention because they matter enough on the shop floor to have received a name.

For the analyst, the state of an object is a reduction of all its properties to a few relevant ones. Of all one could possibly want to know about a spare part, the analyst may retain only the number of units in stock. Likewise, the only item of interest about a switch may be whether it is on. About a machine, it may be its temperature.

An object may be perceived differently by various users, and each perspective requires its own description, while all must be compatible. The same machine may be viewed only as "up" or "down" by maintenance, as "processing batch 125" by production control (which requires it to be "up"), and as being at 1125°C by an engineer.

5.1.2. State Descriptions

All states of an object may not require the same data. In the preceding example, the name of a batch is considered part of the state of the machine processing it, but such a name is obviously not required when the machine is empty, so that, in general, the description of an object's state is not a vector of variables but a list that changes with the state.

When it is possible to label the states of an object, clarity is best served by adjectives or the progressive form of verbs. "Moving," for example, would be an appropriate description of the state of a vehicle, because "the vehicle is moving" is a sentence that sounds natural.

On the other hand, the reader of a diagram with the same state labeled "Move" would have to say "The vehicle is in the 'move' state," which is awkward. There are exceptions where the use of an active verb to describe a state is established usage: a traffic light is referred to a being in the "stop" or "go" state.

The state space of an object is the set of values its state can take. The state space of an on/off switch has just two elements (Figure 5.1). A queue, whose state is the number of units waiting, has an infinite number of states which can be indexed — that is, a countable state space (Figure 5.2).

The state space in both cases is discrete, and thus changes are modeled as jumps from one state to another. The switch goes from "On" to "Off" without spending any time in between. The state of a queue with n units goes to n+1 when a new unit arrives; there is no time at which it is n+0.5.

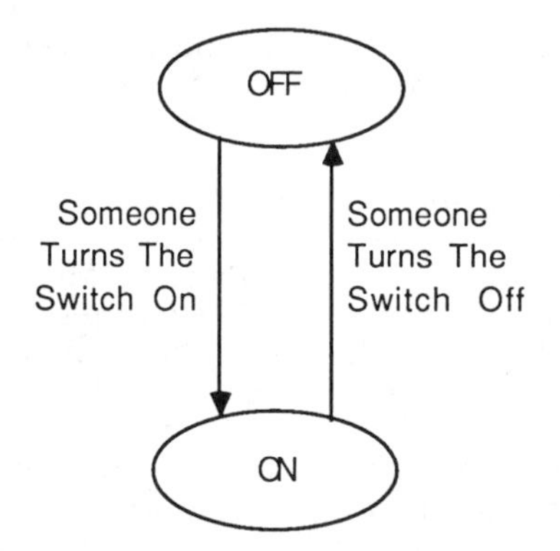

Figure 5.1. Finite State Space: A Switch

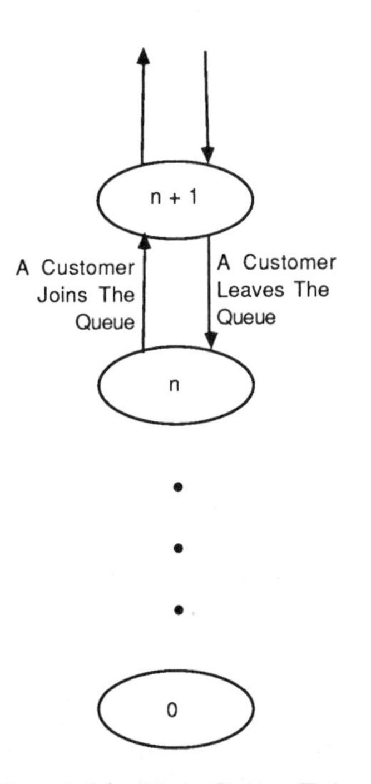

Figure 5.2. A Countable State Space Example: A Queue

The reality thus modeled is that transitions between states last a negligible time with respect to the times between transitions. If we characterize both the state space of an object and all the events causing it to change, we have a complete model of the object.

Figure 5.1 also shows all there is to the graphic conventions used in this book for state transition diagrams. Oval bubbles are used for visual differentiation from data flow diagrams and represent states. Arrows are events with an origin state and a destination state.

If the state of a furnace is its temperature, then its state space is a continuum and changes cannot be modeled as in the preceding. On its way from temperature T_1 to temperature T_2, the furnace does go through every intermediate value. Modeling this furnace theoretically requires knowing at every instant what its temperature is. In practice, it would mean sampling it at a high frequency and is an activity which is much more typical of a programmable controller than a factory host system. How can we simplify this model down to an appropriate level?

Figure 5.3 shows how the furnace can be reduced to a finite state model in a meaningful fashion: turning the heater on instantaneously changes the furnace from "cold" to "heating up." When it reaches the target temperature T_2, its state changes again from "heating up" to "hot." A complete diagram would have to include more states, such as "cooling down."

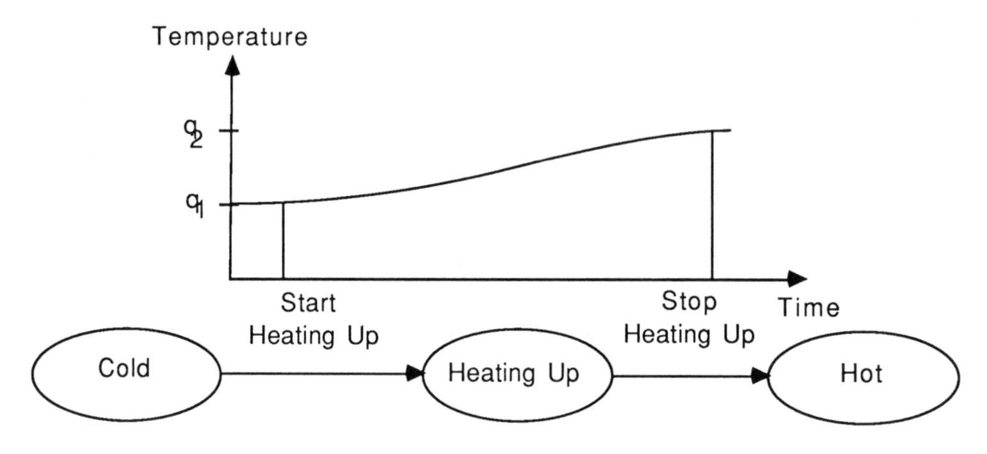

Figure 5.3. Reduction to a Discrete State Space

5.1.3. Events

We call an *event* anything that causes a change in an object with a discrete state space. In the model, these events are instantaneous, and therefore the reality they represent must be of negligible duration. On a shop floor, five hours to heat up a furnace is not a negligible duration, the split second it takes to launch the heating cycle is.

In another context, McMenamin and Palmer [McM84] classified events as "external" if resulting from the actions of outside agents, or "temporal" if caused by the passing of time. These concepts are relevant to state transition modeling, but they must be supplemented with events of internal origin and events induced by history.

External Events

"Operator presses start button" is an event changing the state of a machine from the outside. The operator is not part of the machine and therefore is an external agent. "Batch joins queue" is also an external event, because, until it occurs, the batch is not part of the queue.

On the other hand, "Batch leaves queue" is an event that originates within the queue. McMenamin and Palmer propose to describe external events with sentences of the form:

<external agent> <action verb> <direct object>.

"Object" is here used in the grammatical sense, and is not necessarily a reference to the object being modeled. In "Operator presses start button," the direct object of the verb is "start button," but the object being modeled is a machine.

Temporal Events

"Preventive maintenance becomes due" is an event that happens because a scheduled time is reached. It changes the state of the affected machine from "fit for use" to "preventive maintenance due." Events caused by the passing of time are generated by all scheduling activities. A general format for temporal event names could be:

"It is time to perform" <Action>.

Unlike other types of events, temporal events can be anticipated. Schedules are lists of pending temporal events which must be maintained if anyone is to know that such an event happened. By looking at schedules regularly, it is possible to know when future events will happen and react accordingly.

Internal Events

As an event, a machine failure is neither temporal nor external. After an investigation, an engineer may find that it was caused by vibrations in another machine, and thus eventually assign it an external cause. But if we are modeling the operation of a shop floor, we have to consider what the failure is perceived as when it happens, and that is not as an event caused by an obvious external agent. In this sense events such as "the fuse blows," or "sensor detects a phosphine leak," do not fit in the same pattern as "operator completes loading" or "customer cancels order."

Events Induced by History

If a machine must be brought down because the number of loads processed through it since the last calibration reaches 50, we have an event happening as a planned result of the object's history. Figures 5.4 and 5.5 show how it can be made to fit in the pattern of an external event by defining one state for each value of the cumulative load count. Formally correct, this approach is more a software developer's view than a user's.

5.1.4. State Transition Models and the Markov Property

Temporal and history-induced events have in common that their occurrence is tied to the past. When a temporal event becomes due, it is because an earlier scheduling decision. Likewise, a history-induced event happens as a result of a particular sequence of earlier events.

Models with only external and internal events, on the other hand, have the desirable property that, once the current state of an object is known, the transitions it may undergo are independent of its past. All that matters to start processing on a machine is that it be up and available. How it got to be in that state is irrelevant.

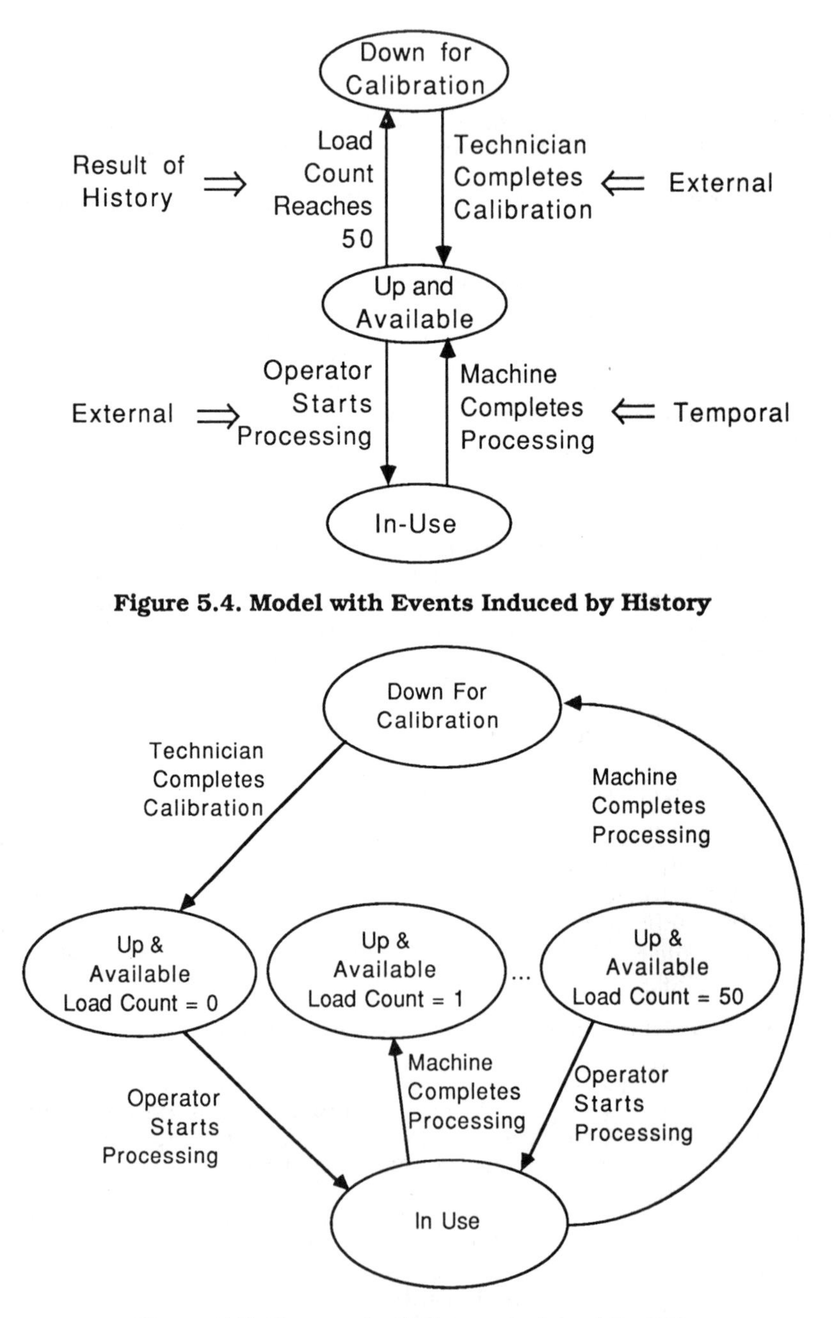

Figure 5.4. Model with Events Induced by History

Figure 5.5. Removal of History-Induced Events

This feature, that conditionally on the present, the future does not depend on the past, is called the Markov property and is the basis of many results in probability theory, some of which are relevant to the statistical analysis of object histories. In systems analysis, it affects model complexity. The preceding diagrams show the price to be paid for removing a history-induced event from a model: the number of states is increased by an order of magnitude.

The number of states and transitions relevant to an object model can easily exceed the representation capabilities of diagrams. Unfortunately, a model that cannot be drawn cannot be communicated to and reviewed by users. Proceeding bottom-up, the analyst can lump together states with enough in common to be given a meaningful name. This approach, however, leads to a usable hierarchy only if all the states merged into one do most of their transitions among themselves.

5.2. REVIEWING A STATE TRANSITION DIAGRAM

If the author of a diagram doesn't know its purpose, the reader won't either. As shown in the examples, there is no such thing as *the* model of an object, but there are as many models as there are reasons to build them.

States must always be labeled, but there are cases where events leading to a state are so obvious that labels can safely be omitted. The reason for omitting them, however, must be that they are obvious, and not that the author can't come up with a reasonable one.

Especially in documents with other types of diagrams, labels must be worded so as to remove any doubt in the reader's mind that bubbles are states and arrows events. A state is something the object is in; an event is a change. Event and state labels that fail to convey this confuse the reader.

5.2.1. Sources and Sinks

Particularly suspicious in a state transition diagram are *sources*, or states with transitions out of them but none into them, and *sinks*, or states one can get into but not out of. There are legitimate reasons for having such states in a model, but even then, their presence may restrict the analysis of historical data.

The most frequent reason for having a source or a sink in a state transition diagram is that some events were forgotten. A user told the analyst how lots go on hold, but did not say where they go from there. This error, detected by a formal review of a drawing, would become apparent much more painfully with the model embedded in software.

When first introduced on the shop floor, a new tool is in an initial state it never returns to afterwards, which is therefore a source. After the tool has been worn out, it is eventually retired, and that state is a sink. This source and this sink are not errors, but relevant states of the object (Figure 5.6).

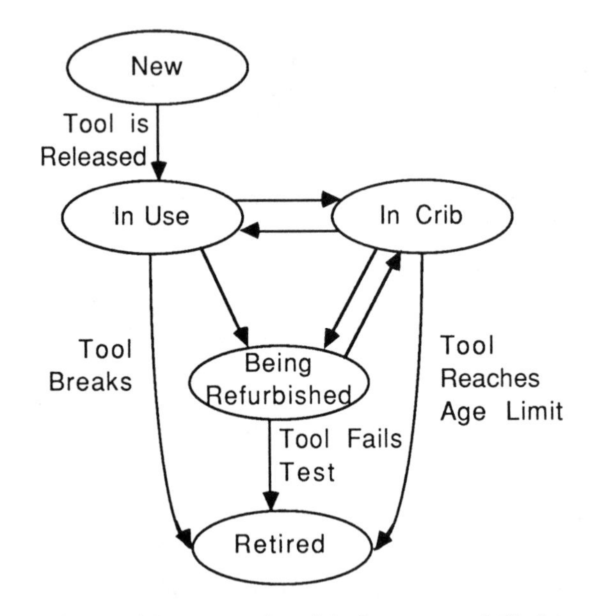

Figure 5.6. Example with Source and Sink

One frequent application of state transition models is the analysis of the use of a resource. What fraction of its time does a machine spend up and idle, in use, and in preventive maintenance? Finding the answer to this question is such a generic requirement that one would expect a single module to calculate these statistics for any object and any time interval for which its history is maintained.

Let us use it on all tools to which the model in the preceding diagram applies. Since all tools are eventually retired and never leave that state thereafter, the fraction of the tool's history spent in that state increases constantly. This renders the report useless.

Figure 5.7 shows how to solve this problem by structuring the model so that the time statistics can be calculated only on substates of "In-Operation." The higher-level model, with the source and the sink, can then be used to calculate tool life statistics from the intervals between the release and retirement timestamps.

5.2.2. Reducibility

A model with a state or group of states it can leave but not return to, it is said to be reducible. If there is such a group, then the rest of the states form another group which the object can enter but not leave. In other words, hierarchical modeling can reduce this system to two states with one a source and the other one a sink.

If a model is irreducible, the average fraction of time spent in each state is an informative statistic. If it is reducible, the focus should shift to the time it takes for objects to go from source to sink.

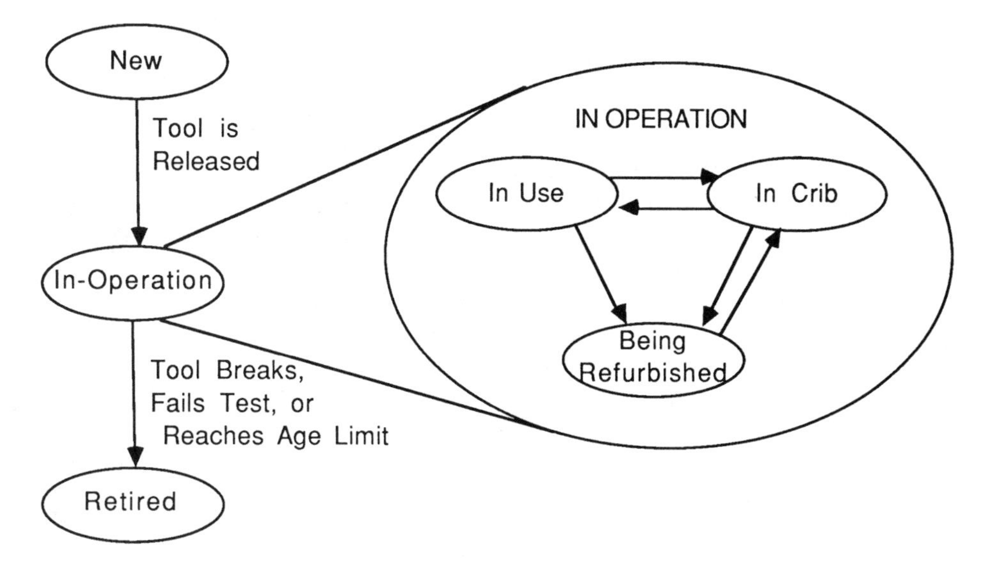

Figure 5.7. Hierarchical Model

5.3. APPLICATION EXAMPLES

5.3.1. A Vehicle

Let us assume that we want to represent a vehicle used on the shop floor. It may be a forklift driven by a person, an automatic guided vehicle, or even a conveyor belt. To keep the example simple, let us make a few assumptions. We could have a more complex, more realistic example, but it would add nothing to the discussion.

Assumptions about the vehicle:

- It travels around a fixed, convex loop.
- It stops at stations along the loop to load or unload supplies.
- It carries one batch of one type of part at a time.
- The number of parts in a transported batch may vary.

Modeling Location by Coordinates

By choosing any reference point inside the loop, we can represent the exact position of the vehicle by its azimuth from that point, as in Figure 5.8. This is not a state transition model as just discussed, and before investing in a real-time tracking station, we might ask ourselves what it is we are trying to accomplish.

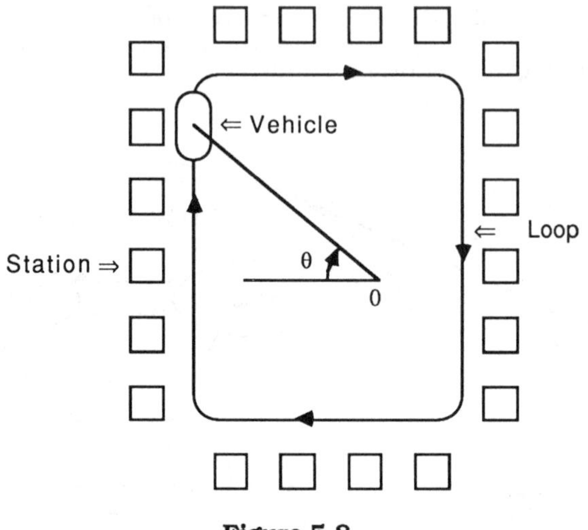

Figure 5.8.

Modeling Location by Station

Discussion with factory personnel might reveal for example that the vehicles only require attention when one of the following events occurs:

- The vehicle reaches a station full.
- The vehicle finishes unloading.
- The vehicle finishes loading and leaves a station.

Between these events, the vehicle takes care of itself, or is taken care of by its driver. The system needs to know when it is done unloading, so that it can order it to load something else. When it is fully loaded, it must be told where to go, and since its speed is known, its arrival can be scheduled. If the vehicle does not arrive close enough to the scheduled time, it means that something has happened that requires a special intervention.

If this outline corresponds to a real need of the factory, then the vehicle lends itself to state transition modeling. It is still a complex model, requiring elaborate tracking, but it is built with a clear application in mind.

Modeling Activity Types

The preceding model may still be overkill if all the users care to know is how much time the vehicle spends loading, unloading, and in transit. This would be the case, for example, if there were a proposal to replace the vehicle with one that travels faster between stations but takes just as long to load or unload. The value of this improvement depends on the fraction of the time the vehicle is moving. In this case, the model can be further simplified.

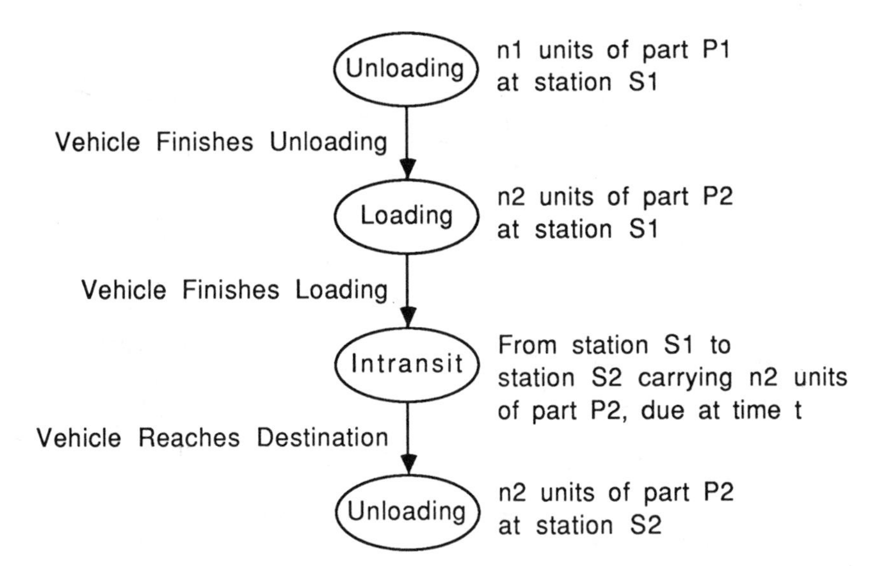

Figure 5.9. First Vehicle State Transition Diagram

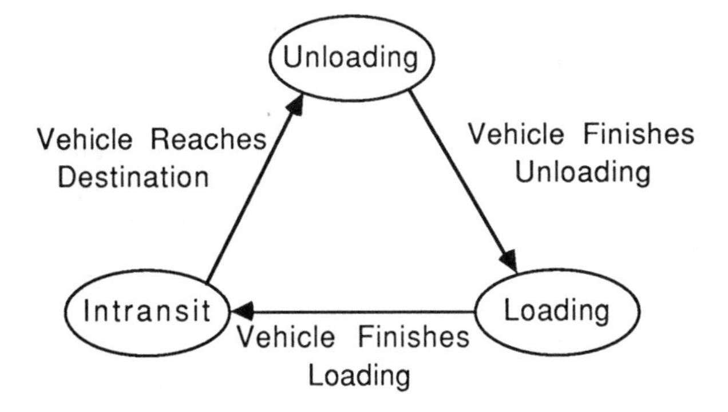

Figure 5.10. Second Vehicle State Transition Diagram

5.3.2. Engineering Changes

Figure 5.11 shows the administration of engineering changes in an assembly factory that was struggling to bring a new product to market. At the time, the product still underwent about 50 engineering changes of various magnitudes every month.

The administrative system was so clogged that it took two to four months for changes to go from initial submittal to implementation, and management hoped that computerization would speed up the process. The first user response to Figure 5.11 was that it provided, for the first time, a clear view of the approval process.

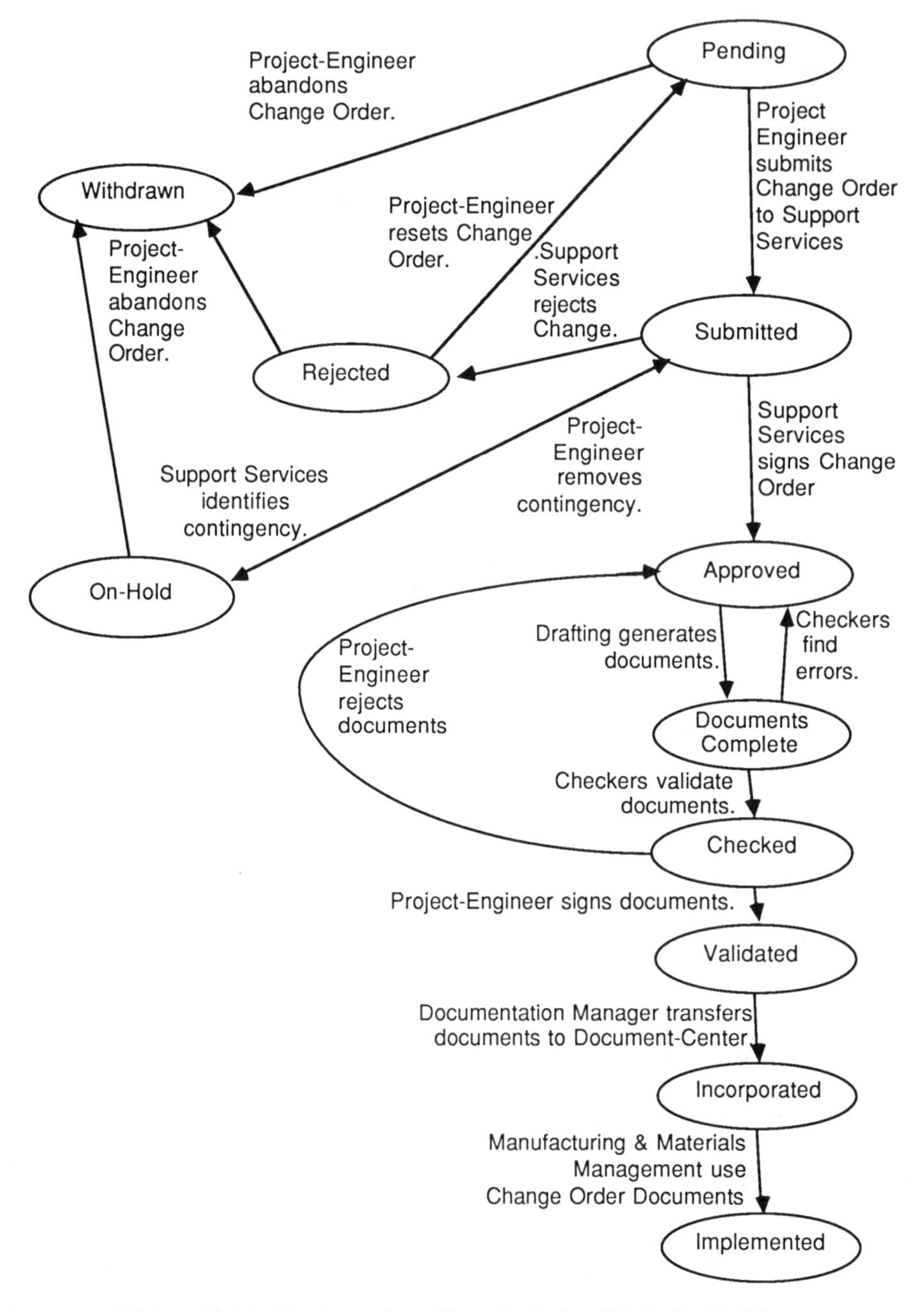

Figure 5.11. Engineering Change Order State Transitions

In one page, it provided more information than the 15-page management policy it was condensed from. The process of deriving the diagram from the policy uncovered ambiguities that had to be resolved by consulting the original authors. Those ambiguities had been overlooked because of the lack of effective review tools.

The managers of the factory were not interested in relaxing control over the release of specifications to the shop floor, even though the current level of control was largely an illusion. Since it took months to get anything approved, the need to actually build products caused the formal system to be frequently bypassed.

Tracking the approval process on a computer could help make the system work in several ways. First, it could enable anyone concerned to locate where a change was by a simple query. Second, priority lists could draw the attention of the engineers involved to the materials to review. Finally, it could report where in the approval process the documents tended to wait longest.

6

MODELING THE FACTORY'S MEMORY

The memory of the system is modeled by sets and relations, and represented graphically by entity-relationship diagrams. A CIM system must remember what processes, equipment, people, and materials are in the factory. Dedicated analysis tools serve this purpose. They are introduced here using the formalism of set theory. The goal is not to provide a place to store records but one from which they can be retrieved. The rules of set theory establish when two sets are equal, and how to build relations by subset selection on tuples.

Users enter lists of elements for finite sets such as factory equipment, decision rules for infinite sets such as potential parameter values, and archival policies to check the unbounded growth of history sets. A mapping is a relation assigning a unique element of a set to each element of another, and is viewed either as a transformation rule or a lookup table. Mappings are used to model unique attributes of factory objects, such as the purchase date of a machine. Queries on a set of relations are constructed using the operations of relational algebra.

Entity-relationship diagrams graphically show the relations between explicitly defined components of the model, while attributes in data dictionary definitions assign values from sets outside the model. Only application designers can modify these structures. Within them, access is granted to factory users selectively, to protect manufacturing specifications.

6.1. MODELING THE SYSTEM MEMORY

6.1.1 What the System Remembers

A CIM system does not receive all its inputs from the outside for every transaction. It must remember what products, routes, and operations are allowed in the factory, who works on what equipment, and where the work in process is, so that this data does not have to be repeatedly entered and references can be validated. To support engineering analysis, it must also remember the history of these structures, of parts, and of machines.

With their focus on transformations, data flow diagrams fail to highlight the structure of the system memory, and special, dedicated analysis tools have to be used for this purpose. The results are variously called

1. "Retained Data Model," by De Marco [DeM84], meaning a model of retained data, not a retained model of data.
2. "Essential Memory Model," by McMenamin and Palmer [McM84], where "essential" means needed to perform the activities the system was built for.
3. "Database Design," by C.J. Date [Dat81], a name which puts the emphasis of the art of using commercially available database management systems to implement the model.

Our discussion of system memory models touches fundamental issues that the reader of a functional specification need not be aware of but the author should be. *Objects, entities, types, aggregates* and many other terms are used in the systems analysis literature to designate sets, with definitions that are mostly circular: a set is defined as an aggregate, which in turn is defined as a set. Frequently, a relation will defined as an *association*, but the latter term will remain undefined.

To break this vicious circle, we use a two-fold approach. First we describe the operations allowed on sets like the rules of a game — that is, semantically; then we introduce the pragmatics of sets through factory examples. Doing this, we find out that the reality of a shop floor cannot be reduced to a model based on these concepts without occasionally bending some of the rules we state.

When dealing with factory users, the notations of mathematical set theory are neither necessary nor useful, but they are in the description of memory modeling concepts to analysts. They will be used as a shorthand in the following pages, in the construction of higher level tools. After serving its purpose, this mathematical scaffolding will be torn down.

6.1.2. Modeling for Retrieval

A product called "Write-Only Memory" may be the oldest joke in the semiconductor industry. Yet there is an overwhelming tendency in CIM systems to design databases for easy update, with the predictable result that queries are difficult. To buck this trend, the analyst must overcome resistance both from users and from software developers.

Most factory managers and engineers have never had a useful database at their disposal and consequently have no idea what it can do for them. "Record keeping" is perceived as a job requirement, and receives attention as such. On the other hand, when faced with a process breakdown, most engineers respond by running experiments, not by analyzing data from the history database because they have never had one to work with.

A CIM database is updated every time a modeled event occurs on the shop floor, which may be once a second. In the same factory, elaborate queries may be made only twice an hour. The updates must furthermore be fast enough not to slow down production. All these factors cause software developers to make retrieval an afterthought.

6.2. SETS AND MEMORY MODELS

6.2.1. From Sets to Databases

Foundations

It is not common for books on systems analysis to base a discussion of memory modeling on its set theoretic foundations, and the reader may wonder why it should be necessary. There are, in fact, several reasons behind this choice.

First, the subject itself is more difficult than data flow or state transition modeling, in the sense that design errors are more difficult to uncover. By building up to the concepts of memory modeling from first principles, we hope to make violations more apparent and their consequences easier to anticipate.

Second, the relational model is an overwhelming presence in the database field and inescapable for the analyst. This model is an application of set theory, from which it emerges naturally. Logicians and mathematicians labored for decades to find a concept of set that did not quickly lead to paradoxes, and we want to piggyback our efforts on theirs.

Finally, the relational model is not without shortcomings, but these tend to be overlooked. Its basic constructs are awkward for the representation of *sequences* and particularly of lists of entries with varying structures. Some of these difficulties are shown in Appendix A of Chapter 7.

Database design can be viewed as related to set theory as circuit design is to logic. Logicians develop new and esoteric ways of proving theorems, while circuit designers combine large numbers of "ANDs," "ORs," and "NOTs" into elaborate systems without needing to know more than the mechanics of these simple operations. Likewise, set theoreticians worry about the consistency of mathematics, while database designers use only the basic, naive set operations to construct pragmatic models.

Our goal is to build modeling tools for factory structures, with machines, materials, and processes, not to examine fine points of set theory. For this reason, we use loosely worded, natural language renditions of the fundamental rules of set manipulations.

The rules are given in the following statements (in italics are concepts that are explained further next):

1. Two *sets* with the same *elements* are equal.
2. Given a set, we can form a *subset* of it by selecting only those elements which have a given *property*.
3. Given two sets, we can form a new set having the original sets as elements.
4. Given a set, we can take the *union* of all sets that are elements of it.
5. We can form a set of all the subsets of another set.
6. There are *infinite* sets.
7. The image of a set by a *mapping* is again a set.
8. Given a set of nonempty pairwise *disjoint* sets, we can form a set with exactly one element drawn from each of the sets.

Engineers in a factory draw lists of machines qualified to perform various tasks. 1 gives a meaning to "the equality of two sets of machines" and gives us a criterion to establish whether it holds. The engineers who drew the lists wrote the machine names in a certain order in list A and in a different order in list B. For example, the engineers may have sorted the machines by decreasing speed for different tasks. The interpretation of A and B as sets implies that this order is irrelevant. If A and B contain the same machines, they are equal *as sets.*

Generalizing the preceding comment, we will to designate a set by the list of its elements, with the understanding that

1. The order of the elements in the list does not matter.
2. Duplicate entries have no effect.

A second issue in real situations is whether the two sets of machines currently happen to coincide, or always should. In the first case, the sets can be changed independently and cease to be equal. In the second case, they cannot. The question for the analyst is whether there are technical reasons for a machine qualified for one task to *always* be qualified for the other.

Pragmatic Neutrality

Sets are neutral with respect to meaning. It is the people who build sets who choose the elements so that they map to external realities. For example, the set Machines = {Furnace1, Furnace2, Aligner1, Spinner} can be set up so that each of its elements is the name of a machine. It is also not difficult to imagine situations in which it is useful to establish whether a name is in this set.

For our purposes, if membership in S can be tested, then it is well defined. There is no implication in the concept of a set that it should make sense to aggregate its elements. S = {Banana, Habeas Corpus, KXSMd3} is a well-defined set even though, pragmatically, there may be nothing that is referred to by "KXSMd3" and there may be no circumstance in which one might want to check whether anything is a member of S.

Database Atoms

The memory model of a system contains a large number of elements of sets that are not sets themselves. Furnace-1 may be an element of the Machines set, but the model contains nothing identified as an element of Furnace-1. We will call such elements *atoms*.

Properties

In 2, appear the concepts of subset and property. A is said to be a subset of B if and only if all elements of A are also elements of B. If we can write a logical formula involving the elements of B and other sets, then it defines a property P(x) which an element B may or may not have. 2. allows us to refer to the set A of the elements of B which satisfy this formula.

The necessity of the subset selection operation is clear. It must be possible to select from among parts in process those that are at a location, or all the machines that are up and available, or the lots of a given product on hold for engineers. One of the key factors in the usefulness of a model is the range of properties it enables users to base selections on.

To evaluate whether an element of B possesses a given property, we need more than a name: we must associate it with other quantities. We do not have all the theoretical tools to do this yet, but we will shortly.

Tuples

3 extends to more than two sets. From sets called Machines, Processes, and Employees, we can form a new set called

Facility = {Machines, Processes, Employees}

From retrieving the elements of this set, we can find out the machines in the facility, the processes running in it, and who works there, but we cannot find out, for example, which machine is used for a particular process.

To do this, we need the concept of an ordered tuple. The notation for a tuple of length n is of the form $(a_1,\ldots,a_n)$ and two tuples $(a_1,\ldots,a_n)$ and $(b_1,\ldots,b_n)$ are equal if and only if $a_1=b_1,\ldots,a_n=b_n$. Now if employee e can perform process p on machine m, this can be represented by the triple (e,m,p). This is a triple and not a set, because, as sets, {e, m, p} = {m, e, p} and the positions of the symbols would not make it possible to identify which ones are the machine, the employee, and the process.

The set of all possible pairs of elements drawn from two sets is called the product of the two sets. The product of Processes and Machines is the set Machines × Processes of all possible pairs of Machines and Processes. Since not every process is performed on every machine, Machines × Processes contains extraneous pairs.

To represent useful information, we need to apply 2 and take the subset of those pairs within Machines × Processes with the property that the machine can carry out the process. Such a subset is the fundamental building block of the factory model and is what relations are made of.

Unions and Intersections

4. allows us to build a set called the union S of the elements of a family of sets. If a factory assembles a number of different products, each has a set of components purchased from the outside. The operation defined by 4 allows us to derive the set of all components purchased by the factory from the family of product component sets.

The intersection T of the family of sets is defined using 2 as having the elements common to all the sets in the family. In the same example of product components, this operation enables us to identify components that are common to a family of products.

Subsets and Enumeration

According to 5, we can form a set of all the subsets B of a set A. A computer assembly factory builds components such as CPUs, memory boards, disk drives, and printers. There is a large number of possible sets of these components, only a small minority of which are "configurations" that can be assembled into a working computer. A possible function for a CIM system in this factory is to automatically verify the technical validity of customer orders — that is, to decide whether the set of components in an order is in the set of valid configurations.

In this and other similar circumstances, it is necessary to form sets of parts and view them as units, which together comprise a set that is included in that of all possible groups of parts.

Infinite Sets, Factory Models, and Database Size

6 decrees the existence of infinite sets. The model of a factory in the memory of a CIM system contains finite sets, built by user assignment: someone explicitly records all machine names and there is a finite number of them. On the other hand, the tolerance interval within which an engineering parameter must be held is infinite, and so is the set of all past and future lots of parts to be made within the factory.

In the case of a tolerance, the problem is to decide whether a particular measured value is in the interval, and, to do this, it suffices for the system memory to contain the interval's lower and upper bounds.

The case of a population of lots is more problematic. When a new lot is started, it must be given a unique name. With a fixed number of characters, whatever the naming convention is, the system will eventually run out of names. Furthermore, the accumulation of lot histories causes the system memory to eventually exceed any preset size limitation.

The managers of real systems have limited space and work around this impossible constraint by only retaining a rolling window of history and "archiving" the old data. Once a lot has been archived long enough, then its name can presumably be reused without causing any confusion.

Choice

In a factory, the following sets are disjoint:

- Available machines qualified for a process.
- Available fixtures that can be attached to any of the machines to perform the process.
- Available operators trained to execute it.

If all these sets are nonempty, then drawing one element from each of them and allocating it to a tool set is a necessary step in launching the process. This is described by 8 as a fundamental operation.

6.2.2. Relations

For now, we are dealing strictly with names: we can say that M_1 is in the set of Machines, but, to discuss when it was bought, we first have to define a way to assign it a purchase date. This is done next.

7 is a statement about a particularly useful type of relation called a mapping. Before discussing it, let us review the concept of relation in general. A relation on n sets $A_1,..., A_n$ is a is a subset R of the $A_1 \times ... \times A_n$ of all tuples formed with elements drawn from $A_1,..., A_n$. A relation contains only some of the tuples formed with elements taken from the *domains* $A_1,..., A_n$.

A factory model usually contains a relation which can be called "Is-Qualified-For" containing all (Machine, Process) pairs such that the operation can be performed on the machine.

Aside from relations such as Reports-To that are user-defined, there are relations such as $\leq$, =, and $\neq$ that are built in, in the sense that computers already know how to test whether a pair (x,y) is in their graph: users do not have to enter a list of all pairs of integers (x, y) such that $x \leq y$.

A relation R over domains $A_1, \ldots, A_n$ is often loosely referred to as a "table." The domain names $A_1, \ldots, A_n$ become "column headers" and the tuples in R become rows, as follows:

A1	...	An
x_1	...	x_n
...	...	...

This confusion is harmless, provided one remembers that the only relevant property of a tuple $(x_1, \ldots, x_n)$ in a relation is membership in R. The following tables represent the same relation, because set membership is affected neither by order nor duplication.

Machines	Operations
Furnace1	Diffusion
Aligner 1	Masking
Aligner 1	Masking

Machines	Operations
Aligner 1	Masking
Furnace1	Diffusion

Thus relations can be displayed in tables, but not uniquely — that is, several tables can represent the same relation. In practice, for housekeeping reasons, duplicates are deleted when relations are displayed as tables. An electronic spreadsheet is a different kind of table, in that it is a two-dimensional array and row order matters. "Build a new table from the first five rows" is a meaningful instruction to a spreadsheet program but not to a relational database manager, because the latter does not understand the concept of "the first five rows."

As another example, highlighting a difficulty, let us consider the representation of the purchase date and the supplier of a machine. This will be a relation defined on the sets of Machines, Past-Dates, and Suppliers. In the table format, it will look as follows:

Machines	Past-Dates	Suppliers
Aligner 1	6/12/87	X
...	...	...

However, to the reader, "Past-Dates" is ambiguous. It could be the date of any past event on the machine. To be clear, this label would have to be "Purchase-Date," but then the information that any past date is a potential purchase date would be lost. This problem is resolved in Chapter 7, using the concept of a mapping.

6.2.3. Mappings

Definition

The number of domains of a relation is called its *arity*. If its arity is two, it is called a binary relation. Much of mathematics is dedicated to the study of a special case of binary relation called "mapping," "functional dependency," or "many-to-one relation."

A mapping is a relation on two domains A and B such that each element of A appears exactly once on the left side of a pair in G. Another way to say it is that a mapping associates to each element of A a unique element of B. In traditional business organizations, each employee has one and only one supervisor who is also an employee. The organization chart can be modeled as a mapping called "Reports-To" with the same set of employees for its two domains.

Rule versus Lookup

Mathematics differs from systems analysis in that it deals mostly with relations and mappings in terms of *rules*. $f(x)=x^2$, as a function of x, is thought of as a rule to obtain the square of a number. The *function subprograms* of many programming languages view mappings from this same perspective.

On the other hand, there is no rule to associate a supervisor to an employee; the only way to find the answer is to look it up in the organization chart — that is, the graph of the Reports-To relation. This difference in perspective is illustrated in Figure 6.1.

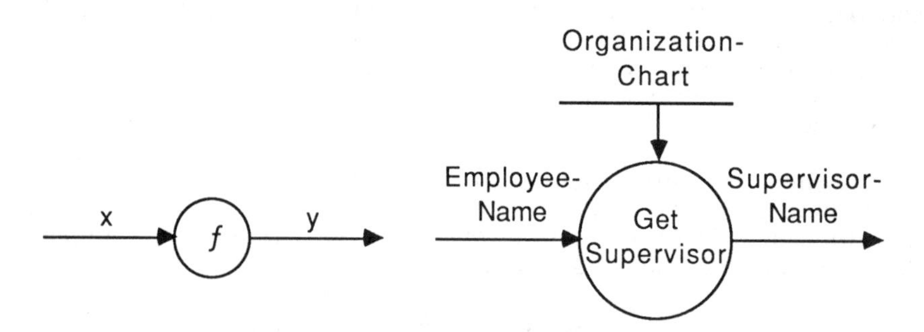

Figure 6.1. Mappings as Transformation Rules vs. Table Lookup

Many-to-Many Relationships

Complex relations can be reduced to mappings, but the resulting model may be obscure to a user. If several operations can be done on the same machine, and several machines can do the same operations, this will be naturally thought of as a *many-to-many* relationship — that is, one in which there is no restriction on the number of times the element of any domain appears in the relation. The reduction of this relation to mappings would only be confusing to users and, if it must be done, should be left to software developers.

6.2.4. Relational Algebra

The reason why the concept of a relation is so central is that whatever a software module retrieves from the system memory can be described as a relation. E. F. Codd [Cod72] defined a set of operations on relations, such that every complex query can be executed by a sequence of those operations. A treatment of relational algebra can be found in [Dat81] or other texts on database theory.

Commercial database management systems (DBMS) implement relational algebra to varying degrees, but this is not the subject of this chapter. The analyst should not be constrained by a particular DBMS, and should be able to assume full availability of the algebra, postponing compromises until an unconstrained model can serve as a reference against which to make them. A brief description of the fundamental operations is given in Appendix A.

The popularity of relational DBMSs is due to the retrieval power of relational algebra. However, the use of this power is predicated on the availability of well-defined, meaningful relations and the design of such a structure is constrained by the concepts just discussed, as we shall see in following examples.

Confusion is being created by some software companies selling "relational" DBMS for personal computers. The review of a such a product in a trade magazine contained the following statement: "Links provide the pathways for retrievals (...). However, if you decide, after the database has been designed, to retrieve data in some way not provided by an existing link, the database structure will have to be modified."

This contradicts the claim that the DBMS is relational. The only way relations are linked in the relational model is by matching data values, and getting rid of the fixed access paths required in older, cruder data models is one of its goals. The suppliers of the preceding DBMS borrow the popularity of the relational model without supplying it.

The reviewer admits further in the article that another product "sticks closer to what theorists call a relational database," with the implication that not too much attention really needs to be paid to these people's fixations. Unfortunately, the advantages that are the reason for the relational model's popularity cannot be obtained from products that don't implement it.

The communication of ideas about the relational model is jammed by the diffusion of bowdlerized versions, which puts its practical application to manufacturing and other fields in jeopardy.

6.2.5. Need for Graphic Notations

A set may be involved in two types of relations that in practice require different conventions. First, there are relations between sets the user models as individual components (or "objects" in the sense of Chapter 9) of the factory, such as Machines and Operations. Then there are mappings from one of those sets (or the graph of a relation) to ranges that are not themselves independent components of the factory. One example is assigning a purchase date to a machine.

For the reader to see at a glance all the relationships between Parts, Machines, People, and Processes that can be set up within the system, we need a graphic convention to represent in one single picture all the relations a given set is involved in. For this purpose, we shall use entity-relationship diagrams. The sets only contain names. To represent components of the factory, we must map these names to quantities variously referred to as "attributes," "properties," "state variables," or "fields." In the machine purchase date example of Section 6.2.2, the column label can be clarified by viewing "Purchase-Date" not as the name of a set but as a *mapping* from the set of Machines to that of Past-Dates.

The tuple of a set and all its attributes is frequently called a *record type*, while a member of that set with all its attribute values is called a *record occurrence*. The list of fields in the record determines how its occurrences can be queried. In a structured specification, these record structures are documented in a *data dictionary*. The mappings defined on a relation will be referred to as attributes of the relation, or fields of the records associated with the relation. If the vocabulary of tables is used, then records become "tables," fields become "columns," and occurrences become "rows." The construction of data dictionaries is discussed in Chapter 7.

When looking at an entity-relationship diagram, no harm is done in confusing sets with the associated record types from the data dictionary because their names match and the purpose of a name is to make such a confusion possible. In daily life, no distinction is needed between the name "John Smith" and the person it refers to.

6.3. GRAPHICS

6.3.1. Entity-Relationship Diagrams

Figure 6.2 is an example of the graphic convention used in mathematics to represent mappings between several sets in one single picture. Several restrictions make this notation unsuitable for systems analysis without some modifications:

- We need to represent not only mappings but general relations.
- The diagrams can get confusing when set names are phrases rather than single letters.
- We need to represent set inclusion.

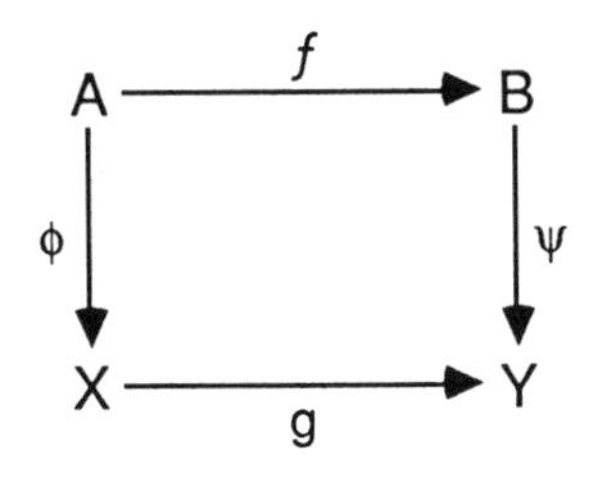

Figure 6.2. Mathematical Diagram

Figure 6.3 shows first the mapping f on domains A and B expressed as an entity-relationship diagram. The set names are boxed to enhance name separation, and the name of the mapping is enclosed in a diamond to distinguish it from a set name. "N" and "1" indicate that f is a many-to-one relationship. In the entity-relationship model, this diagram does not indicate whether *every* element of A must have an image in B, and thus the true domain of f could be a subset of A.

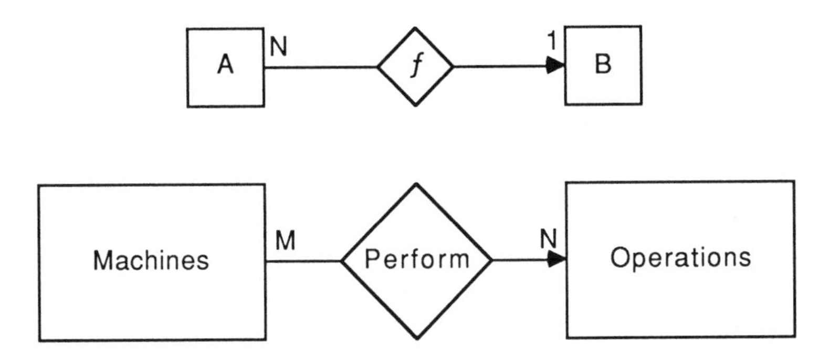

Figure 6.3. Entity-Relationship Diagrams (Binary)

The entity-relationship diagrams in this book are nonstandard in having an arrowhead to indicate a direction. The usefulness of this extra symbol appears in the diagram representing the relationship between machines and operations. It can be read as "Machines perform Operations," and the arrow tells the reader what the subject and the object of the sentence are.

If the relation were named "Require," the arrow would be pointing in the opposite direction. "M" and "N" indicate that this is a many-to-many relationship — that is, one machine can perform several operations and an operation can be done on several machines.

6.3.2. Diagrams for Relations of Arity >2

Most of the relations we need to express in diagrams are fortunately binary. Figure 6.4 is a ternary relation modeling the frequent case where the execution of an operation in a machine requires special attachments. The diagram indicates that there is a relation in the model where (Operation, Machine, Tool) triples can be stored and retrieved.

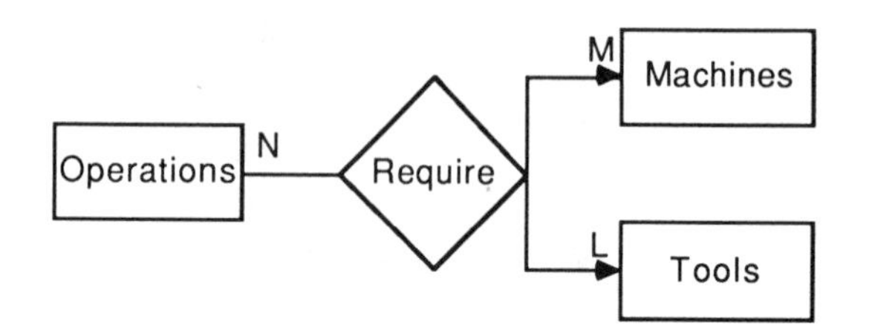

Figure 6.4. Entity-Relationship Diagrams (n-ary)

When an operation is thus linked to several machines, the diagram does not say whether the machines are alternates or the operation requires all of them simultaneously. The same remark applies to tools. Such semantic issues must be resolved for the model to be complete. Those are beyond the scope of the diagram.

6.3.3. Representing Set Inclusion

In another departure from standard entity-relationship diagrams, we represent a subset by a box included in the superset. This is a natural way to model this special relationship. An example is given in Figure 6.5. Operators and Engineers are both Employees.

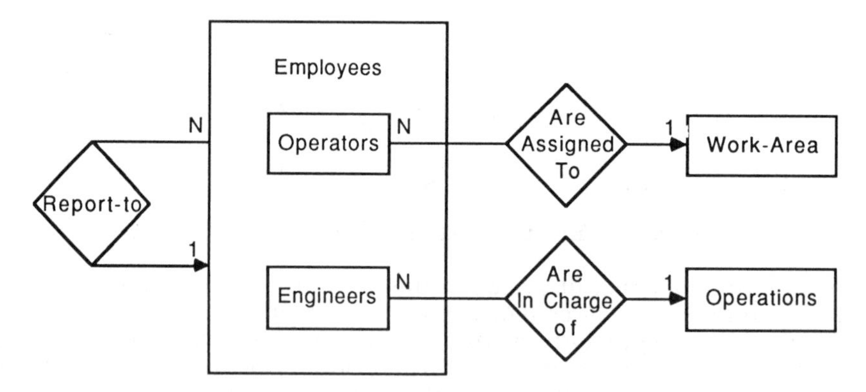

Figure 6.5. Diagram with Set Inclusion

The relationship between Operators and Work-Areas does not involve other Employees, and therefore the arrow representing it starts at the edge of the "Operators" box. On the other hand "Employees report to Employees" is a relationship involving all categories of Employees (except the chief executive officer), and therefore the arrow modeling it starts and ends at the edge of the Employees box.

6.4. SECURITY

The preceding discussions contain references to "the user." In reality, the memory model of a computer system in a factory is accessed for retrieval and update by managers, engineers, operators, equipment controllers, and external, networked computer systems. Utter chaos would ensue if all these different types of users were given full access to the whole model.

The definitions of the relations form a schema that is the domain of the application software designer and out of bounds to all users. There are two reasons for this. First, users rarely have the technical skills to build a usable schema. Second, they access the database through application software specified in the data flow model and predicated on the schema. If a user deleted a relation from the schema, the application would stop working.

One implication of this restriction is that the relations users can update must be built exclusively on sets of atoms. In other words, the value of a field in a relation cannot be the name of another relation. Appendix A of Chapter 7 shows an example in which this restriction is painful to work around.

Given a schema, the task of filling it with the factory's structure of products, routes, operations, machines, and employees is controlled, if not carried out, by a small group of users responsible for database administration. Unrestricted update access to these structures by engineers may be desirable in a pilot line, but would not be acceptable to the production manager of a high volume line.

Retrieval access to factory data is also restricted to preserve company secrets, so that even managers and engineers may manipulate only small subsets of the factory status and history. Finally operators only see instructions to carry out their tasks and prompts to log the results.

Security can be excessive. For example, restricting an operator to access the system only through an assigned, physical terminal does not really protect against abuse any better than using passwords and granting selective access in the system's memory. It also imposes the assignment of physical ports to know which terminal an operator is using, and this would preclude the use of local area networks.

Appendix A — RELATIONAL ALGEBRA OPERATIONS

We discuss here briefly the four fundamental operations of relational algebra with which queries can be built. The reader is referred to Date [Dat81] for details.

6.A.1. Selection

A selection from a relation is the result of a query of the form "Display all machines qualified to perform operation X." In the vocabulary of tables, the result of a selection is a relation with the same columns but only the rows meeting a specified condition. Selection from a relation produces a relation included in it and with only tuples such that some of their coordinates satisfy another relation.

6.A.2. Projection

In the vocabulary of tables, projection consists of retaining only some of the columns. "Display all operations for which there is a qualified machine" is a query that would be answered by projecting the Machine-Operations relation onto its second domain.

6.A.3. Join

If we have (1) a relation called Machine-Operator associating operators with equipment they have been trained on and (2) another relation called Machine-Operation saying what operations can be done in what machines, we may want to use this to relate operators to operations. A join will do that.

We shall only define the equijoin of two relations with a common domain. It is possible to extend this definition to other types of join and to more than two relations. Let us assume that the domain the two relations have in common is C. By eventually rearranging the order of the domains, we can get the two relations are R on domains C, $A_1, \ldots, A_n$ and S on C, $B_1, \ldots, B_m$. The join of the two relations is a relation T on C, $A_1, \ldots, A_n, B_1, \ldots, B_m$ where the tuples in T are formed by concatenating the tuples in R and S with matching C values.

If (Furnace-1, Joe Smith) is in Machine-Operator and (Furnace-1, Gate-Oxide) is in Machine-Operation, then (Furnace-1, Joe Smith, Gate-Oxide) will be in the equijoin of the two relations. A projection of the join onto the last two domains will then contain (Joe Smith, Gate-Oxide).

6.A.4. Division

If we have a set of operations, and we want to know the list of machines that can individually do *all* the operations in the set, we will get the result by dividing the Machine-Operation relation by the unary relation defined by the desired set of operations.

7

DATA DICTIONARIES IN MANUFACTURING

The detailed structure of data flows, factory objects, and memory entities is given in a formal data dictionary, along with comments and natural language definitions of atomic elements. Data dictionary definitions break down a record into smaller units, in turn explained by text, using De Marco's conventions.

Textual definitions must be aimed at specific readers, and are only needed when names are not clear enough. In data flows, fields containing complex, mathematical functions of other data require both conceptual and operational definitions to explain (1) what they are intended to mean, and (2) how they are calculated.

The established terminology of a factory should be followed, unless there are strong reasons not too, such as complaints by employees about its obscurity. De Marco's data dictionary conventions are limited in that they do not provide means to express variant or recursive structures needed in manufacturing to represent state dependent lists of variables or bills of materials.

7.1. DEFINING ATTRIBUTES IN A DATA DICTIONARY

7.1.1. Purpose of Data Dictionaries

In data flow diagrams, data flows are identified only by their names. Likewise, the sets and relations shown graphically in entity relationship diagrams are the skeleton of the system memory. The data dictionary serves to flesh out both of these models by filling out the details.

It is useful to know who an employee works for, the name of the process a part is made by, or the machines that are qualified for an operation, but, in all those cases, we could not possibly run a factory with only this information. To each set element or relation tuple we must be able to assign descriptions and numeric or logical parameters.

7.1.2. De Marco's Conventions

This book, not being a specification, does not contain one large data dictionary, but analysts must be able to write and review one. De Marco's definition conventions are used throughout the text, wherever they clarify meaning.

Before proceeding, let us review them, in De Marco's words:

=	means "is equivalent to"
+	means "and"
[...]	means "either-or," that is, "select one of the options enclosed in the brackets"
{...}	means "iterations of the component enclosed"
(...)	means "the enclosed component is optional"
...	means "enclosed is a textual explanation of the entry"

The "{ }" and "( )" notations are shared with set theory, but the meanings differ. In the rest of the book, unless explicitly stated otherwise, they will always be used in the De Marco sense.

As we have said in Chapter 6, a record type is a tuple of the form

$$A = \underline{\text{Identifier}} + f_1 + \dots + f_k$$

where A is a set and $f_1, \dots, f_k$ are mappings called "attributes" defined on A, with ranges $B_1, \dots, B_k$ *outside* the system model. A may be the graph of a relation we have defined, and the list of attributes may be empty.

Example:

Machine = <u>Machine-Name</u> + Description + Maker + Date-of-Purchase

In a record occurrence, Machine-Name is replaced by an element of the set of Machines, and the attributes take on the value assigned to the machine.

Example:

Machine-Name	=	Furnace-1
Description	=	Diffusion furnace for 4-in and 6-in Wafers
Maker	=	XYZ Semiconductor Equipment, Inc.
Date-of-Purchase	=	12/15/89

The "-Name" field identifies each occurrence as an element of the set that is the domain of the attributes. The field (or sequence of fields) playing this role in a record definition is underlined, and, in the relational model, is called the primary key.

"-Name" is one of many possible suffixes that can be attached to these fields. Alternatives include "-Key," "-ID," and "-Code." Any will do, but the analyst should choose one and stick to it throughout the data dictionary, because a reader who sees several suffixes in one document will mistakenly assume them to have different meanings.

Although, theoretically, mapping names should be unique, it is allowable to use the same in several definitions as long as no harmful confusion arises. "Description," for example, can be used to designate attributes of machines, operations, and parts without any detrimental effect.

On the other hand, an attribute usually cannot be confused with its range without loss of clarity. "Date-of-Purchase" takes its values in the set of all valid dates, but if the definition said "Date" instead of "Date-of-Purchase," the reader would have no way to know what the field was *intended* to mean. It would be as likely to be the date of last calibration as the date of purchase.

If an attribute has a small, finite range, then the whole set of allowed values can be explicitly shown in the data dictionary in one of the two following forms:

Machine-State = [Ready | In-Use | In-PM | Down]

or

$$\text{Machine-State} = \begin{bmatrix} \text{Ready} \\ \text{In-Use} \\ \text{In-PM} \\ \text{Down} \end{bmatrix}$$

Now, we recall that a data flow is a highway along which packets of data of known composition flow. If an arrow in a data flow diagram is labeled

"Machine," the reader of the specification can look up the composition of such a packet in the dictionary.

If the system memory model contains a relation called "Factory-Machines," containing many occurrences of the "Machine" record, its structure can be expressed in the data dictionary in the following form:

Factory-Machines = {Machine}

Here the brackets designate a possibly empty sequence of the enclosed elements. Thus "{Machine}" means "a list of occurrences of the Machine record type." In the preceding chapter, brackets to enclose a list the elements of a set were used above only as a tool to introduce other concepts. From this chapter onwards, brackets will always have the meaning of iteration within a data dictionary.

The discussions of normal forms and data models by C.J. Date [Dat81] and [Dat83] show that relations in which all fields are mappings defined on the primary key have desirable properties. For example, if the Machine record were defined as follows:

Machine = Machine-Name + ... + Location + Floor-Covering

with "Location" designating the room in which a machine is placed. Then "Floor-Covering," be it linoleum or carpeting, is only associated with the Machine through its Location. In other words, Floor-Covering is an attribute of the Location, not of the Machine. One obvious problem with this structure is that a location cannot be assigned a Floor-Covering unless there is a Machine in it. A less obvious but similar example is given in Appendix A.

Also, since mappings assign an image to each element of the source set, every field is filled in every occurrence of every record. In other words, there are no null values, which is essential for relational algebra to be usable. Null values have to be interpreted somehow in order to perform a selection or a join, but, depending on circumstances, their meanings may vary.

To take examples from daily life, the absence of a party affiliation for a citizen means "none," but a null value for a man's maiden name means "not applicable." Among engineering parameter measurements, a null value tends to indicate instrument malfunction or operator negligence. In a selection or a join, should these null values match with everything or with nothing? This problem is discussed extensively by C.J. Date [Dat83].

However, to a degree, the analyst can and should take liberties with these principles for the sake of communication with users, and leave the resolution of the preceding problems to software designers. We will examine the case of optional attributes and the use of a relation in lieu of a field.

7.1.3. Optional Attributes

Optional attributes are included between parentheses in the definition of the record type for the whole set. If some, but not all batches have a work order number, then the data dictionary definition for a batch could be as follows:

Batch = Batch-Name + Quantity + (Work-Order-Number)

To avoid null values, two structures would have to be defined, as follows:

Batch = Batch-Name + Quantity

Batch-Work-Order = Batch-Name + Work-Order-Number

From the point of view of user communication, the tradeoff is between more simple definitions and fewer definitions with the added complexity of optional fields.

7.1.4. Relations as Attributes

If the system must remember information about each part within a batch, this can be expressed by the following definition:

Batch = Batch-Name + Batch-Attributes + {Part-Name}

Strictly speaking, this definition establishes a many-to-one relation between parts and batches which should be expressed by an entity-relationship diagram. If it is modeled both there and in the data dictionary, there is a danger that subsequent editing of both will render the global description inconsistent.

If the relationship is expressed only in the data dictionary and users are comfortable with its form, then the analyst may decide to leave it in. However, to conform with the relational model, the software designer will split it as follows:

Batch = Batch-Name + Batch-Attributes

Part-Batch = {Part-Name + Batch-Name}

7.1.5. Textual Explanations

When the list of attributes of a record type is not sufficiently explicit, it can be supplemented by textual explanations, set off by '*' marks on both sides, as in:

Lot = * Set of parts processed together. *
= Lot-Name + Product + Part-Count +...

7.1.6. Redundancies

If the data dictionary definition for "Lot" lists "Product-Name" as an attribute, it says that each lot maps to a product. This fact should already be known from the entity-relationship diagrams, and therefore, strictly speaking, should not be repeated here.

The following definitions follow this rule:

Product = {Product-Name + Description + Status + Revision-Number}
Lot = {Lot-Name + Due-Date + Status + Speed}
Order = {Order-Name + Quantity + Unit + Time-Received + Due-Time}

No matter what database system is used to implement the system, the attributes in these definitions will be in the corresponding records. On the other hand, the link between orders and products will be implemented by making Product-Name an attribute of Order only in a relational DBMS. Other methods would be used in hierarchical or network databases.

7.2. DEFINITIONS

7.2.1. The Reader's Point of View

Users read definitions to get information they need about a word and don't already have. To write good definitions, the analyst must know who will read them and for what purpose. The reader may wonder how to recognize the defined object when encountering it in the system, what operations are allowed on it, or how it maps to the factory.

Quoting Aristotle, De Marco [DeM78] says that definitions should say what kind of a thing the defined object is and how it differs from other objects of the same kind. The attributes of a data dictionary entry may be complex structures, themselves defined elsewhere in the dictionary. The definition of an entry in terms of atoms is obtained by looking at several definitions.

The question then arises of how atomic attributes should be defined. The definitions of higher-level structures are syntactic, simply describing the entries as made up of smaller pieces. The discussions of meaning are in the comments. We shall review two circumstances in which the definition of atoms generates special problems:

- When the atom is a primitive, such as "Time-Received" for an Order.
- When it is a complex mathematical function of other terms, such as a manufacturing yield.

7.2.2. Defining Primitives

Is there a need to define "Time-Received"? In other words, what is there that the reader might conceivably not know and that a definition could clarify? If there is nothing, the analyst is well advised not to waste time coming up with a definition for the sake of meeting a formal requirement.

In the case of "Due-Date," the choice of the word "date" carries implications worth explaining, namely that no unit of time smaller than a day will be considered. "Due-Time," or "Time-Due," although a less common phrase, would leave open the option to make parts due at 4:35 pm on 2/17/89. The relation of the due date to the factory's work calendar may also need clarification. The definition could say, for example, that the parts are due by the end of the last worked shift on or before the due date.

7.2.3. Defining Inferred Results

Special difficulties arise in the definition of outputs shown to users that are summaries of inputs coming from the factory floor. We shall explore this problem through the example of the "Yield" of a product through an operation in a time period.

The data dictionary definition of a yield could be as follows:

Yield-Report = Time-Period + {Product + Operation + Yield}
* Calculated for product P, operation Op and period i by the following formula:

$$\text{Yield}(P,Op,i) = \frac{\text{Outs}(P,Op,i)}{\text{Ins}(P,Op,i)}$$

where:

- Outs(P,Op,i) is the number of parts of product P leaving operation Op during period i, and
- Ins(P,Op,i) is the number of parts that were in the batches of product P that left operation Op in period i when they first arrived at the operation. *

This definition is operational in that it explains clearly how to calculate a number. If the analyst is challenged to explain why this formula was chosen, the most likely first answer is: "That's just how we define it. If you want it calculated otherwise, we'll accommodate you."

No one would say that about temperature. "If you want to call 'temperature' not the thermometer reading but the square of its logarithm, that's what it'll be." It sounds wrong, because "temperature" is not only a word but a reference to a physical concept. The numbers placed in a temperature field will be used, for example, to predict the flow of heat between two parts of a solid using well-known equations. By redefining temperature, the analyst would be jeopardizing the pragmatics of the system.

Yield calculations in a semiconductor factory are the subject of spirited discussions between production supervisors, to whom it is a performance measure, process engineers, to whom it is a diagnostic tool, and planners, who must take it into account in forecasting shipments. The analyst's motivation in accepting any formula the users will agree to is to stay out of this political minefield.

However, the fact that these discussions take place shows that, to users, yields are more than the outputs of formulae. If they didn't think that there was a parameter of the operation called yield existing independently of any formula, they could not possibly care what formula was used. To them, it is not just a number, but a physical quantity like temperature.

To find out what a yield truly is, the analyst must examine what it is used for. "If the yield is 80%," says the planner, "then I know that, on the average, I must release 125 parts to the operation for every 100 I want out of it. And since the yield varies from batch to batch, I know that I must have a buffer stock after the operation to compensate."

The process engineer's perspective is different: "I look at the yield every day. If it moves up and down by one or two points around the same value, I don't worry about it. If it drops five points overnight, I know something's wrong and I have to intervene. If it moves up one point and stays there, it tells me we've done something right..."

The planner wants to infer the future from the past. The engineer wants to know whether period-to-period variations point to a real change. From these statements, the picture of an underlying yield should begin to emerge as the time-dependent probability of success in processing a part of the product through the operation.

The yield formula then takes on the meaning of an *estimator* of this underlying probability. By specifying the probabilistic model more fully, it is then possible to rate different estimators on the basis of various statistical criteria, such as whether or not they are biased or how large their estimation variances are.

The definition of yield as probability of success is conceptual but by itself insufficient, because it is not operational: it does not provide a means of calculating a number. The conceptual dimension of the definition provides the reader with an intended meaning, and therefore with objective criteria for evaluating how good a particular formula is.

7.2.4. Established Usage

The terminology in use in a factory should usually be carried over into the system specification. The reasons for respecting it are clear: a system that speaks a language users already know is easier to implement and get accepted than one that forces them to learn new terms.

There are, however, circumstances when it may be preferable to change it. For example,

1. A factory deviates from standards followed by other factories of the same company, with which it might exchange technology.

2. Names contain encoded information that is now available elsewhere in the system.

3. The vocabulary in use is so obscure that many operators don't understand it.

Software designers bypass this issue by making all names user-definable. The limits to this approach are reached when the system terms become so user-specific that communication with the software supplier breaks down. If all screen prompts are user-defined, questions and problem reports may become difficult to formulate because the support organization must first remap the new names to the original ones.

Each manufacturing organization has a special vocabulary developed frequently for nontechnical reasons. Nissan calls its kanbans "workplates" to avoid the perception that it is emulating rival Toyota. Departments performing the same functions may be called "quality control," "quality assurance," or "product assurance," the names being switched to exorcise the memory of a former manager or to create a perception of change. The demand from a system that it exclusively use the vocabulary of one organization can be met only at the cost of making it inapplicable elsewhere.

In manual systems, product names tend to contain encoded information about the product and serve as retrieval keys for folders in file cabinets. For example, the first two characters may point to a manufacturing facility, the following three refer to a process technology, and so on, resulting in names like "WXBIP3571." In a CIM system, the sole purpose of a name is unique identification. No other information needs to be encoded in it, because the database management system, not a clerk, takes care of retrieving it.

There is no technical reason not to use euphonic, easily remembered names. A CIM system should not prevent users from following their traditional naming conventions, but analysts should be wary of imposing them on the system.

Most information encoding schemes are harmless, the only objection to them being that they introduce redundancies, and hence a risk of inconsistency. Some schemes are worse, such as the use the first part of a machine name to represent its location on the shop floor. When the machine is moved, the choice is between letting it have a name referring to the wrong location or breaking the continuity of its history by changing its name.

The administrative tools of factory management are sometimes made deliberately obscure. In a semiconductor factory, an engineer designed a form for operators to report on defective parts, and did not call it "Defect

Report" but "Discrepant Material Report" on the grounds that it sounded more "professional." The name was thereafter abbreviated to "DMR."

The operators in that factory were mostly not native speakers of English, and had had limited schooling in their own countries. Most did not know the word "discrepant." Nobody knew what "DMR" meant. It would be a mistake on the part of an analyst to call the equivalent of this form in a CIM system by the same name.

Manufacturing systems analysts are pressured by fads in their choice of words. Once MIS becomes passé, a simple name change brings the factory to the age of CIM. Similar transformations turn menu help into "rule-based expert systems" and reorder point inventory management into Just-In-Time.

In some organizations, compliance with fads makes the difference between funded and cancelled projects. In preferring fashionable to standard terms for political marketing, the analyst takes the risk of confusing users, overselling, and of having the system itself become unfashionable long before it has outlived its usefulness. On the other hand, those are problems the analyst would like to have, whose proposals are turned down for being unfashionable.

7.3. LIMITATIONS OF THE DE MARCO CONVENTIONS

7.3.1. Variant Structures

The state of the vehicle from Section 3 of Chapter 5 is an item the system must remember. How can we create a data dictionary definition for it? One might be tempted by the following:

$$\text{Vehicle-State} = \left[\begin{array}{l} \text{"Loading"+ At-Station + Load-Description} \\ \text{"Intransit" + Origin-Station + Destination-Station} \\ \qquad\qquad\quad \text{+ Load-Description} \\ \text{"Unloading" + At-Station + Load-Description} \\ \text{"Waiting" + At-Station} \end{array}\right]$$

This definition, however, violates the data dictionary conventions, which do not allow constants, and there is nothing we can do about it. To identify the state names, we would also have a separate definition as follows:

$$\text{Vehicle-State-Name} = \left[\begin{array}{l} \text{"Loading"} \\ \text{"Intransit"} \\ \text{"Unloading"} \\ \text{"Waiting"} \end{array}\right]$$

But we couldn't possibly write

$$\text{Vehicle-State} = \underline{\text{Vehicle-State-Name}} + \text{Vehicle-State-Parameters}$$

because the list of state parameters depends on the state name. We should be able to read the definition as:

> "IF the Vehicle-State-Name is 'Intransit'
> THEN the vehicle has Origin- and Destination-Stations, and a Load-Description..."

Niklaus Wirth [Wir76] would call this a variant record type. Not only did the requirement for modeling such a structure arise naturally from the analysis of the vehicle, but it also appears necessary to give the end user control over what the allowed states are. Our list does not include "Down," and that is an immediately obvious shortcoming, but any fixed list is bound to be found insufficient by some users.

In other words, the true requirement is not for a particular state-transition model for vehicles, but for the ability to set up such models. The curious reader will see in Appendix B that implementing variant types while giving users control over the definitions is no mean feat.

7.3.2. Bills of Materials and Recursion

Models of manufacturing bills of materials are another noteworthy example. A bill of materials can be represented by a "Goes-Into" relation relating an assembly and its immediate components, as shown in Figure 7.1, with the following additional rules:

- A part never goes into itself.
- If a part doesn't go into anything, it is an end-product.
- If a part has no other parts going into it, it is a raw material or a purchased component.

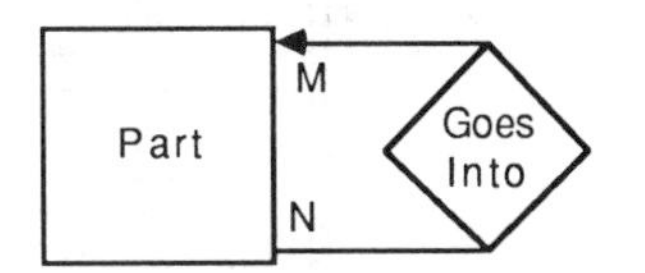

Figure 7.1. Goes-Into Entity-Relationship Diagram

This can be expressed in the following data dictionary entries:

Goes-Into = {Component + Assembly + Quantity}
Assembly = Valid-Part-Name
Component = Valid-Part-Name

The Goes-Into relation is a simple model of the structure of an assembled product, but far removed from the intuitive perception of it. To make a shippable product, one needs a list of components assembled in the final operation, which may be purchased from the outside or be subassemblies with their own list of components. The bill of materials structure of a product is a tree, as shown in Figure 7.2 for a bicycle.

Figure 7.2 shows, down to a certain level, the components of a bicycle in an intuitive manner. It illustrates a bill of materials as a list made of purchased components and smaller bills of materials. This concept can be abstracted into a data dictionary definition for a bill of materials as follows:

$$\text{Bill-of-Materials}=\left\{\begin{bmatrix}\text{Purchased-Components}\\ \text{Bill-of-Materials}\end{bmatrix}+\text{Quantity}\right\}$$

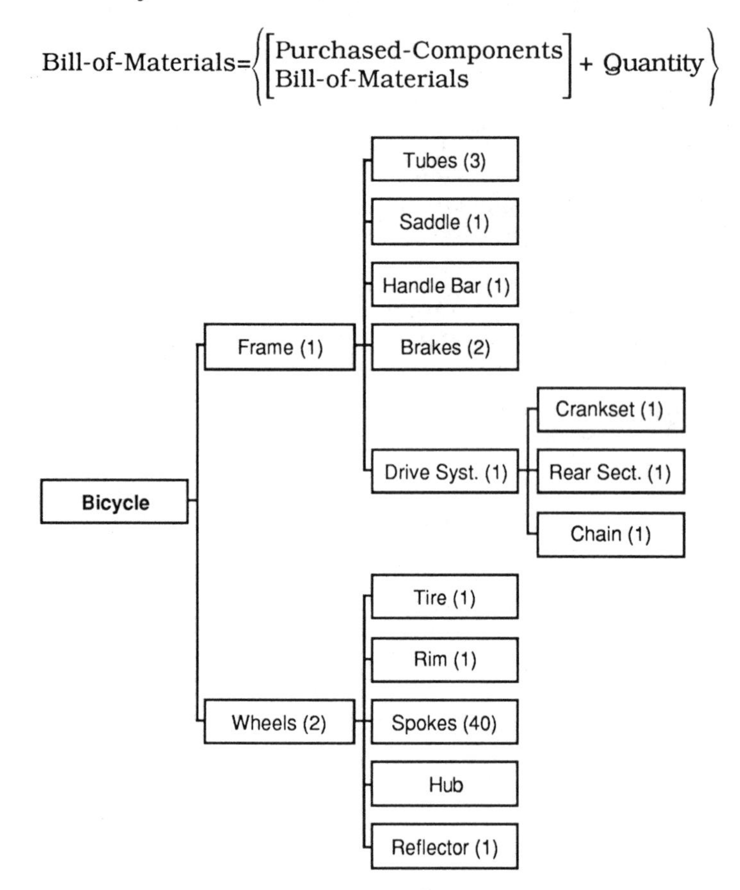

Figure 7.2. Incomplete Bicycle of Materials of a Bicycle

This definition contains a feature called recursion, not allowed by De Marco and generally familiar only to software engineers: a bill of materials is defined partially in terms of itself. Bills of materials are one of many cases of recursion which arise naturally in modeling factories. The Goes-Into table is a list of the branches of the bill of materials, and the tree can be constructed from it as in Figure 7.3.

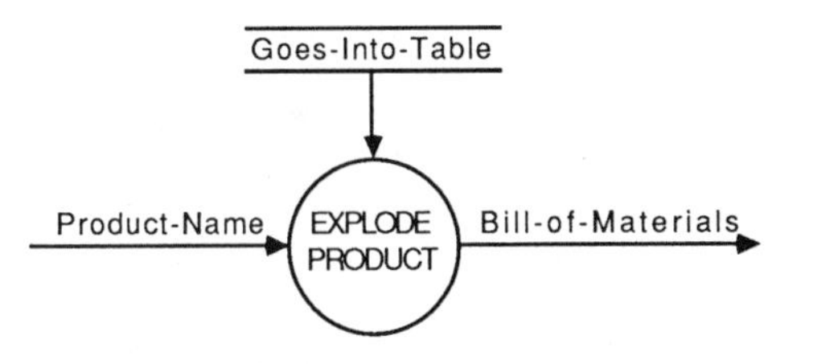

Figure 7.3. Exploding a Goes-Into Relation into a Bill of Materials

Appendix A — AN EXCURSION INTO SOFTWARE DESIGN

So far, we have looked into the use of tools by analysts working with users. On the other side of the fence are development engineers who must design and code software to meet the requirements specified by analysts. In this appendix, we examine the special case of querying the state of a machine to give the reader a feel for the kind of decisions a software designer must make.

As in the preceding, we have a vehicle on the shop floor, traveling along a fixed route between stations where it loads or unloads various cargos. We want the vehicle modeled in the system memory in such a way that we can find out what state it is in. We want the state description provided in the answer to include all the state variables. Since we don't have infinitely fast computers and vehicle states are queried often, we must also take performance constraints into consideration.

We would like states and state variables to be user-defined. If possible, we would also like the database structure to make it impossible to store mistaken, incomplete, or inconsistent data. Finally, it would desirable to be able to make other, ad hoc queries on vehicle states using relational algebra.

Now let us tackle this problem.

7.A.1. Variant Types in PASCAL

In the PASCAL programming language, it is possible to define a record type for vehicle states as follows:

```
Type Name = Packed Array[1..64] of Char;
Type VehicleState =
   Case State : (loading, unloading, intransit) of
         loading:     (CurrentStation: Name;
                        LoadType: Name;
                        Quantity: Integer);
         unloading:   (CurrentStation: Name;
                        LoadType: Name;
                        Quantity: Integer);
         intransit:   (OriginStation: Name;
                        DestinationStation: Name;
                        LoadType: Name;
                        Quantity: Integer);
```

Although we can use this feature to build application programs with the ability to model the different variable lists associated with each state, it does not solve our problem for several reasons. First this structure exists while the program is running, but it is not a place where semipermanent data can be stored. Second, the list of allowed states is "hardcoded in" — that is, written into the PASCAL program source and not editable by users.

7.A.2. First "Relational" Structure

To solve these problems, let us examine how a relational database could be set up to store vehicle states. The first method that comes to mind is to define a table with the union of all fields that could possibly be relevant to a vehicle state; for example,

Vehicle-State = {Vehicle-Name + Vehicle-State-Name + (Current-Station) +
(Origin-Station + Destination-Station) +
(Load-Type + Quantity)}

By searching this table for the record identified by a vehicle name, we can retrieve its complete state. However, the list of states and state variables is still not user-editable, and there is worse: the table is not a true relation, and relational algebra cannot be used on it because of the optional fields. This structure also offers no guarantee of integrity, in that it is possible to load it with nonsensical data, such as a destination station assigned to a vehicle that is unloading.

This does not mean that integrity cannot be maintained, but only that it isn't maintained by the database structure and must be by the application software. The reader may also notice that structures with many null values are wasteful of storage space, which is a consideration, albeit of secondary importance compared to the usability of relational algebra.

Reusing a field with different meanings depending on the state is a "space saving" method so disastrous that it would be best ignored. However, since it is actually used, it must be discussed. The following table is an adaptation to our example of an input form supplied by a maker of numerically controlled machine tools. It is a throwback to the days when modeling concepts were constrained by the use of punchcards.

Loading	1	At-Station		Load-Type	Quantity
Unloading	2	At-Station		Load-Type	Quantity
Intransit	3	Origin-Station	Dest.-Station	Load-Type	Quantity

Vehicle	State				
V1	2	S1		L1	100
V2	1	S2		L2	50
V3	3	S3	S4	L4	200

To understand one row of this table, the reader must first look up the state number in the header to find out the meanings of the fields for the corresponding state. Imposing such a form on users is asking for mistakes. Using it internally wreaks havoc with relational algebra.

7.A.3. Variants in PROLOG or LISP

Let us consider the following alternative to our flawed table:

Vehicle-State = {Vehicle-Name + Vehicle-State-Name}

$$\text{Vehicle-State-Name} = \begin{bmatrix}\text{Intransit}\\ \text{Loading}\\ \text{Unloading}\end{bmatrix}$$

Intransit = {Vehicle-Name + Origin-Station + Destination-Station + Load-Type + Quantity}

Loading = {Vehicle-Name + Current-Station + Load-Type + Quantity}
Unloading = {Vehicle-Name + Current-Station + Load-Type + Quantity}

This structure would avoid null values, and offer some guarantee of integrity by providing no place to store irrelevant variables. The state of a vehicle would be queried by retrieving two records:

1. Look up the Vehicle-State-Name in Vehicle-State;
2. Take Vehicle-State-Name to be the name of another relation in which to look up the state variables.

Unfortunately, as discussed in Chapter 6, commercial relational database systems usually do not allow the value of a field to be the name of a relation. These systems require field values to be atoms, and the preceding structure fails on this count. Allowing some end users to edit state definitions would be tantamount to letting them alter the database schema, a privilege application designers are reluctant to give.

This structure can be implemented in languages such as PROLOG or LISP, because they allow the value of a variable to be the name of another. Using these languages today, it is possible to build prototypes, but not a system that can be concurrently updated from a hundred terminals.

7.A.4. Variants with No Redundancy

The preceding approach still relies on the application software to ensure that the state of a vehicle is recorded consistently in Vehicle-State and in the relation identified by Vehicle-State-Name. A vehicle could, for example, be listed in Vehicle-State as "unloading," but appear in the Intransit table.

Such inconsistencies are possible because the name of the vehicle state is stored in two places. To avoid this, we could resort to the following approach:

Valid-Vehicle-States = {Vehicle-Name + State-Name}

$$\text{State-Name} = \begin{bmatrix} \text{Intransit} \\ \text{Loading} \\ \text{Unloading} \\ \ldots \end{bmatrix}$$

Intransit = {Vehicle-Name + Origin-Station + Destination-Station +
Load-Type + Quantity}

Loading = {Vehicle-Name + Current-Station + Load-Type + Quantity}

Unloading = {Vehicle-Name + Current-Station + Load-Type + Quantity}

Valid-Vehicle-States does *not* point to the state the vehicle is in but lists all the states it could be in. The state it is in is indicated only by the presence of the vehicle in the relation defined for that state. The state of a vehicle would be queried as follows:

1. Get a list of all the valid states for the vehicle;
2. Search all the relations defined for the valid states, for the unique vehicle record.

With three states, this would require retrieving three records from Valid-Vehicle-States and then searching up to three other relations. With imperfect technology, this may be too high a performance price to pay for the elimination of redundancy.

7.A.5. Second Relational Structure

To stay within the constraints of relational database systems and yet allow users to edit state definitions, we could use the following structure:

Vehicle-State = {Vehicle-Name + State-Name + Variable-Name +
Variable-Value}

By retrieving all records matching a given vehicle name, we would get its complete state. This structure, however, does nothing to protect integrity; it is up to the application software to ensure that the pairs of variable names and values associated with a vehicle actually describe relevant state variables.

Aside from the potentially large number of records to retrieve just to query a vehicle state, this structure also makes it impossible to represent a vehicle state with an empty list of variables. When there are state variables, it forces recording the state name once for each variable, thus affording an opportunity for inconsistencies.

The root of these problems is that Variable-Value assigns a value to a (Vehicle-Name, Variable-Name) pair, whereas State-Name is functionally dependent on Vehicle-Name only. Date [Dat81] would say that the Vehicle-State relation is not in "second normal form." We could resolve this by changing the model to

Vehicle-State = {Vehicle-Name + State-Name}
Vehicle-State-Variables = {Vehicle-Name + Variable-Name + Variable-Value}

Then the query would require searching two tables.

7.A.6. What Is Actually Done

We have examined several structures we could use to model the state of a vehicle and found problems with all of them. We have encountered null values, hardcoded states, and queries that are too complicated for frequent use. For systems to be actually built, something has to give, and the designer's dilemma is to figure out what will hurt least. Systems in operation do have tables full of null values, or hardcoded state definitions, or a combination of both.

8

MODELING PROCEDURES AND DECISIONS

The inner workings of processors at the bottom level of a data flow explosion are explained in a "structured natural language" limited to imperative clauses listed in sequence or indented to represent selection or iteration. Complex decisions are modeled separately, by trees if the decisions are coupled, or by tables if all combinations of conditions are possible. Structured analysis does not provide algorithms, only ways to communicate them.

8.1. MODELING PROCEDURES

Data flow diagrams model a system as concurrent processes with no apparent coordinating structure. The hierarchical explosion of the system is similar to the partition of a factory into facilities, then of each facility into shops, then of each shop into work centers, and so on down to the level of individual machines. The units recognized at all intermediate levels can perform several tasks concurrently, but, at the bottom level, are machines which can do only one thing at a time.

When one reaches that level in the hierarchical explosion of the "data factory" that a software system is, the analyst needs to switch focus to the description of how an individual processor works. It is then and only then that control flow considerations take priority.

As a modeling method, structured analysis is based on the principle of "divide and conquer." Chapter 4 covered the division of a system into networks of smaller ones. We now proceed to "conquer" these pieces.

8.2. STRUCTURED NATURAL LANGUAGE

De Marco calls his main tool for this stage of the analysis *structured English*. *Structured natural language* is preferred here because it describes the same technique as well in Japanese or German as in English.

The processors found at the bottom level of data flow explosion carry out their tasks in the form of operations executed one at a time. Text is an appropriate medium here, because it is read sequentially. A graphic road map is more effective than text to locate a house in a city. On the other hand, communication of a cooking recipe is generally not graphic.

The textual description of a procedure is not a smoothly blended narrative. The operations must be clearly separated and their precedence constraints emphasized. This is done by restricting the sentence structure to imperative clauses, numbering to indicate sequence (see Example 1), and indenting to represent iteration or selection.

Example 1: Following is the description of a simple MOVE-IN, with the data flow structure shown in Figure 8.1.

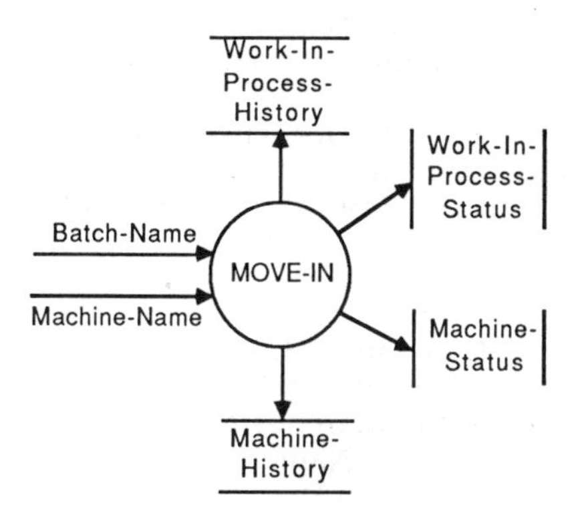

Figure 8.1.

1. Put the Machine in use.
2. Start processing the Batch.
3. Append the Batch-Name and the time to the Machine-History.
4. Append the Machine-Name and the time to the Batch-History.

The MOVE-IN transaction of Example 1 is simplistic, in that it does not even consider the possibility that the machine may be unavailable or that there may be no machine by this name. The sequencing of instructions is intended to represent the way a user would think of the process.

You allocate a machine before you start using it, and therefore step 1 precedes step 2. You only record events in history once they have actually happened, which is why steps 1 and 2 precede 3 and 4. From the point of view of the software developer, in this case, the order truly is immaterial, because the updates are made to the database all at once or not at all.

Example 2 describes how to build a single-level bill of materials for a part. In a relational database system, this module could be implemented exactly as shown here. With other types of software tools, the construction of the component list would be done as in Example 2'. Example 2 is the way a user would describe it; Example 2' is how a traditional programmer would, and is more difficult for a user to follow because it involves an iteration and a decision.

Example 2:

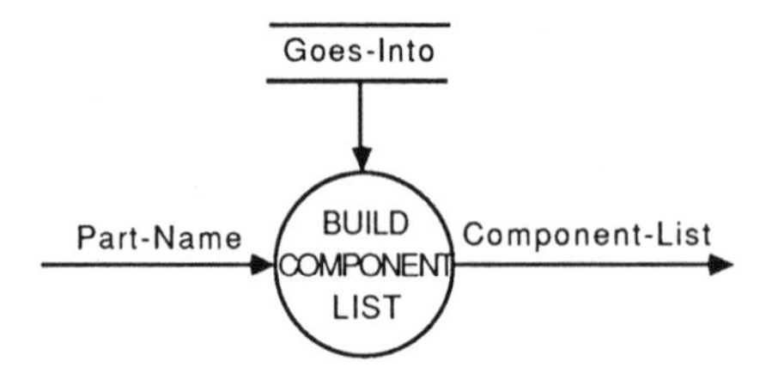

Figure 8.2.

Component-List = {Component-Name + Component-Count}
Goes-Into = {Component-Name + Assembly-Name + Component-Count}

1. Retrieve all pairs in the Goes-Into table with an Assembly-Name matching the input Part-Name.
2. Build the Component-List from the Component-Names in those pairs.

Example 2':

1. Start with an empty Component-List.
2. For each entry in the Goes-Into table, do the following:
   2.1 If the Assembly-Name matches the input Part-Name, append the Component-Name to the Component-List.
   2.2 Otherwise skip it.

Example 3 shows a more elaborate decision structure. A sample of measurements taken on a batch of parts is used to determine whether anything went wrong at the last operation. This diagnosis is made on the basis of the mean and standard deviation of the measurements and the actions taken depend on (1) whether there is any anomaly, and (2), if there is one, of what kind. When the statistics are outside their limits, the issuance of an alarm on one does not preclude the same on the other, and therefore the last two if's have no "else" clauses.

Example 3:

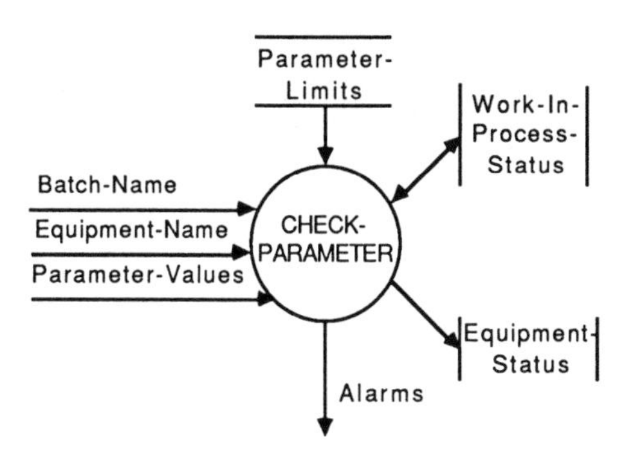

Figure 8.3

Parameter-Values = Parameter-Name + {Values}

1. Look up the Operation the Batch is at in the Work-In-Process-Status.
2. Retrieve from the Parameter-Limits the limits defined for the Parameter mean and standard deviation at the Operation .
3. Summarize the parameter Values of the Batch into mean and standard deviation.
4. If both mean and standard deviation are within limits, do the following:
   - 4.1 Set the state of the Batch to 'Pass.'
   - 4.2 Set the equipment state to 'Available.'

   Otherwise
   - 4.3 Put the Batch on hold.
   - 4.4 Bring the Equipment down.
   - 4.5 If the mean is out of bounds, issue an Alarm saying so.
   - 4.6 If the standard deviation is out of bounds, issue an Alarm saying so.

Although still manageable, Example 3 shows how quickly the complexity of decisions can rise beyond the capacity of structured natural language to represent them clearly. In such cases, the analyst should switch to decision trees or decision tables.

8.3. DECISION TREES

Figure 8.4 shows an inventory valuation procedure. The captions of the tree can be described by the following definitions:

Inventory	=	* material for which only the total number of parts is relevant. *
Lot	=	* Set of parts processed and tracked together. *
Route	=	* Sequence of fabrication operations. *
Base-Cost	=	Product-Name + Amount
	=	* Cost of the raw materials of a product.*
Standard-Cost	=	Product-Name + Operation-Name + Amount

= * Cost assigned to a part in process for a product by assigning an amount c(i) to operation i, without considering the expenses actually incurred in processing the part to this point. If the base cost is B, the standard cost of a part at the n-th operation of its route is

$$C = B + \sum_{i=1}^{n} c(i) \; *$$

Off-Route-Cost = Product-Name + Amount
= * Cost assigned to a part in process for a product while not undergoing a processing sequence — for example, stored between two routes. *

If a lot of parts is not on any route and never has been, but off route costs exist for this product at this operation, the tree shows that accounting uses them. Otherwise, accounting uses the base cost.

8.4. DECISION TABLES

Glenford Myers [Mye79] devotes several pages to the generation of decision tables from textual descriptions for the purpose of generating software test cases. This process can be bypassed by specifying the behavior of a module by a decision table to begin with, when applicable.

A decision tree is appropriate in the case of standard cost assignment to parts because the conditions are strongly coupled. Depending on the answer given at one node, the next questions will be different. If material is in a lot, the next question is whether it is on-route, off-route, or generated by a split. If it is in inventory, the next question is whether off-route costs exist.

On the other hand, what should be done if the due time for preventive maintenance on a machine depends on independent conditions with no hierarchy. Is the machine in use? Are we in regular working hours? Is a qualified technician available? A decision tree for such a problem is not particularly enlightening because too many logical combinations of answers are possible. The decision table is also a tool to ensure that all combinations of answers have been envisioned.

The first, trivial table in Example 4 shows that a machine is brought down for preventive maintenance (PM) only when it is due and the machine is not in use. The second table is less trivial and shows that preventive maintenance work is started if the machine is down, a technician is available and either it is regular work time or there is an overtime budget.

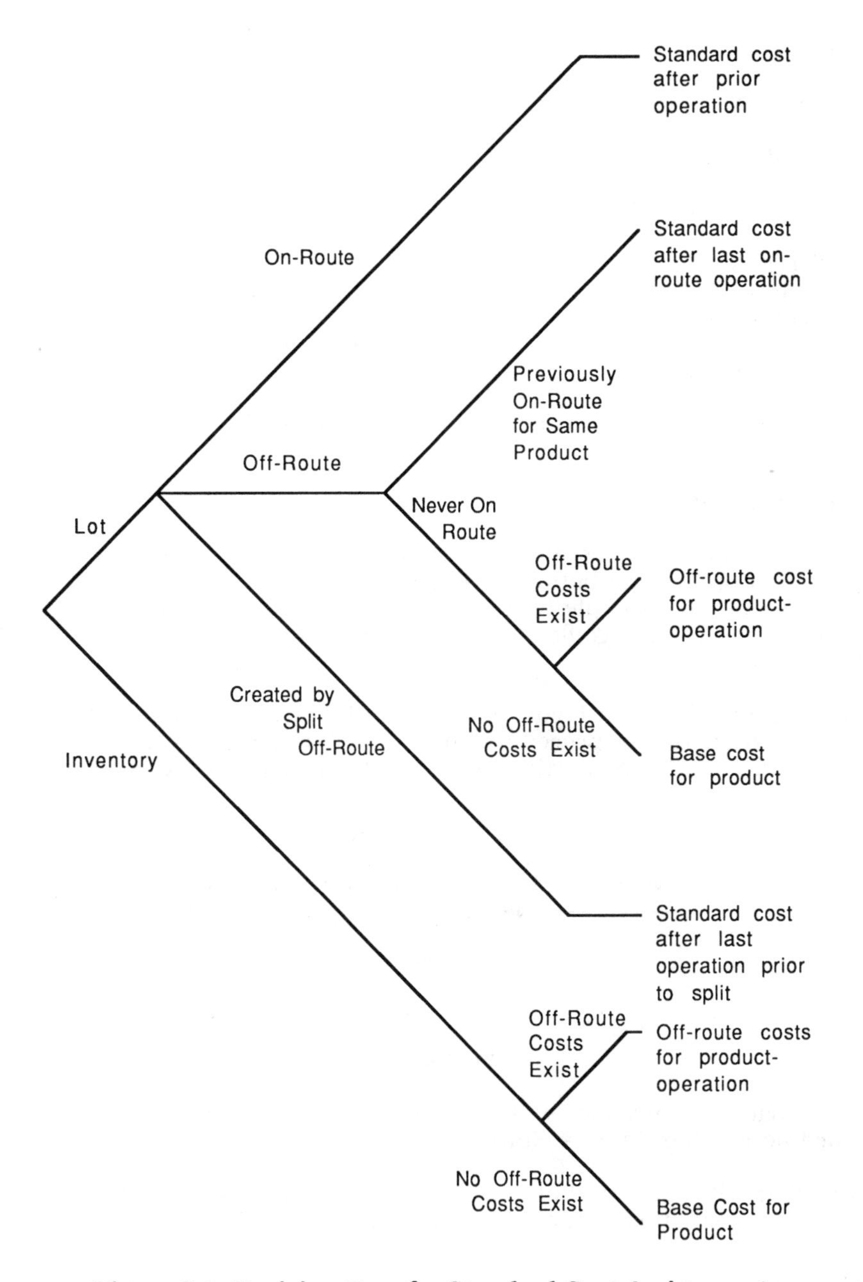

Figure 8.4. Decision Tree for Standard Cost Assignment

Example 4:

CONDITIONS 1	Y = Yes, N = No			
PM due	Y	N	Y	N
Machine in use	Y	Y	N	N
ACTION 1				
Bring down for PM	N	N	Y	N

CONDITIONS 2																
Machine down for PM	Y	N	Y	N	Y	N	Y	N	Y	N	Y	N	Y	N	Y	N
Technician available	Y	Y	N	N	Y	Y	N	N	Y	Y	N	N	Y	Y	N	N
Regular work time	Y	Y	Y	Y	N	N	N	N	Y	Y	Y	Y	N	N	N	N
Overtime budget ≥0	Y	Y	Y	Y	Y	Y	Y	Y	N	N	N	N	N	N	N	N
ACTION 2																
Start PM	Y	N	N	N	Y	N	N	N	Y	N	N	N	N	N	N	N

8.5. ALGORITHMS

The preceding examples all describe trivial algorithms. In some cases, the tasks carried out by the bottom level processors of a data flow explosion are not that simple. In Example 5, it is to schedule n jobs through 1 machine so as to minimize the average flow time of the jobs, subject to the constraint that all due times be met.

This activity, shown in Figure 8.5, cannot be divided into smaller ones. Yet the specification of the procedure to turn a Work-Load into a Schedule is nontrivial and predicated on the knowledge of an appropriate algorithm.

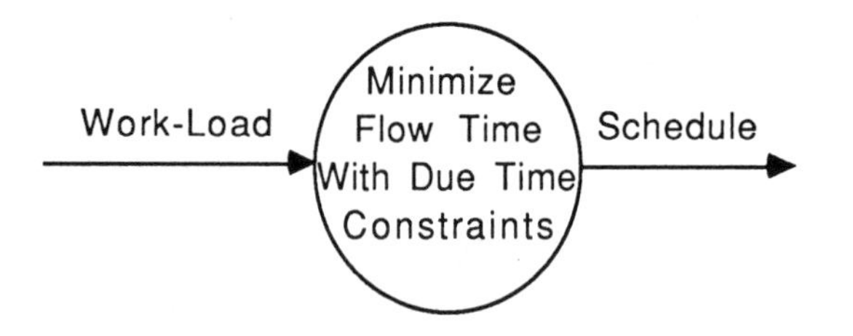

Figure 8.5.

Work-Load = {Job-Name + Processing-Time + Due-Time}
Schedule = {Job-Name + Start-Time}

French [Fre82] shows that this problem is solved by Smith's algorithm, which is described in structured natural language as follows:

Example 5:

1. Assign initial values to the following variables:
   1.1. Let time t=0 be the beginning of machine operation.
   1.2. Let t be the total machine time needed to process all the jobs.
   1.3. Let U be the set of all unscheduled jobs (initially all jobs).

1.4. Let V be the (initially empty) sequence of scheduled jobs.

2. Until all jobs are in V, do the following:
   2.1. Build the set $W \subseteq U$ of all jobs with Due-Time later than t.
   2.2. Find the job $J \in W$ with the longest Processing-Time p.
   2.3. Set $t = t - p$.
   2.4. Remove J from U.
   2.5. Append job J to the front of V.

3. In V, set the Start-Time of all the jobs to be the sum of the Processing-Times of the previous jobs.

Structured analysis does not provide algorithms, only ways to express them. Step 2.1 assumes that there is at least one job with due time later than t. This assumption is valid on work loads for which all due dates can be met, and Smith's algorithm is only applicable to those.

Structured analysis leads the analyst to separate the function of validating work loads from that of executing the algorithm itself, and thus to provide uncluttered descriptions of both.

9

ESSENTIAL SYSTEMS ANALYSIS IN MANUFACTURING

Essential systems analysis is a strategy to build system models with the tools of structured analysis by characterizing planned responses to events and partitioning memory by objects. The perfect technology assumption serves to distinguish true functional requirements from features needed only to accommodate available hardware.

Activities are classified as fundamental or custodial, depending on whether they directly produce outputs for end users or update the system memory. Administrative activities simplify the specification of fundamental activities by assuming all input validation tasks. Interactions with external entities are carried out through a ring of interface activities built around the logical core of the system.

9.1. THE ESSENTIAL MODEL OF A WORK CELL

McMenamin and Palmer [McM84] characterize software systems as "planned response systems." From their perspective, specifying a system is equivalent to saying how it should respond to certain external or temporal events, in the sense of Chapter 5. We shall examine this approach through an example.

9.1.1. Event Partitioning

"Work cell controllers" are a recurring topic in the CIM literature, but their descriptions are usually focused on hardware and communication software, while the functions they are supposed to perform are taken for granted. Let us list events a work cell controller should respond to:

PART MOVES

Material handler moves parts into work cell.
Another work area requests parts.
Planner downloads move schedule to work cell.
Planner orders inventory transfer to another work cell.
Equipment completes processing on a batch of parts.
Operator or engineer manually moves parts out of work cell.

ALARMS

Equipment sensor issues alarm.
Statistical anomaly is detected in readings from equipment sensor.
Environmental sensor issues alarm.
Statistical anomaly is detected in readings from environmental sensor.
Statistical anomaly is detected in measurements on work in process.

EQUIPMENT MAINTENANCE

Employee reports equipment trouble.
Maintenance employee completes repair.
Preventive maintenance becomes due on a machine.

ACTIVITY START-UP AND SHUTDOWN

Work cell starts up.
Work cell shuts down.

This list may be incomplete and relevant in one factory but not in another. Such a list is normally drawn in a free flowing brainstorming session and is initially devoid of structure. It is up to the analyst, afterwards, to find relevant headings under which to group entries. Aside from enhancing readability, these headings give clues about what processors the work cell controller bubble should be exploded into in data flow modeling.

The first event is "Material handler moves parts into work cell." In the data flow model, this identifies an entity, external to the work cell controller and informing it of the delivery of incoming parts. The analyst then finds out what messages the work cell controller should issue to this and other external entities in response (see Figure 9.1).

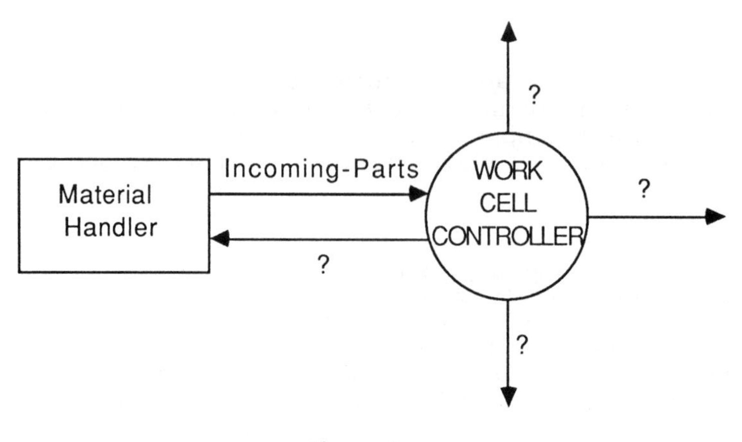

Figure 9.1.

9.1.2. Object Partitioning

As discussed in Chapter 5, an "object" is identified by a noun phrase referring to a user concern. In Chapter 6, we emphasized the semantic neutrality of concepts such as sets and relations. Object partitioning is a memory model building strategy that is based on identifying sets with objects. (See [McM84] for a discussion of such alternative strategies as "private component files.")

Following is a list of objects within a work cell, which might have come out of user discussions:

- Equipment
- Routing segments
- Employees and a unique supervisor
- Power, gas and liquid outlets
- Work calendar
- Environmental parameter distributions

Armed with this list, the analyst now has to find out the relations between objects that need to be modeled, as in Figure 9.2, and list all the object attributes in data dictionary definitions.

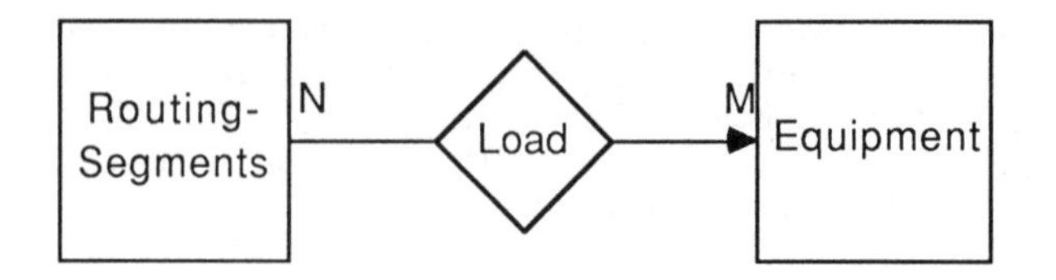

Figure 9.2.

9.1.3. Integration

Once, using event and object partitioning, the functions of a work cell controller have been specified, the analyst also needs to consider its position within the environment of the factory as a whole. In particular, manufacturing organizations should be expected to restructure their factories several times within the life of a work-cell controller.

The software system should be designed to minimize the amount of editing work generated by such activities and the opportunities for introducing errors and inconsistencies.

It must be possible to assign to or remove objects from a work cell, and several additional capabilities are also needed for more sweeping changes, such as

- Creating a new work cell.
- Deleting an empty work cell.
- Merging existing work cells.
- Splitting an existing work cell.
- Turning facilities into work cells of a larger one.

The reader will note that these activities are not planned responses to events, but changes in the memory model. They are examples of "custodial activities," another concept of essential systems analysis, and discussed next.

Each relationship between work cells and other objects defines keys for data consolidation. The minimum use of such keys is for reporting on specifications, schedules, status, or history. However, it should also be possible to use these keys for supervisory control.

Example:

An air-blower malfunction might cause a pressure anomaly in all the work cells it serves. This anomaly may be too small to trigger an alarm in any of the individual work cells, but its presence in all the work cells related to the blower could be detected by a pressure average and an alarm could be generated at that level.

9.1.4. Conclusions on Work Cell Modeling

The complete specification of a work cell controller is not attempted here. The point of the preceding discussion is to show how it could be done in a specific case, and the basis for some of the software design decisions is already apparent.

Work cells can be modeled in several ways, ranging from making "Work-Cell-Name" an attribute of other objects to dedicating a computer to each work cell. The software model should be sufficiently rich to perform the intended functions of the controller, but analysts should be aware of the rigidity introduced by dedicated hardware.

Work Cells as Attributes of Other Objects

The representation of work cells in a system originally built without this concept usually is as fields within other records. Work-Cell-Name is made an attribute of Employees, Equipment, Process-Operations, and Lot-Status. Such work cells can be used as consolidation keys in reports.

However, cross-references can only be generated by scanning the lists of Employees, Equipment, Process-Operation and Lot-Status. There is no way to represent an empty work cell nor to distinguish an error in work cell assignment from the creation of a new one.

Work Cells as Independent Data Objects

In this approach work cells are defined independently of other entities, which are then mapped to them. Work cell definitions can be under document control, work-cell assignments can be validated, and cross-reference tables generated. Such work cells are purely data elements, and rely on functional software to change their state.

Work Cells in Object-Oriented Programming

Here, work cells are software objects emulating the behavior of a dedicated computer. Active work-cell control is achieved without hardware coupling. Such a work cell is thus an element of data bundled with the software to manipulate it.

Dedicated Hardware

The most frequent perception in factories is that, somehow, a work cell controller must be a piece of hardware. It is usually depicted as a box with cables connecting it to process equipment. Each work cell has a dedicated computer.

The computer system structure is coupled with the work cell structure at the time of installation. Thus any attempt to move walls or otherwise redefine work cells may result in workload imbalances between the computers. Access to all work cell computers from all terminals may be difficult, and load sharing between computers in case of malfunction is impossible.

9.2. EXTRACTING THE ESSENCE OF A SYSTEM

In the application of event and object partitioning to the specification of a system, the analyst is faced with decisions about how truly functional a requirement is and the need to classify activities. The bubbles of the data flow diagram are all processors, but tend to cluster naturally in types, the identification of which simplifies the task of the software designer.

9.2.1. The Perfect Technology Assumption

Although it is a frequently stated goal of analysts to focus on what a system should do, as opposed to how it does it, the distinction is in practice difficult to make, and Figure 9.3 shows why. The "Daily-Production-Extract" is what GENERATE DAILY EXTRACT produces, but it can also be viewed as how the system produces its daily "Production-Report." De Marco distinguishes between "logical" and "physical" features, but gives no easily applicable, objective criterion to decide which is which.

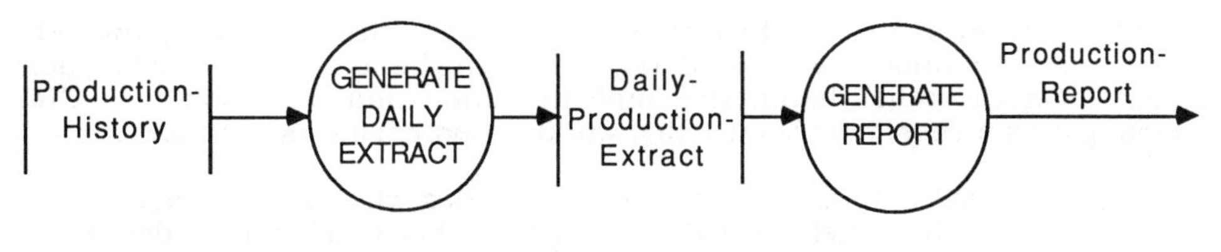

Figure 9.3.

McMenamin and Palmer suggest that the analyst should *temporarily* assume the availability of a perfectly reliable, infinitely fast computer with no memory limitations, and then ask whether, with such a computer, a requirement would still exist.

The perfect technology assumption and its purpose are frequently misunderstood. It brings forth comments like "This is the real world. We don't have perfect technology." The point of the perfect technology assumption is not to act as if it were satisfied, but to serve as a thinking aid to distinguish a feature that truly represents what a system is *for* from one that is needed only to accommodate the idiosyncrasies of available hardware.

In the previous example, "GENERATE DAILY EXTRACT" fails the test: why would someone with perfect technology go through the intermediate stage of the extract in order to generate a report? On the other hand, the availability of a perfect computer would not affect the need for production reports. Therefore, the model of the truly functional requirements is given in Figure 9.4.

Figure 9.4.

9.2.2. Fundamental and Custodial Activities

Once a data flow model has been pruned of extraneous, nonfunctional requirements using the perfect technology assumption, the analyst sees various patterns emerge. McMenamin and Palmer distinguish between "fundamental activities," which are the reasons for a system's existence, from "custodial activities," which maintain the memory of the system so that the fundamental activities can take place.

Telling a work center how many parts of a certain type it should process in the next week is a fundamental activity of a production scheduling system. Defining routings, bills of materials, and work centers is a custodial activity: it does not produce directly useful outputs but the preceding fundamental activity cannot be carried out unless it has been done.

Users buy systems for their fundamental activities, which legitimately receive more attention than custodial activities in the analysis and evaluation process. On the other hand, the implementation and operation of a CIM system places a considerable custodial workload on engineers and managers.

A well-designed subsystem of custodial activities will not win praise for its authors, but will be largely unnoticed. On the other hand, a poor design in this area will generate user frustration.

9.2.3. Administration

Under this heading, McMenamin and Palmer group all input validation activities. These may be needed to avoid recording that an event happened on a February 30, to enforce syntax rules such as the absence of spaces inside a name, or to ensure that an input name matches an entry in a table. Administrative activities do not need to be included in data flow models until fundamental activities are agreed upon.

In Chapter 8, the description of Smith's algorithm —minimizing the flow time of n jobs on 1 machine subject to due date constraints— was based on the assumption that all due dates could be met. The inclusion of an error handling mechanism would have cluttered the description, but such a mechanism is necessary if the system is to operate reliably.

Figure 9.5 shows the data flow structure such an administrative activity could have. Feasibility is established by scheduling the jobs in order of increasing due time. The issuance of diagnostic messages is treated separately because it may be done in different ways.

9.2.4. Logical Core and Interface Ring

It is not enough for a software system to produce logically useful outputs. It must also be able to communicate them to external entities. For this purpose, the logical core of the system must be surrounded by an interface ring, as shown in Figure 9.6.

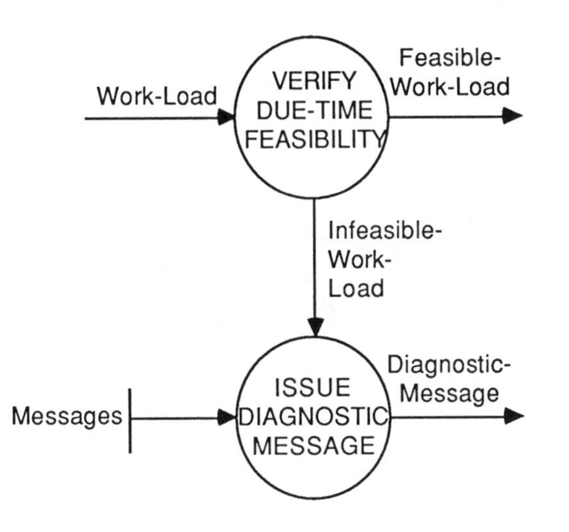

Figure 9.5.

The functions of the interface software are determined by the nature of the external entities more than by the type of data to be passed. The external entities a CIM system must communicate with fall into the four broad categories of (1) human users, (2) transport systems, (3) processing equipment, and (4) other computer systems.

Communication with human beings is carried out through video terminals for engineering and managerial work, hand-held terminals and bar code readers for shop floor transactions, and printers. Systems analysis of this area is user interface design.

If the transport system is driven by people, the interface is a special case of human communication. If the transport system is automatic, then its controller must communicate with the CIM system, and the main issues are the existence of ports and of a common language, or protocol.

Communication with process equipment involves downloading instructions from the CIM system and uploading results, both in terms of throughput and parametric measurements. The analysis issues are identical to the case of transport systems.

Other computer systems connected to a CIM system may include (1) the CIM systems of customers or suppliers, and (2) computer-aided engineering workstations. The communication issues are in principle the same as for equipment and transport systems, but the practical difficulties are lessened by the existence of widely used standards.

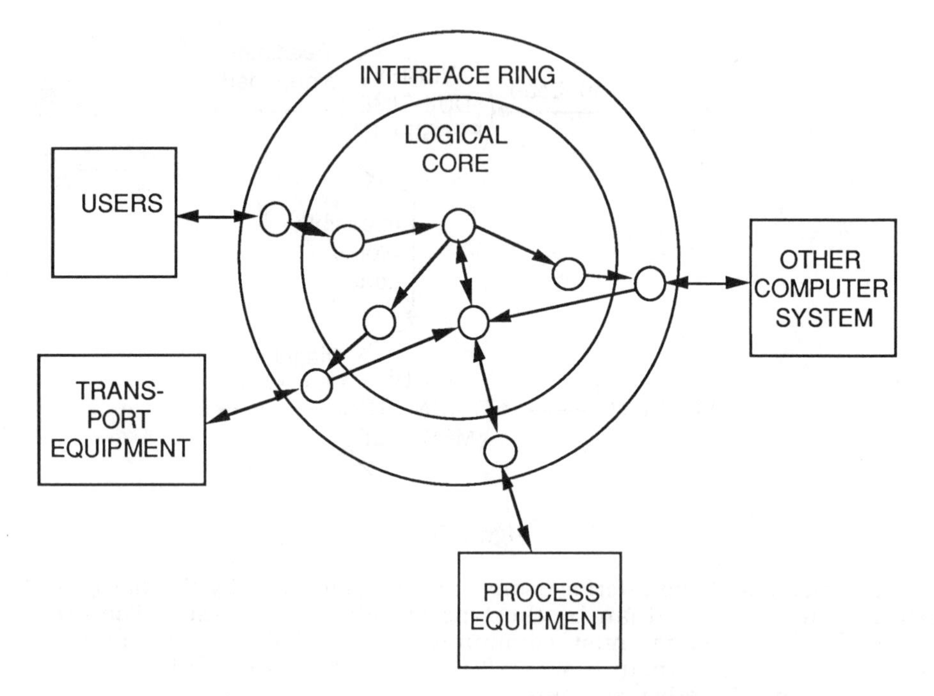

Figure 9.6. Logical Core and Interface Ring

10

MATERIAL FLOW MODELING

In material flow models, processes are viewed as networks of states or operations. Representation by pictograms is too specific for most purposes, and is replaced by the generic ASME symbols for operations, inspections, transport and storage. These symbols are traditionally used for graphic coding of specifications and industrial engineering analysis. We expand the approach to hierarchical modeling and use it to characterize general flow patterns.

10.1. MATERIAL FLOWS IN FACTORIES

In the course of normal operations, a factory buys raw materials, holds work in process, and sells finished goods. In exceptional circumstances, it may also buy or sell machines or real estate but such investment decisions are not the concern of this chapter. Making materials flow — transforming raw materials into finished goods — is the fundamental activity of the factory: it is the purpose for which it is built. The factory exists because its owners decided to make, not buy, the goods.

If a factory already has a CIM system, its memory must contain a model of material flows. The construction of this model is what, in Chapter 2, we called systems analysis in the second sense. On the other hand, if a factory is evaluating its CIM needs, then the *ability* to model its material flow pattern is a requirement of the CIM system.

Before introducing general tools, we shall examine the issues of material flow modeling through the examples of (1) slicing a silicon monocrystal into wafers for integrated circuit manufacturing, and (2) building an MOS transistor using the planar process. (The reader is *not* assumed familiar with these technologies.)

Figure 10.1 shows pictograms of the successive operations needed to turn a six-foot long silicon monocrystal into thousands of "wafers" that are six inches in diameter, 0.3 mm thick, have a polished surface and a known crystal orientation. The monocrystal is the output of a "crystal pulling" process. It initially has tapered ends and a rough, wavy surface.

The first step is to cut off the ends and grind the crystal down to a cylinder of the desired diameter. A narrow stripe on the resulting cylinder is then marked for orientation and etched. One of a variety of saws is used to slice it up, and the remaining operations shown in Figure 10.1 result in clean, polished wafers.

The preceding description, borrowed from [VLS82], emphasizes processing— that is, it shows what is done to parts and how it is done. From the pictograms in Figure 10.1, the reader can understand the procedures followed, and even imagine the appearance of the equipment.

Such an explicit representation of procedures implies commitment to a technology. The slicing operation, for example, is shown as performed with a rotary blade. Alternative means are listed in [VLS82], such as a 100,000-psi water jet or a bandsaw with a diamond coated blade, but the diagram excludes them. Such a specific diagram may be useful for training operators, but we also need tools to represent the fact that slicing takes place, regardless of how it is done.

Another approach to the graphic representation of manufacturing processes is shown in Figure 10.2, which is taken from an integrated circuit engineering textbook and outlines the making of a field-effect transistor in MOS (Metal-Oxide-Semiconductor) technology. A reader familiar with semiconductor processing could skip the following paragraphs, but in so doing, might miss the essential difference with Figure 10.1: nowhere is any mention made of how the steps could be executed.

Some of the captions in the textbook original, such as "Source and Drain Diffusion" or "Aluminum Sputtering" were in fact more technology-specific than in Figure 10.2. The change to "Generate Source and Drain" and "Deposit Aluminum" is meant to emphasize the absence of technological commitment.

A field-effect transistor regulates current flow between a "Source" and a "Drain" by applying a voltage to the "Gate" area separating them. In MOS technology, the source and drain are created by introduction of impurities in a substrate of doped silicon, the gate is isolated by a film of silicon dioxide (S_iO_2), and the interconnections are made of aluminum.

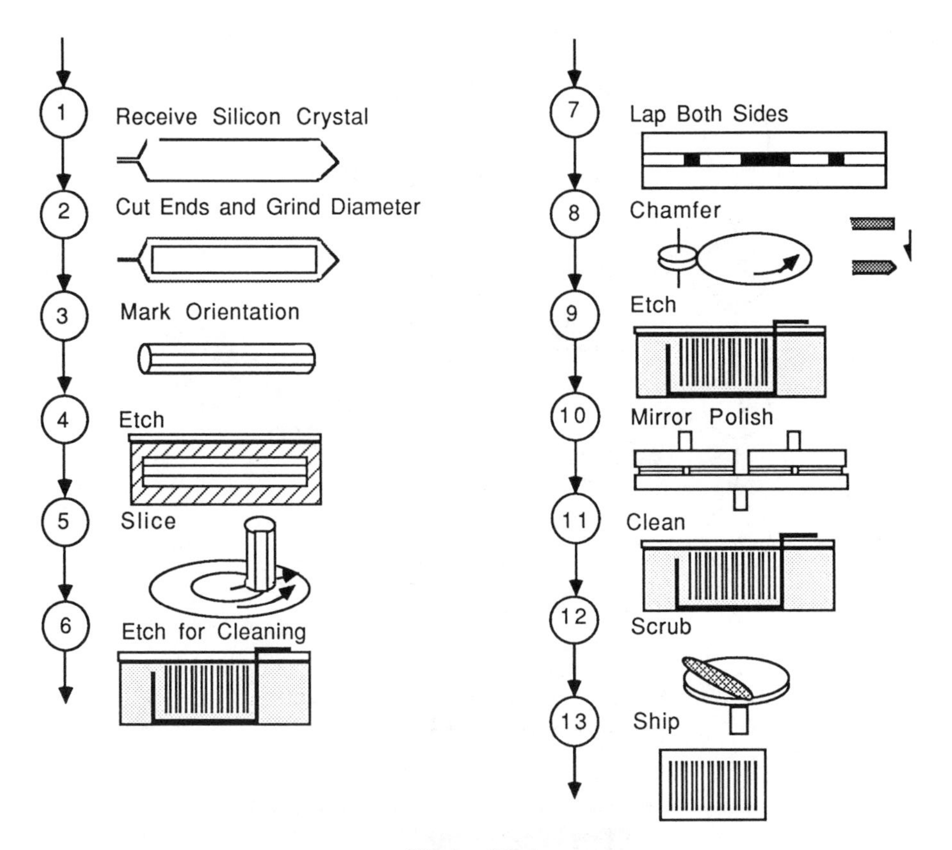

Figure 10.1.

Figure 10.2 shows the transistor being made by repeated applications of (1) thin film growth, (2) patterning, and (3) doping the substrate in selected areas. The pictograms of Figure 10.2 are not drawn to scale. If the wafer thickness is 0.3 mm and the initial oxide is 1000 Ang, a drawing to scale would show a thickness ratio of 3×10^{-5}.

An "n-type" wafer is initially doped with impurities giving it free negative charge carriers. Conversely, p-type impurities produce positive charge carriers. Once those are introduced in sufficiently large quantities in the source and drain areas, these become "p-type," in that positive charges become majority carriers. The p-type areas are isolated from each other by being embedded in an n-type substrate.

For the gate to work, the oxide covering it must be much thinner (on the order of 100 Å) than the oxide used to isolate the source and the drain from the metal film. Finally the connections of this transistor to the outside are made by opening contact holes, depositing a layer of aluminum, and etching away line separations.

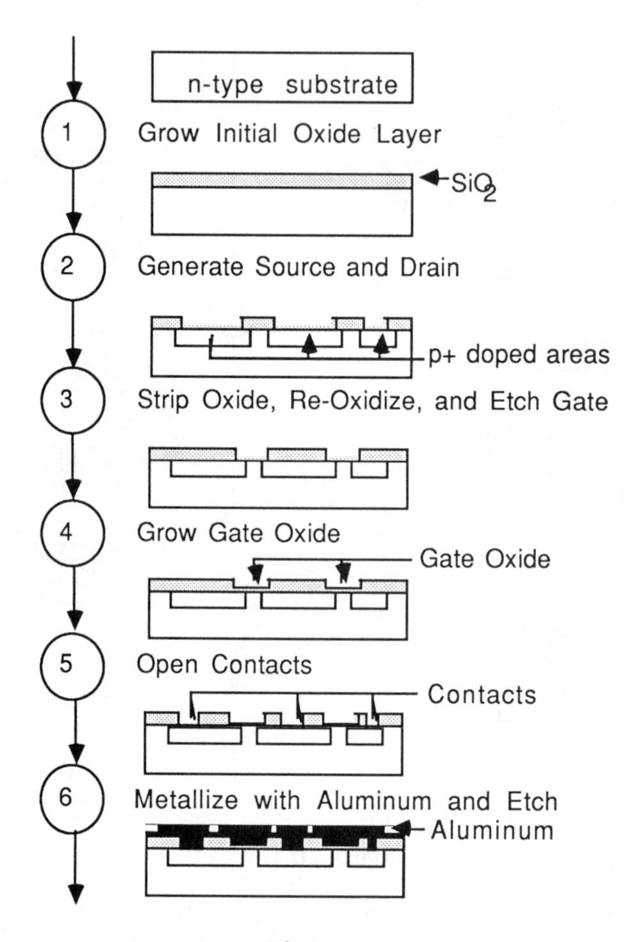

Figure 10.2.

Where Figure 10.1 showed transformations, Figure 10.2 shows a list of states without committing to any method for effecting the transitions between the states. The advantages of a representation that leaves open implementation choices are similar in this context to what they are in software. This model is portable between factories with different equipment configurations, and in particular from a pilot line to a high volume facility.

In the example, the source and drain regions can be created by diffusion of gaseous impurities into the wafer in a furnace, or by shooting accelerated ions into it (ion implantation). Etching can be done dry, in a plasma, or wet, in acid solutions. Finally patterning can be done by projection of a mask onto the whole wafer at once (projection alignment), or by optical reduction onto a section of the wafer and stepping. These processes are not interchangeable, but their domains of applicability overlap.

Pictograms were used in the preceding examples as the most obvious graphic representation for process specialists. However, when communicating with anyone else, the pictograms have to be explained, and the advantage of immediate understanding vanishes.

In this book, manufacturing and its computerization are discussed in generic terms. The intention is to introduce concepts the reader can apply to any factory. Pictograms, however, are specific to a manufacturing activity. If the concept of representing a process flow by a list of states is introduced by a semiconductor example, the reader might think that it has no relevance to metal working.

In circumstances such as operator training, there is no need to be generic. On the other hand, some aspects of a material flow pattern must be communicated to experts in such fields as production scheduling. To these experts, the process specific information in pictograms is largely irrelevant but must be explained anyway if no generic model is available. In this situation, the pictograms become an obstacle to — rather than an instrument of — communication.

Since, for pictograms to be effective, they must have some verisimilitude, drawing them becomes time consuming and expensive. A simple, more abstract representation reduces this workload. Figures 10.1 and 10.2 were drawn from the perspective of process integration— that is, the concatenation of several operations for making a complex part. This requires some knowledge of the component operations, but not to the depth required of a process engineer specialized in one of them. To build a complete MOS process, one needs to know something on how to grow oxide, how to dope silicon, and how to draw patterns on wafers, but not as much on each process as a thin films, diffusion, or photolithography engineer.

The process integration perspective is insufficient for CIM. On the one hand, we must also be able to model the details of process engineering; on the other hand, we need to represent physical flows as seen by production managers, including such elements as inventory buffers, inspections, and transport systems.

10.2. THE ASME SYMBOLS AND THEIR VARIANTS

This book uses the symbols proposed by ASME (American Society of Manufacturing Engineers) for material flow modeling. Over other conventions developed by computer scientists, the ASME standard has the advantage of having existed since before computers were invented and having been taught to generations of industrial engineers. It is a ready-made communication tool.

Through the decades, the usage of the ASME symbols has become loose. Here we tighten it for two reasons. First, we want to maximize the expressive power of the language. Second, we want to build models that can be loaded into CIM systems. Both of these goals require clarity and precision.

Many mistakenly assume loose usage to provide greater freedom in modeling. While the absence of strict rules allows the expression of more different ideas, it restricts their communication because, on the receiving end, the reader is left to infer the author's meaning from messages with multiple interpretations.

The second motivation is even stronger: it is impossible to automate vagueness. For each type of symbol to be translated into data structures, it must have an unambiguous meaning.

Having been designed to model material flows, the ASME language has most of the pieces needed to represent them. We describe some additions made in the Japanese industrial standards (JIS), and also borrow the concept of hierarchical modeling from structured analysis.

10.2.1. Vocabulary

Operation

This symbol represents a transformation of input materials into different output materials. It should be used when and only when a transformation takes place; but such activities as packing for shipment qualify as "transformations." The material leaving an operation is not of the same type as the material entering it and is not interchangeable with it.

Inspection

Material going through an inspection step undergoes no intentional transformation. It is tested against a set of criteria, and leaves on different material flows depending on the results. Inspection steps are also collection points for engineering parameters.

Queue

In the ASME conventions, this symbol designates an allowed accumulation of material in general. In this book its use is restricted to the following cases:

1. Material to be processed through a resource that is not immediately available.
2. Material placed on hold until disposed of by management.
3. Material waiting for more identical material to join it and form a batch that can be loaded into a piece of equipment.

Waiting

In ASME conventions, this symbol is synonymous with the previous one. We shall use it only for material held by shortages of other materials. For example, if A's and B's must be assembled together, an early arrival of A's may be thus held until enough B's are available.

Transport

This symbol represents the explicit use of a transport system between two operations. The distinction between a transport step and a material flow is explained next. There is a time at which material is loaded onto a transport system and a time at which it leaves it. The materials undergo no intentional transformation while underway, but may suffer handling damage.

Material Flow

Theoretically, input and output material flows should always be labeled. However, in long fabrication sequences, it is not necessary. If the input to the first operation in the sequence is part P, then the flow out of operation O is implicitly understood to be "Part P after Operation O." Spelling it out would make the diagrams needlessly busy.

In the ASME conventions, material flows are indicated by lines without arrowheads, with the direction of flow implicitly understood to be top to bottom. To maintain clarity without any orientation constraints, we always use arrowheads. A material flow symbol indicates a transfer of materials, but, unlike a transport step, it does not model the transfer activity itself. The reasons for not modeling a transfer as a transport step may be

- The machines are physically linked so that parts flow without outside intervention.
- The transfer is short and requires no critical resource.

Figure 10.3.a shows the timelines of two work centers between which the material flow model shows no transport step. The time interval between departure from A and arrival at B is unaccounted for. If this time is sufficiently short, no harm is done. However, Figure 10.3.a shows it to be as long as the time spent at either A or B.

The model in Figure 10.3.b provides a means of accounting for the time spent in transit. Loading and unloading, however, are not modeled. If the time for these activities were not short with respect to transit time, Figure 10.3.b would have to be refined further.

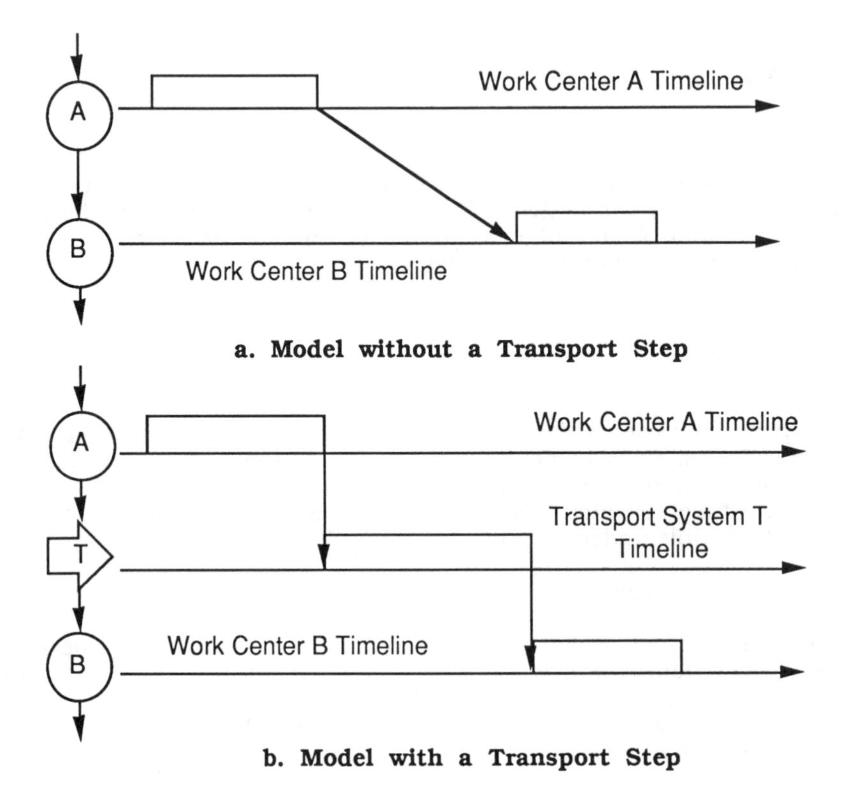

a. Model without a Transport Step

b. Model with a Transport Step

Figure 10.3.

10.2.2. Convergence and Divergence

In some circumstances several material flows point to or from an operation, and we need to give precise meanings to these constructions.

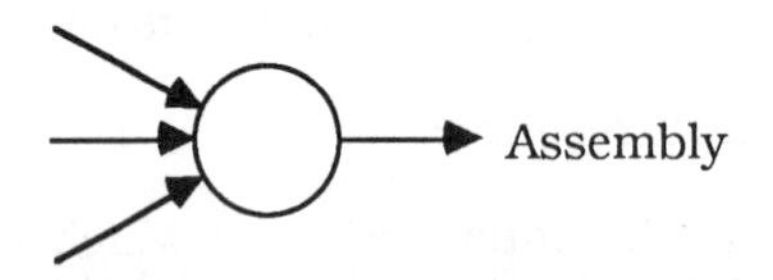

Several material flows entering an operation will be taken to mean flows of components to assembly.

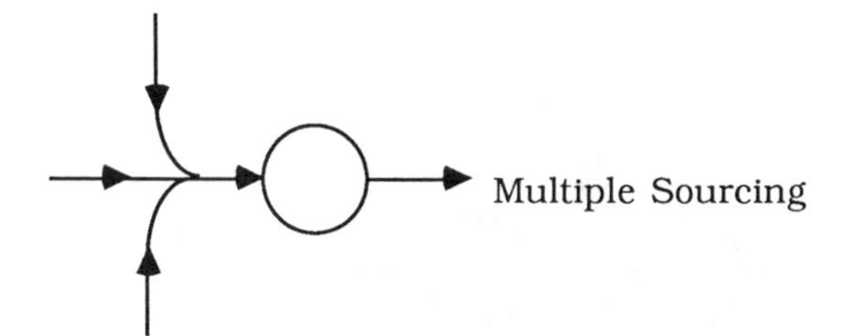

Flows merging before reaching an operation indicate a single type of part coming from multiple sources.

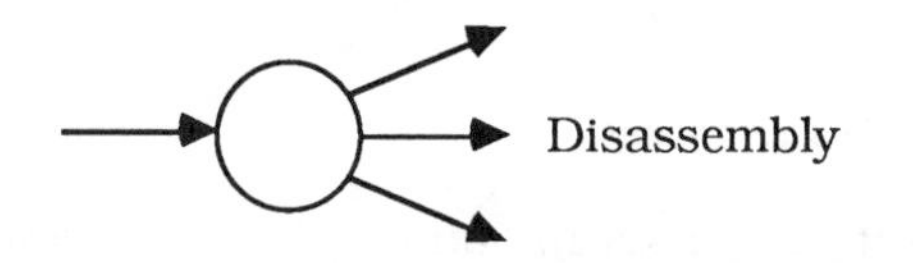

Multiple flows coming out of an operation indicate disassembly — that is, one single type of part is split into several, as an animal in a slaughterhouse.

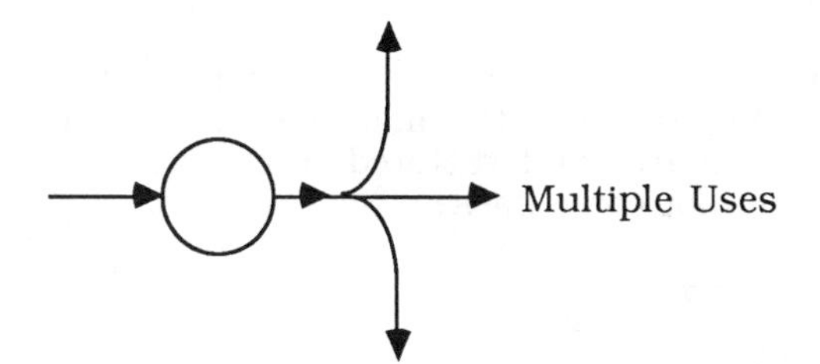

A forked output flow represents multiple uses of a single part type.

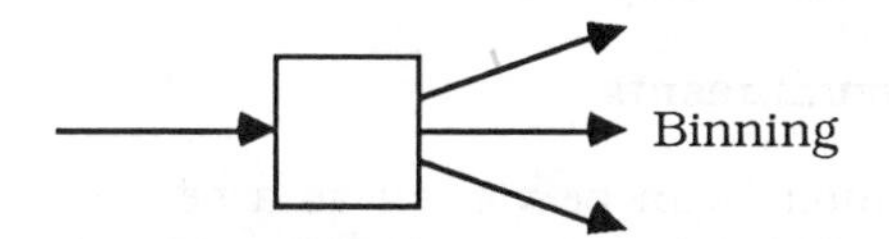

Muliple flows out of an inspection represent a split of the input flow into various grades. The simplest case is the separation of good parts from rejects. A more elaborate procedure may be sorting eggs by size — or memory chips by speed — into several salable products.

10.2.3. JIS Symbols

The ASME graphic vocabulary has been supplemented in Japan with the following (JIS-Z-8208 Standard). The scrap symbol is the only one used in this book.

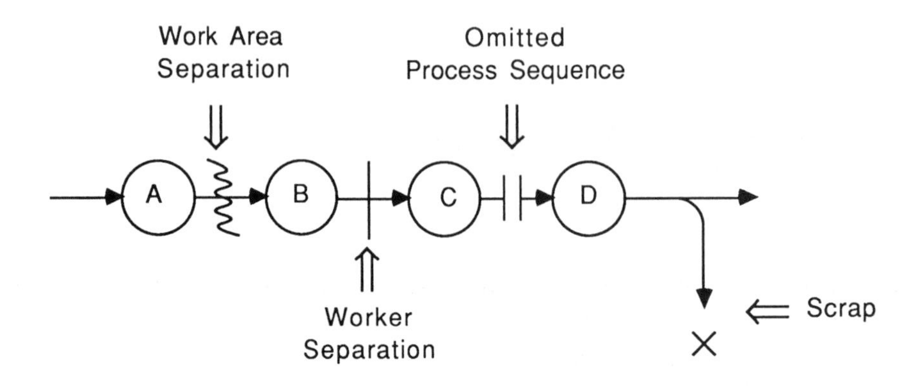

Figure 10.4. JIS Symbols

Users frequently say that they want to "unscrap" material. The problem with making flows to scrap reversible is that another word must then be found to designate irreversible disposal. If a silicon wafer is shattered, or if a faulty casting is melted, it cannot be retrieved. In this book, the JIS scrap symbol is reserved for this case.

A two-stage procedure, where material is marked as scrapped before it actually is, can be modeled as in Figure 10.5. The material can be retrieved from the A buffer by removing the mark, and restored to the normal flow. However, if it goes on to scrap, it cannot come back.

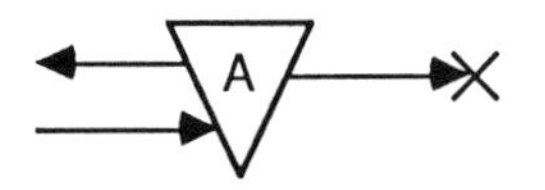

Figure 10.5. Two-Stage Disposal

10.2.4. Representation of Deepening Commitments

The construction of a material flow model can be viewed as a sequence of progressively deeper manufacturing commitments, or a gradual reduction in the number of degrees of freedom left to designers.

List of States

Figure 10.2 determines (1) the physical make-up of a product and (2) a breakdown of its manufacturing process into stages. The resulting transistor is defined not by its electrical characteristics, but by patterns, dopant concentrations, and film thicknesses. The diagram is a commitment to the construction of the circuit in MOS, to the exclusion of other technologies with which the same functional properties might be obtained.

The sequence of states shown in the diagram is one of many possible ways to break the process down into smaller units. Since Figure 10.2 is excerpted from a textbook, its purpose is didactic, and the changes from state to state are the largest that could be explained to the reader. In a manufacturing environment, the breakdown is made so that there are known technical means of carrying out the transitions. Figure 10.6 shows process modeling as a list of states in a generic sense. This model is refined in the following sections.

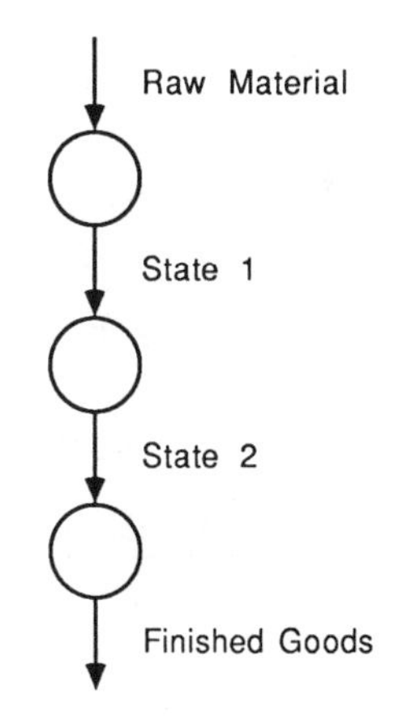

Figure 10.6. List of States

List of Operations

For each transition, the next stage, shown in Figure 10.7, is to choose one out of the many possible ways of implementing it. The transition from a list of states to a list of operations is mostly a switch in point of view. The information content of a detailed list of states is close to the description of a method to effect transitions between states in a less detailed list.

The description of a method may not imply an equipment selection down to a make and model, but it does restrict that selection. Stating that patterning is to be done by photolithography, using positive photoresist and steppers is an intermediate stage between keeping all options open and specifying model names. This stage is useful when the factory is yet to be built or when there are several sites with different equipment under consideration.

Addition of Inspections, Scrap, and Rework

Inspections, scrap, and rework are not intrinsic to the process, in that they are mechanisms to cope with failures. If the process worked perfectly as described in the preceding stage, they would not be needed.

These activities may be diseases, and, as every book on quality says, should be eliminated, but we are discussing tools to model material flows as they are, not as they should be. Inspections may be "counting dead bodies" and reworks may be uneconomical, but both happen in most factories and cannot be ignored (see Figure 10.8).

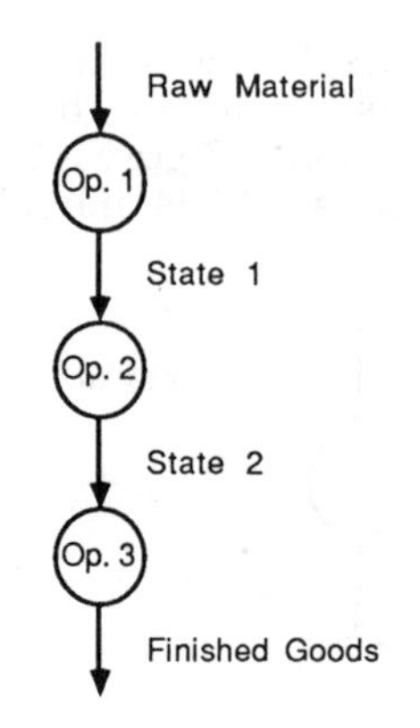

Figure 10.7. List of Operations

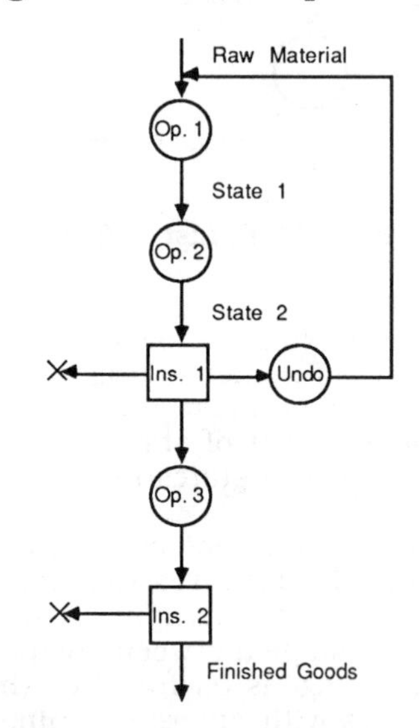

Figure 10.8. Addition of Inspections, Scrap and Rework

Assignment of Equipment and Transport Steps

This stage maps the process to an actual shop floor, specifying work centers and eventually alternative work centers for each operation. Transfers between two physically separated work centers implies resorting to a transport system. Thus the transport steps in the process definition are there as a consequence of equipment assignments. Figure 10.9 shows two work centers A and A' qualified to perform operation 2. The forking and merging material flows show that the same materials are input to and output from A and A'.

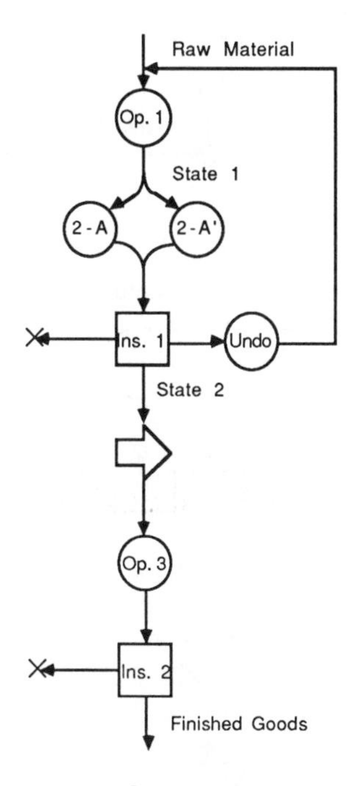

Figure 10.9. Assignment of Equipment and Transport Steps

Location and Sizing of Inventory Buffers

On the other hand, the location and sizing of inventory buffers is not deducible from equipment assignments. It is a management decision, affecting material flows in a pattern such as that shown in Figure 10.10.

10.3. TRADITIONAL USES OF MATERIAL FLOW DIAGRAMS

10.3.1. Graphic Coding of Documents for Identification

Document control organizations traditionally use the ASME symbols as a classification scheme for manufacturing specifications. A routing sheet may be encoded, as in Figure 10.11. The icon in the upper left-hand corner of the title page of every document tells the reader what kind it is.

10.3.2. Process Analysis

The reduction of a manufacturing process to a network of ASME symbols highlights its composition. In particular, it brings out the presence of too many inspections or transport steps, as in Figure 10.12.

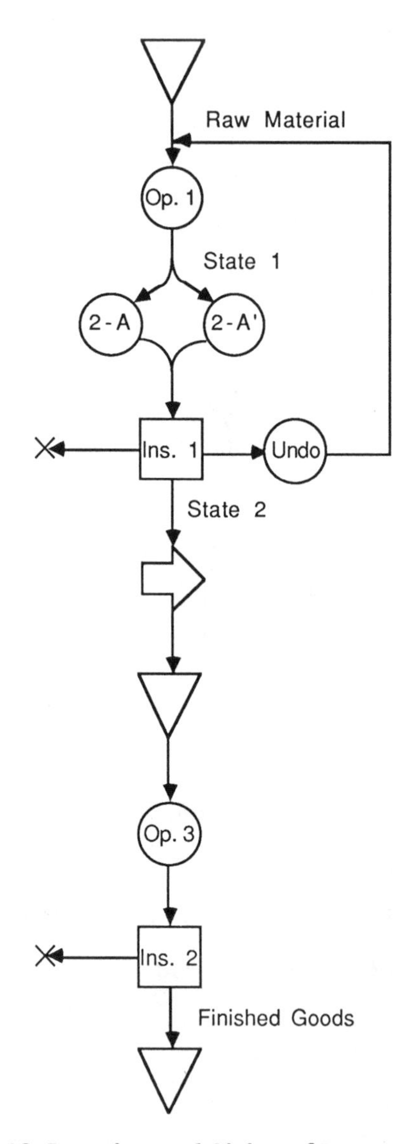

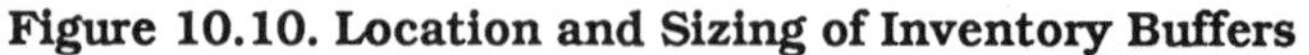

Figure 10.10. Location and Sizing of Inventory Buffers

10.3.3. Physical Flow Analysis

We have seen that the need for transport steps can be deduced from the assignment of work centers to operations simply from work center location codes. But the analysis of physical flow can be carried further by stringing out the ASME network over a blueprint of the shop floor, as in Figure 10.13.

With this type of diagram, the industrial engineer can visualize the physical flow pattern of the shop floor.

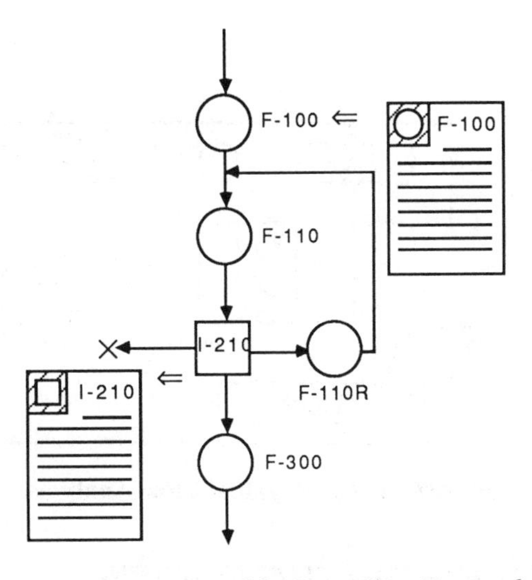

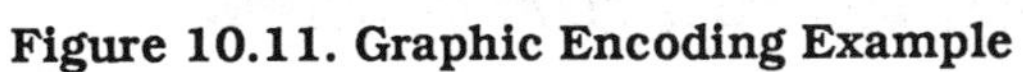
Figure 10.11. Graphic Encoding Example

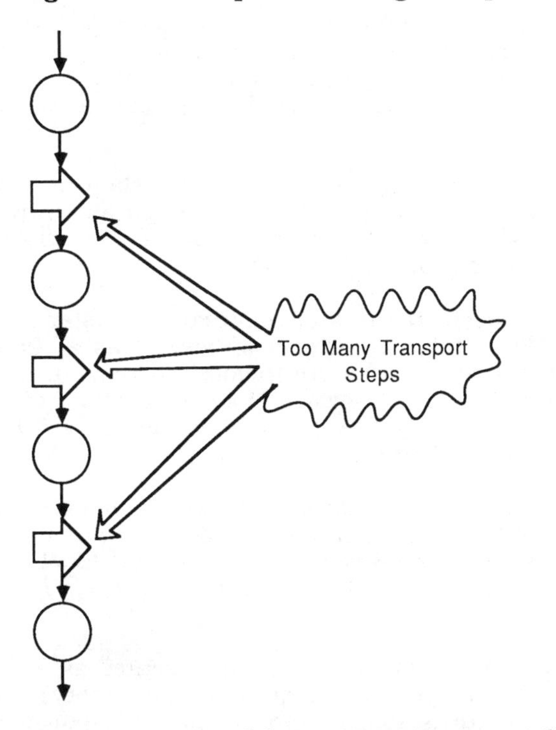

Figure 10.12. Example With Too Many Transport Steps

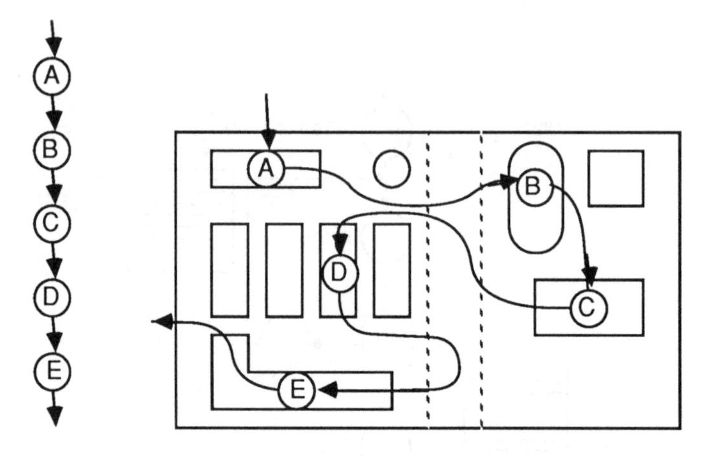

Figure 10.13. Physical Flow Analysis

10.4. HIERARCHICAL MATERIAL FLOW MODELING

In Chapter 3, the external context diagram of a CIM system was obtained by drawing a boundary around it and examining incoming and outgoing data flows. "Shipment," as a label, meant not a set of parts sent to a customer but the content of the attached paperwork. In this chapter, we carry out a similar analysis on the parts themselves.

The hierarchical approach to data flow modeling is portable to material flows. Complete models can be built or presented as multilevel explosions of external context diagrams. As in the case of data flows, this approach contributes to the clarity of the final result.

A common approach in the semiconductor industry is a fixed, two-level decomposition. Wafer fabrication may be viewed as a "Process," which is a sequence of "Operations," which are themselves sequences of "Steps." As the density of integrated circuits grows, so does the number of steps in a process. An LSI process of the early 80s could be made up of 50 operations of five steps each, for a total of 250 steps.

By the end of the decade, VLSI or ULSI will commonly have 500 steps and a complexity crying out for hierarchical modeling. For the same concepts to be applicable to a short metal working process and to VLSI wafer fabrication, we must allow operations to be sequences, or even networks, of smaller operations.

If the whole process of Figure 10.1 is represented as a single operation, as in Figure 10.14, the emphasis is placed on the factory's function as viewed from the outside. It turns silicon monocrystals into polished wafers ready to undergo the planar process of semiconductor manufacturing. Unlike Figure 10.1, it highlights the separation of crystal slicing from circuit manufacturing, and it may make the reader ask why the two processes are not integrated in a single factory.

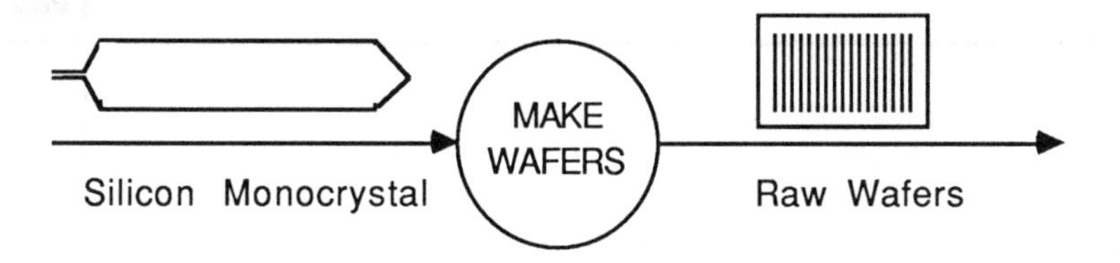

Figure 10.14. Crystal Slicing

The reasons that can be given are reminiscent of those used in partitioning a software system into modules:

- The same wafers can be used by many different semiconductor factories.
- Slicing and polishing is a mechanical process, technically far removed from the solid state physics and photochemical basis of the planar process.
- The quality of the output can be validated— that is, the quality of a polished wafer can be assessed.

On the other hand, once the planar process is started, the functionality of the resulting circuits can only be tested once it is completed.

The MAKE WAFERS bubble of Figure 10.14 can be exploded into a lower level diagram in many ways, one of them being Figure 10.1, which dives straight down to the most detailed level, and another one being Figure 10.15, where an underlying structure of the process is suggested.

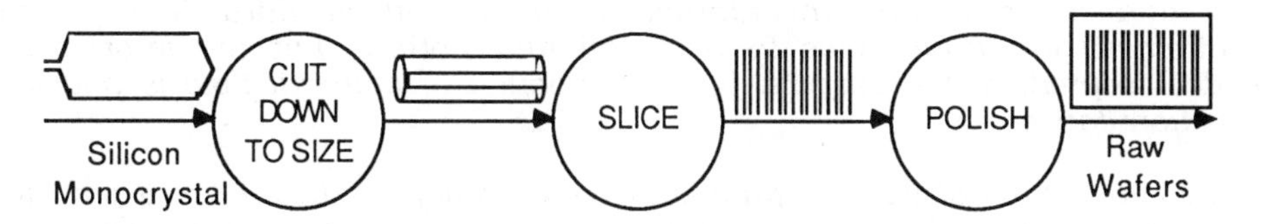

Figure 10.15. First Level Explosion

Figure 10.15 highlights the slicing operation as the keystone of the whole process, with all upstream activities being preparations for it and all downstream activities being finishing. This structure, which resembles the transform model in software design, is also found in other industries, such as die casting. A symptom of it is that the same name is loosely used to describe a process and an operation within it. There is no way the planar process could be modeled after this pattern.

Nontrivial consequences can be derived from this keystone structure. In this type of factory, for example, production scheduling tends to be simple and not a differentiator between competitors. The greatest challenge for the manufacturer is to maintain the equipment needed for the keystone operation and keep tools and fixtures available.

11

MATERIAL FLOW VERSUS DATA FLOW

Depending on how it is used, the same object may be viewed as data or material. Data is read from and written into files; materials are withdrawn from and inserted into inventories. Conservation of data flow is the requirement for input flows to carry all and nothing but the information needed to produce the outputs, but conservation of material flow is physical and quantitative.

A sample material flow diagram is walked through to show how it provides an understanding of a process, and a data flow model of a reorder point materials management system is built to drive the material flows. Its shortcomings are then highlighted by another walkthrough and by an external context diagram.

The ISO-OSI model of data communications is then reviewed for ideas on a similar model for material handling systems. Routing and scheduling concepts cross over from data to materials, but error-correction codes do not. A standard, layered model of material handling would simplify the work of both suppliers and users.

11.1. DIFFERENCES BETWEEN MATERIAL FLOWS AND DATA FLOWS

To a marksman, a gun is a tool and not a piece of data. However, to a detective, a gun is data that is read for clues. As this example shows, the determination of whether an object is data is made from the use of it and not from its physical nature.

Intuitively, there are objects which we do not hesitate to call "data" or "materials." In other cases, the following examples will show that the distinction is not so obvious. When there is a doubt, it should be resolved not from the nature of the flowing objects, but by a decision of which rules best model their use.

The decision is straightforward in the following cases:

- Engineering Parameter Value — When a dimension, a weight, or a resistivity is measured on a part and recorded, then no one would argue that this is anything but data.
- Customer Order — Likewise, orders from customers form a flow of data, whether they are communicated by mail, telephone or otherwise.
- Finished Goods — On the other hand, finished cars, or integrated circuits shipped in response to orders represent a material flow.

Other cases are not so obvious, such as

- Documentation of Military Parts — Along with integrated circuits, U.S. Defense contractors are required to ship part histories including engineering parameters. The engineering records are part of the finished goods. Are they data or materials?

- Software — Software is usually shipped to customers on diskettes or tapes. Material is being sent in a package, but its value is negligible with respect to the data it carries. Is the shipment path a material flow or a data flow?

- Subway Tokens — Subway tokens, as used in New York City, represent a right to ride, which one would normally think of as a data item. But those tokens are also minted pieces of metal circulating between passengers' pockets, collection turnstiles, and sales windows. Do they flow like materials or like data?

Material and data flow diagrams can both be said to "conserve flow," but with different meanings. Material flow conservation is a physical law stating that every particle of matter entering an operation must be accounted for: what goes in stays in until it comes out. The variation in inventory over a period of time is exactly the difference between input and output quantities.

This reasoning is not applicable to data. Figure 11.1 shows a long list of in and out quantities being reduced to a single number. There is no quantitative conservation of flow. Yet the diagram conserves data flow in the sense that the inputs contain all the data needed to produce the output and no more.

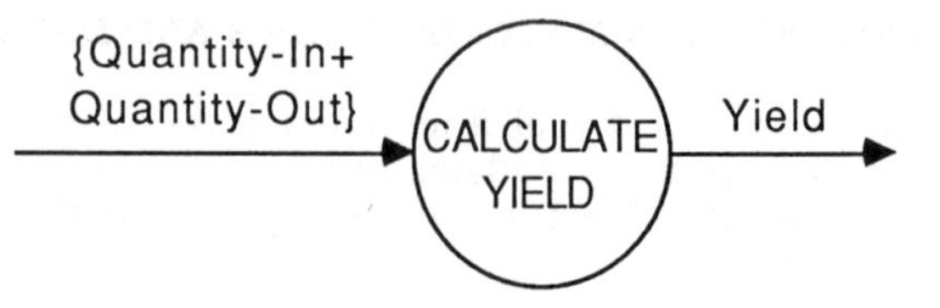

Figure 11.1. Yield Calculation

Figure 11.2 shows the opposite situation, where a few parameters given to an equation solver are all it needs to produce a long table of values.

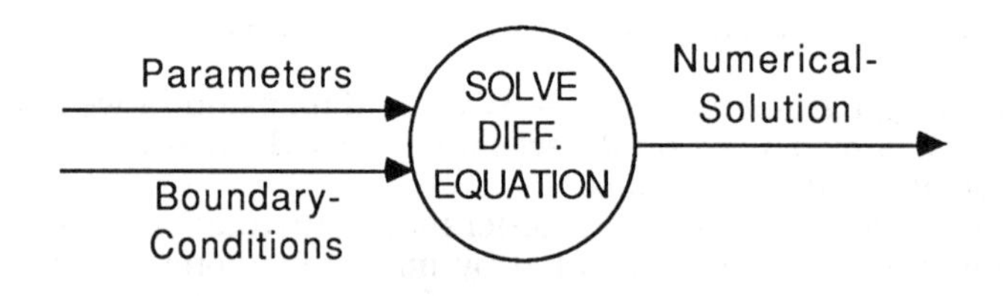

Figure 11.2. Differential Equation Solver

In this example, a handful of input parameters is turned into thousands of values. Here again there is no quantitative conservation of data flow, but there is conservation in the sense that the input flows contain all that is necessary and nothing that isn't.

Obviously, a data processor cannot produce the desired outputs without all the necessary inputs, but one might think that extraneous inputs, on the other hand, do not hurt. In fact, they do. In the case of yield calculations, an extraneous input flow may cause unnecessary data collection and increase the record keeping workload on the shop floor. In the case of the equation solver extraneous inputs would confuse users, who would assume them to be needed and be puzzled as to what values to give them.

Inventories in a material flow diagram map to data stores in a data flow diagram. However, inventories and data stores behave differently, as shown on Figure 11.3. Insertion and removal of a part in the Finished-Goods inventory both translate to updates of the Finished-Goods data store.

An arrow pointing *away* from the inventory therefore corresponds to an arrow pointing *to* the data store. An arrow pointing away from a data store indicates a read operation. Data that has been read still remains in the data store. On the other hand, material that has been removed from an inventory is no longer there.

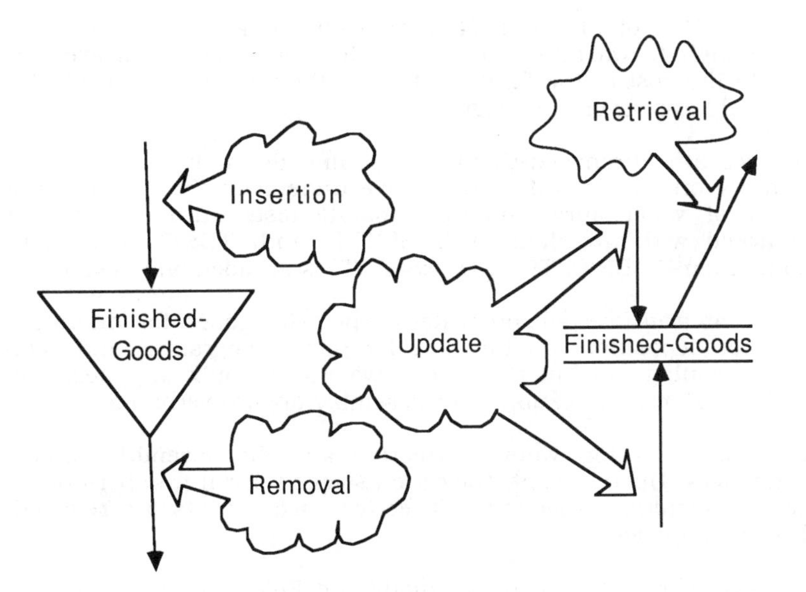

Figure 11.3. Read and Write versus Insert and Withdraw

11.2. FROM MATERIAL FLOW TO DATA FLOW

Material flow diagrams can be reviewed by walkthroughs. If Figure 11.4 represents material flows in any manufacturing activity, then it should be applicable to integrated circuits (ICs) sold in plastic dual in-line packages (PDIP). Let us test this conjecture.

Customer shipments are drawn from a Finished-Goods inventory, as shown on Figure 11.4. However, there are no spare parts to an IC; customer returns are replaced with ICs drawn again from the Finished-Goods inventory, nothing ever travels along the Spare-Parts material flow, and it could be eliminated in this case.

The bill of materials of a finished, assembled IC is as follows:

1. A die[1] of silicon containing the circuit.
2. An aluminum lead-frame.
3. Glue to attach the die to the frame.
4. Gold wire to connect the bonding pads of the die to leads on the frame.
5. Plastic to mold casings around the bonded die.

Die and lead frames are Finished-Parts, but glue, gold wire and plastic are bulk supplies, for which the name of "Finished-Parts" is a poor fit. The

[1] Although the plural should normally be "dice," the semiconductor industry uses "die." "Wafers" and "die" are terms used by every IC maker except Texas Instruments, where they are respectively called "slices" and "bars."

implicit assumption of Figure 11.4 is that the bulk-supply items are not important enough to warrant attention. If in fact these items are a major contributor to the cost of the Finished-Goods, then a separate "Bulk-Supply" inventory should appear on the diagram.

Lead-frames are Purchased-Components, and die arrive at the Finished-Parts inventory on silicon wafers that have undergone the planar process. The last steps of wafer fabrication are automatic tests, during which defective die are marked with ink dots. If "INSPECT" and "TEST" are considered synonymous, then "FABRICATE AND INSPECT" is an adequate description.

The raw materials are silicon wafers, specialty gases for diffusion and chemical vapor deposition, photoresists for photolithography, and acids for etching and cleaning. All but the silicon wafers are bulk supplies, and the preceding discussion of this subject for assembly applies verbatim.

The raw materials for wafer fabrication and die assembly come from external suppliers. Upon receipt, they are assigned serial numbers or lot IDs and undergo incoming inspections. Defective parts or lots are rejected and returned to the suppliers.

Figure 11.5 shows the result of modifying Figure 11.4 to remove the objections raised in the walkthrough, and embodies an alternative model of material flow. Whether this model is an improvement as a basis for a CIM system depends on the following factors:

1. Are bulk supplies considered important enough to justify being represented in the CIM system?

2. Does excluding the possibility of shipping spare parts to customers restrict the applicability of a CIM system?

If the IC factory is an in-house supplier for a computer maker, then that company may want to use the same CIM system in its IC and computer manufacturing facilities. Computer spare parts may have to be shipped to customers, and therefore the Spare-Parts material flow should not be deleted from the diagram.

Material flow diagrams do not show any mechanism to cause materials to move. They provide network models of factories which are independent of how production is scheduled. Material movements in a factory are caused by exchanges of information. A possible approach to materials management is shown in Figure 11.6.

In the special case discussed here, operations 1 to 4 of the material flow diagram, being generic, map exactly to bubbles in the data flow diagrams. The data transformation corresponding to an operation labeled "Load wafers into oxidation furnace" would be handled by a generic "Move-In" bubble in the data flow diagram receiving the detailed information through incoming data flows.

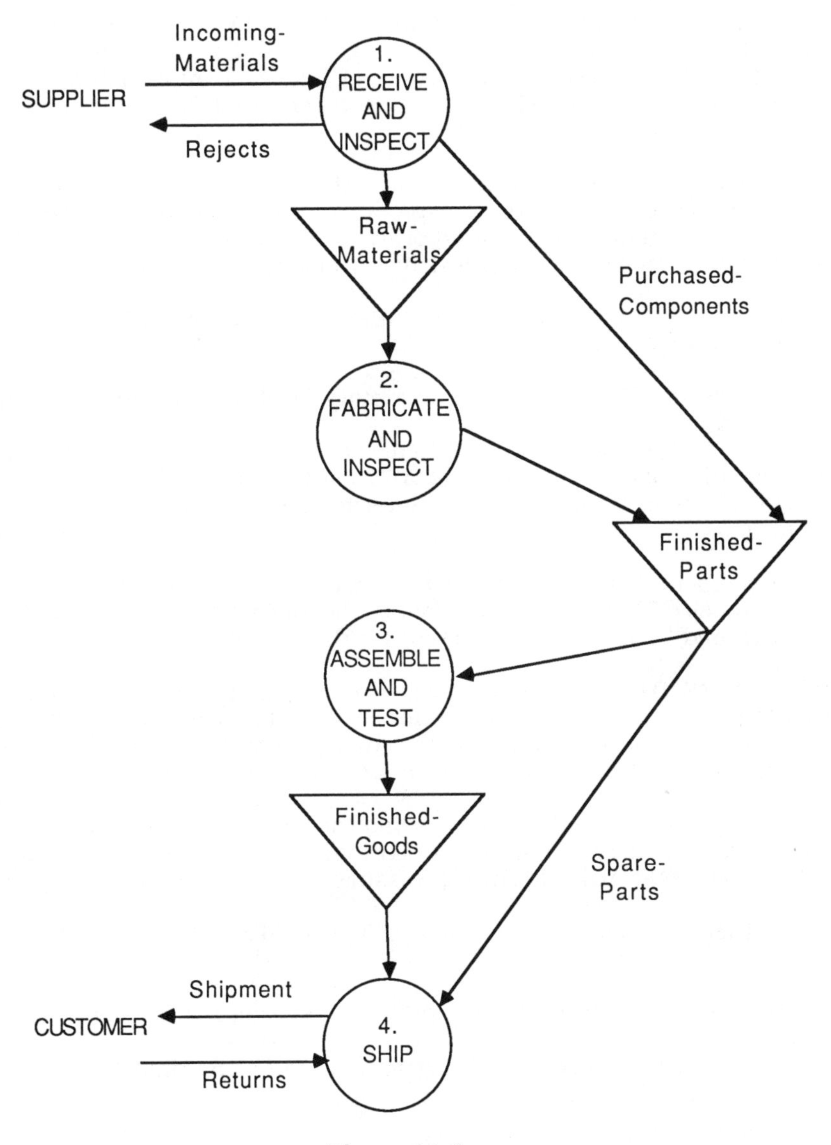
Incoming-
Materials
SUPPLIER
Rejects
1.
RECEIVE
AND
INSPECT
Raw-
Materials
Purchased-
Components
2.
FABRICATE
AND
INSPECT
Finished-
Parts
3.
ASSEMBLE
AND
TEST
Finished-
Goods
Spare-
Parts
Shipment
CUSTOMER
Returns
4.
SHIP

Figure 11.4.

The materials management system of Figure 11.6 is based on a reorder point model of inventory control— that is, material requisitions are triggered by inventory level crossings. The low level, at which replenishment becomes necessary, and the high level, to which inventory is brought back up, are decided by the materials manager.

Customers are external entities, and both a data source and a data sink. A customer order for n parts of type A is communicated to the materials management system, which incorporates it in a Shipment-Schedule, along with other customer orders and a demand forecast for the longer term.

Before making a Shipment according to the Shipment-Schedule, SHIP does the following:

1. It reads the Finished-Goods inventory level and the Reorder-Points for the products involved.
2. If the Shipment causes a Reorder-Point crossing, then a Finished-Goods-Requisition is issued to ASSEMBLE-AND-TEST.
3. The Finished-Goods inventory level for parts of type A is decremented by n.

What happens if the scheduled Shipment exceeds the amount of finished goods in stock? This case is overlooked by the system described in Figure 11.6. It should be modified to account for this possibility.

When ASSEMBLE-AND-TEST receives a Finished-Goods-Requisition, it checks the Finished-Parts inventory for Reorder-Point crossing and issues the appropriate requisitions either for Purchased-Components or for Fabricated-Parts. The possibility of running out of stock is ignored here, too.

FABRICATE-AND-INSPECT works likewise to meet the requirements for fabricated parts. In response to requisitions for Purchased-Components and Raw-Materials, MANAGE-MATERIALS issues Purchase-Orders to suppliers.

Several features should be added to Figure 11.6 to cope with the risk of running out of stock:

1. MANAGE-MATERIALS could spread large orders over time in the Shipment-Schedule.
2. Out-of-Stock alarms could be generated and communicated to Production Control.

Figure 11.6 is so busy that it is no longer clear what information the manufacturing system as a whole exchanges with its environment. Figure 11.7 shows what happens when bubbles 1 through 5 and all the data stores are "imploded" into a single bubble interacting with the external entities.

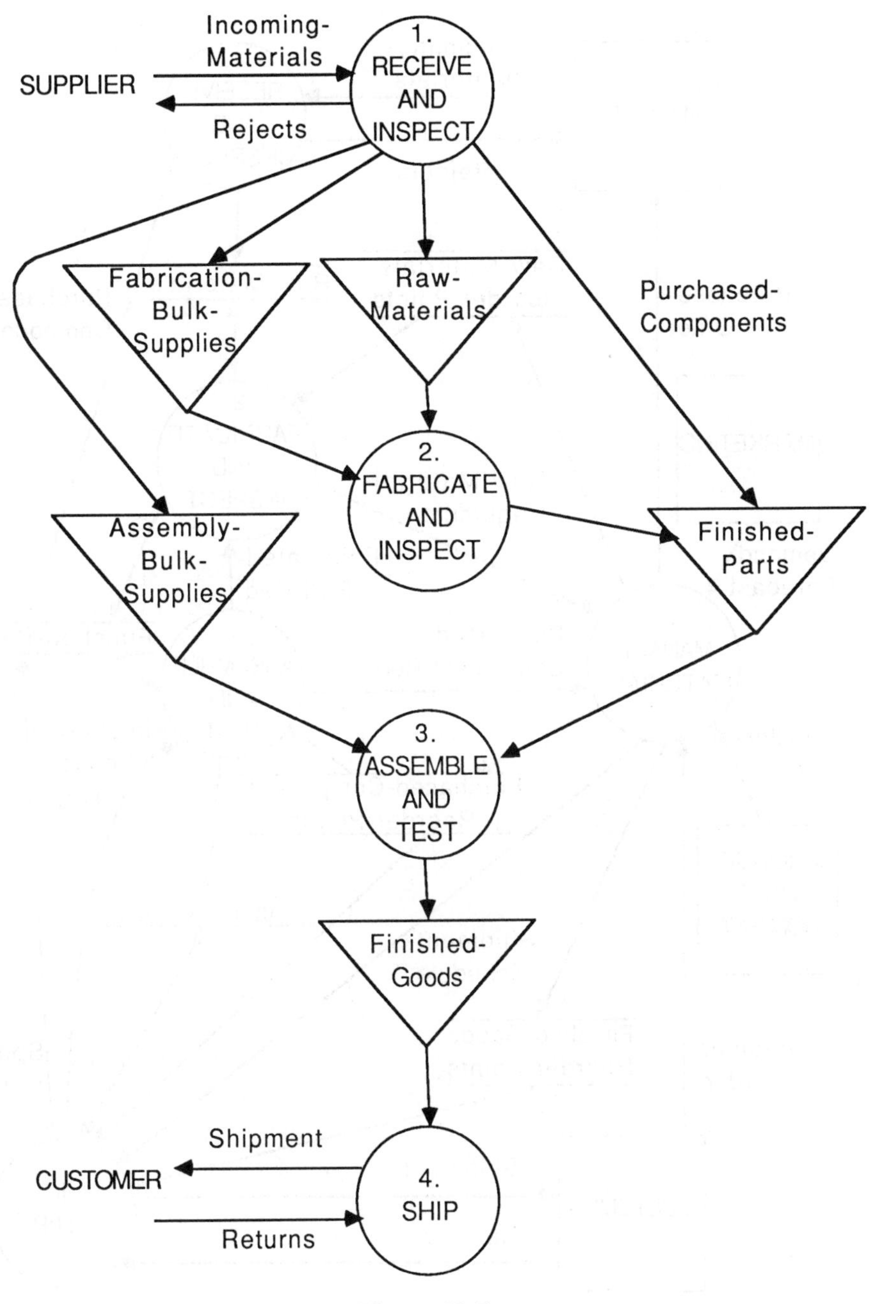
Incoming-
Materials
SUPPLIER
Rejects
1.
RECEIVE
AND
INSPECT
Fabrication-
Bulk-
Supplies
Raw-
Materials
Purchased-
Components
2.
FABRICATE
AND
INSPECT
Assembly-
Bulk-
Supplies
Finished-
Parts
3.
ASSEMBLE
AND
TEST
Finished-
Goods
Shipment
CUSTOMER
4.
SHIP
Returns

Figure 11.5.

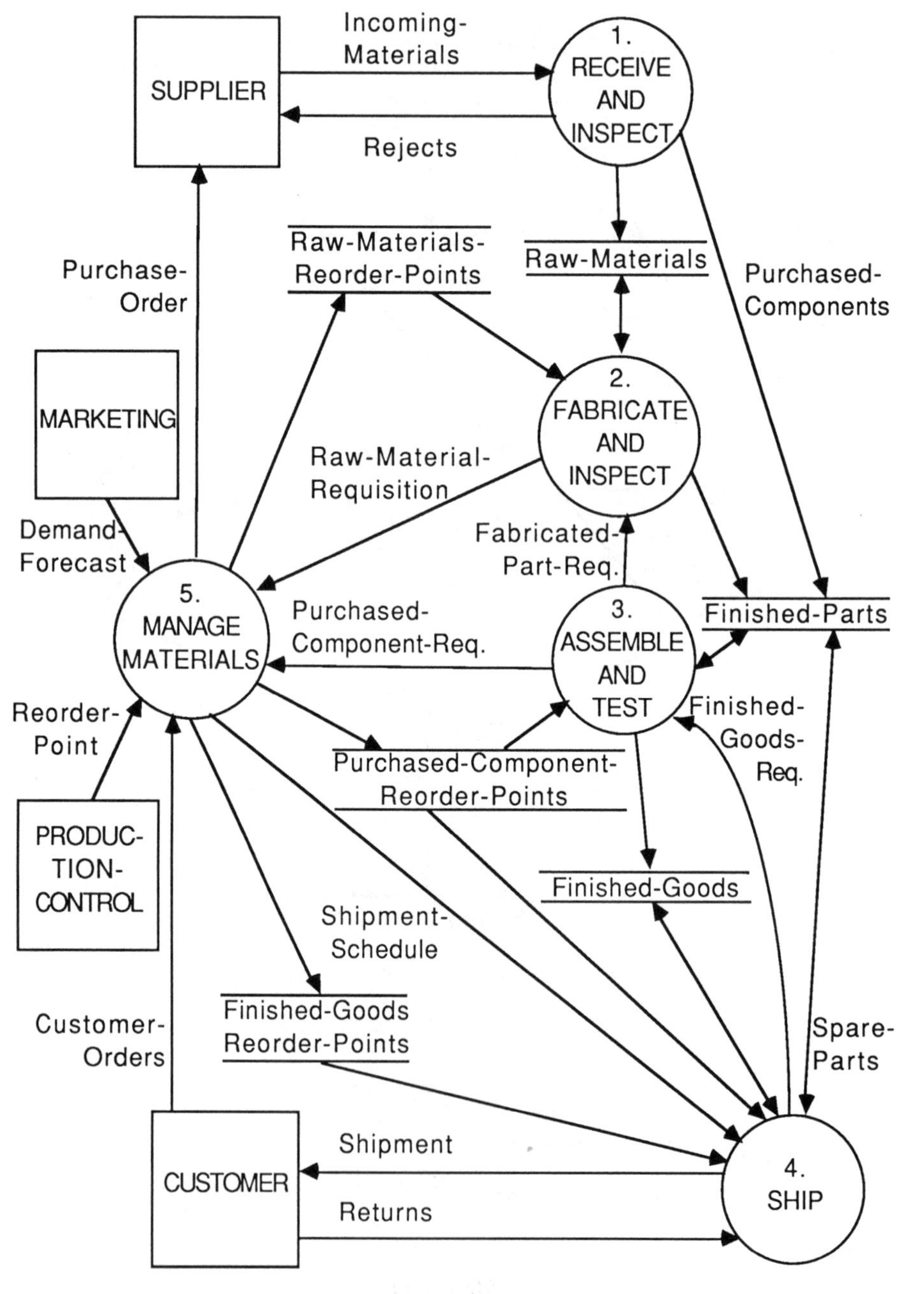
SUPPLIER
Incoming-
Materials
Rejects
1.
RECEIVE
AND
INSPECT
Raw-Materials-
Reorder-Points
Raw-Materials
Purchase-
Order
Purchased-
Components
MARKETING
2.
FABRICATE
AND
INSPECT
Raw-Material-
Requisition
Demand-
Forecast
Fabricated-
Part-Req.
5.
MANAGE
MATERIALS
Purchased-
Component-Req.
3.
ASSEMBLE
AND
TEST
Finished-Parts
Reorder-
Point
Finished-
Goods-
Req.
Purchased-Component-
Reorder-Points
PRODUC-
TION-
CONTROL
Finished-Goods
Shipment-
Schedule
Customer-
Orders
Finished-Goods
Reorder-Points
Spare-
Parts
Shipment
Returns
CUSTOMER
4.
SHIP

Figure 11.6.

The limitations indicated in the subtitle of Figure 11.6 — namely the omission of the data flows corresponding to manufacturing process definitions, manual data entry, and historical analysis — become obvious in Figure 11.7. It is also clear that no more objects could be added to Figure 11.6 without making it unreadable.

11.3. TOWARDS A LAYERED MODEL OF MATERIALS HANDLING

11.3.1. The ISO-OSI Model of Data Communication

The International Standards Organization (ISO) has defined a seven-layer *open system interconnection* (OSI) model, which is being used more and more to describe computer networks. By enabling engineers to understand precisely what they mean when they say that a system belongs in the "transport" or the "presentation" layer, the ISO-OSI model eliminates many opportunities for confusion.

No such standard model exists in the world of material handling and transport systems. As a result, two automatic guided vehicle systems are described in terms that make them sound identical, when in fact one of them travels on a fixed path and must be told which parts to move, while the other one chooses where to go and sequences parts according to some user-specified criterion. An equivalent to the OSI model for materials would clarify such matters.

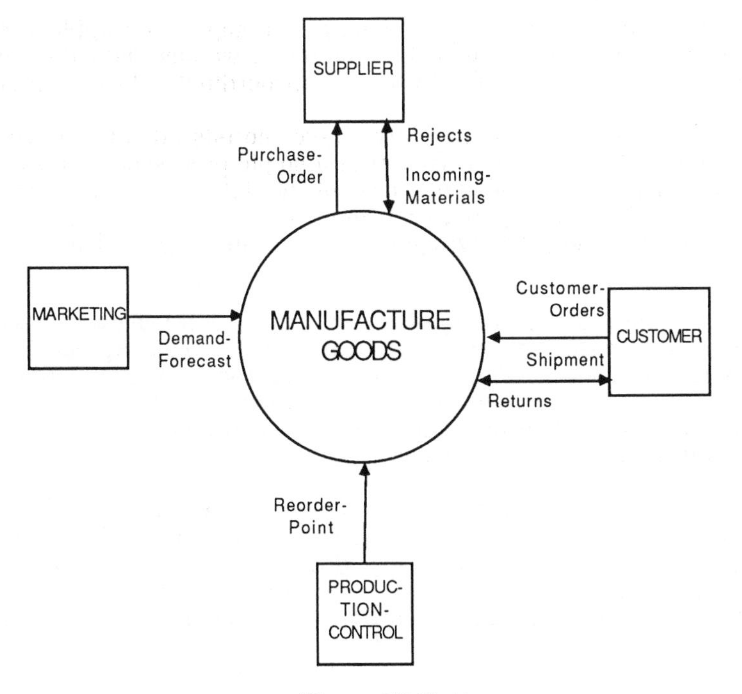

Figure 11.7.

The following summary description of the OSI model is based on Tannenbaum [Tan81]. The systematic comparison of data communications with material handling has two purposes. The first one is to give the software engineer a feel for some critical issues of material handling. The second one is to introduce the manufacturing engineer to the OSI model, on which such standards as General Motors' *manufacturing automation protocol* (MAP) are based.

Figure 11.8 shows why, in contrast with many other books on CIM, we devote so little space to computer communications. Systems analysis is about what applications say to each other, and not about the means by which messages are transmitted. On the other hand, materials handling and transport systems are integral parts of manufacturing systems, and therefore a legitimate object of inquiry.

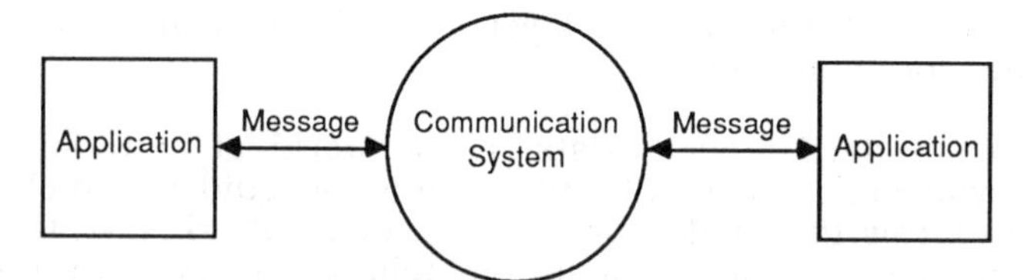

Figure 11.8. Communication Context

Layer 7 — Application

Assuming that computers can exchange messages, the application of this capability remains to be decided. In Chapter 4, we saw data flow models of software systems as towers of networks of uncoordinated logical processors.

Most software designs to date have consisted of simulations or approximations of such structures within a single processor, but a conceivable alternative consists of choosing one level of the data flow model, assigning one processor to each, and mapping each data flow to a communication channel. This approach is called the *data flow architecture*, and is clearly not easy to implement.

A set of rules, called an *application layer protocol* would have to be specified for processors to handle the traffic of messages in the absence of a control structure. More generally, such protocols are associated with any model of distributed computation or distributed databases. The term "protocol" is also used to designate software running on each node of the network and implementing the rules.

The preceding discussion crosses over to material flow systems. Machines, shops, and factories are physical processors exchanging materials through handling and transport systems. The designers of manufacturing systems decide whether to go from iron ore to finished automobiles under a single roof as in Ford's River Rouge plant in the 1930s, or to resort to a network

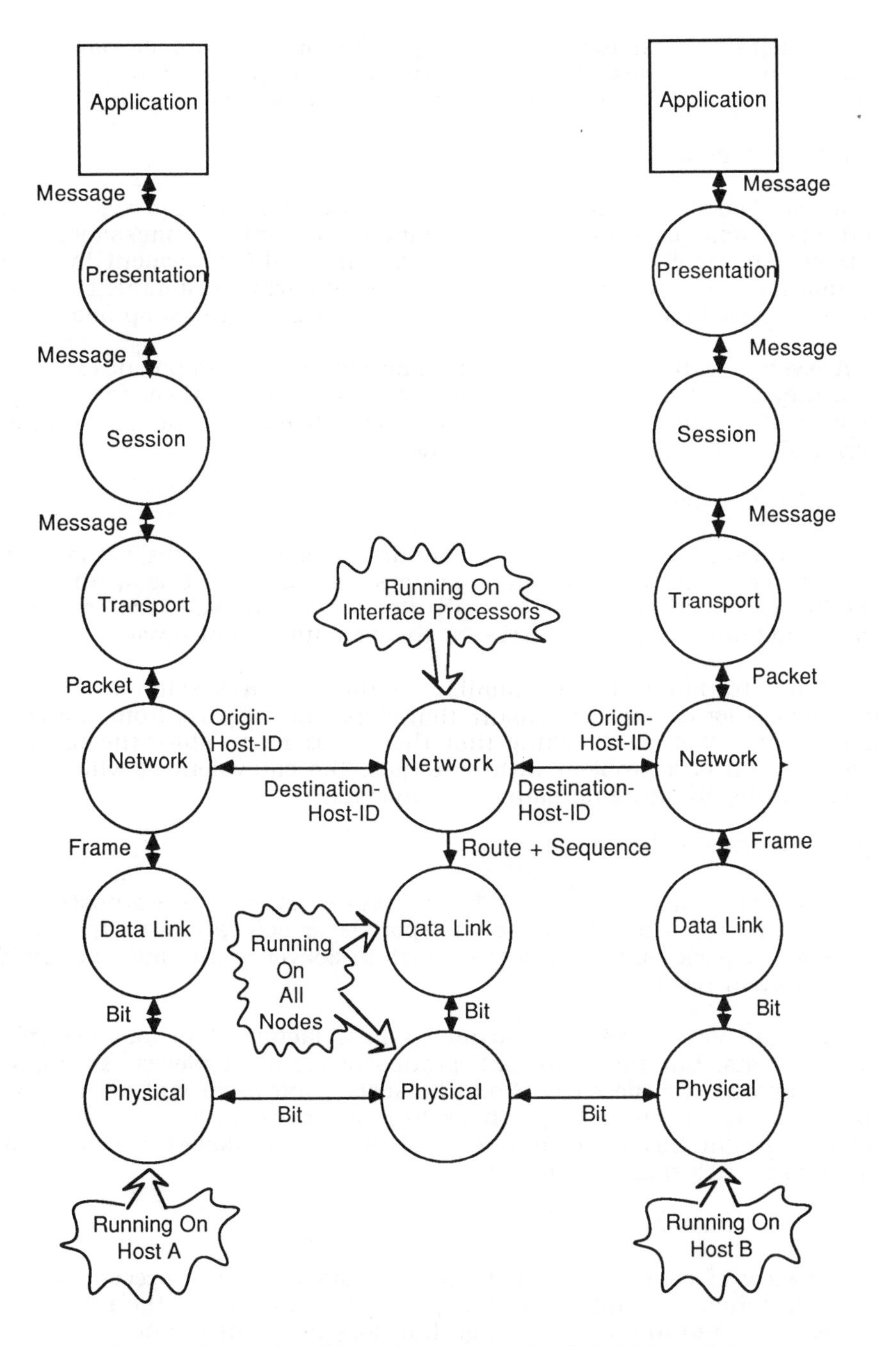

Figure 11.9. The Seven Layers of the ISO-OSI Model

of suppliers. The network design specifies movements of parts between manufacturing nodes, and the network wide production planning and scheduling system takes on the role the application layer protocol.

Layer 6 — Presentation

When needed, the presentation layer of a communication system encrypts and decrypts, compresses and expands, or converts messages between different character codes. In material handling, the equivalent layer would contain all operations consisting of packaging parts for shipment to a remote site and unpacking them into the containers used on the shop floor.

A system with such features would accept lots for shipment exactly in the form they are produced in the origin facility and deliver them ready for use at their destination. A system without these features places a preparation workload on the shipping and receiving organizations.

Layer 5 — Session

The session layer sets up the communication prior to the actual exchange of messages, establishing such matters as whether it is authorized and whether it should go in one or both directions. This software can also have accounting functions to bill users for the communication service.

Administration functions similar to those of a session layer exist for materials — for example, to ensure that shipments of parts from one factory to another are allowed. Checking that there is room to store the parts at the receiving end is a problem with materials, the equivalent of which does not appear in the literature on data communication.

Layer 4 — Transport

Once a session has been established, processes on the two hosts are ready to communicate. The role of the transport layer software is to split messages into smaller packets, rely on lower layers to deliver them, and reassemble the packets upon receipt.

For transport between departments in a factory, lots may be split into smaller units, but these are still groups of identical pieces, as opposed to sentences within a message. From the strict point of view of production, it is not necessary that they be put back together in the original order. Transport between geographically separated factories is more likely to involve grouping lots onto a truck than splitting lots.

Layer 3 — Network

The view of communication between hosts in all the preceding layers is purely in terms of results. The transport layer cares about the integrity of the packets delivered to the destination, but does not want to know how they get there. That function is delegated to the network layer and is carried out through a *communication subnet* — that is, a network of physical processors driven by the network layer software. The network layer communicates between machines as opposed to between processes.

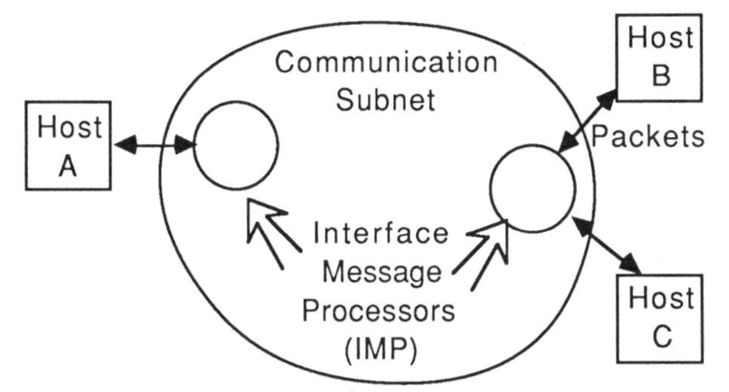

Figure 11.10. Communication Subnet

When the subnet offers several paths between two hosts, the selection is made by the network layer. In many local area networks, packet sequencing, and the resolution of contention between nodes, is more of an issue than routing.

For materials, the "transportation subnet" is the infrastructure of roads, paths, and vehicles through which a transfer is carried out. As in the case of data, only the result matters to the higher levels. An overnight delivery service, for example, is a black box to its users. To them, it doesn't matter which path a parcel follows, provided it arrives intact and on time. The functions within the delivery service which route and sequence parcels comprise its network layer.

Layer 2 — Link

The data link layer is in charge of communication between two *adjacent* nodes. Once a route is chosen and the transfer of a packet is scheduled, the network layer splits it into frames and the data link layer ensures that these frames are correctly transmitted from one node to the next. The receiving node acknowledges the arrival of complete frames, and notifies the sender of a timeout when an expected frame has failed to arrive. If a frame is damaged on arrival, the data link layer software has it resent, and it can also detect and discard duplicate frames.

The acknowledgment and timeout detection functions cross over to the material handling realm. However, preventing damage to a shipment is clearly much more vital with materials, because they cannot be easily duplicated and resent.

Layer 1 — Physical

The physical layer receives bits from the data link layer and is responsible for delivering them to the data link layer of the receiving node. It turns the bits into electrical signals appropriate for transmission, and adds various error detection and correction mechanisms to compensate for the unreliability of the physical transmission lines. It is not obvious that there is any function in material handling systems corresponding to the physical layer of the ISO model.

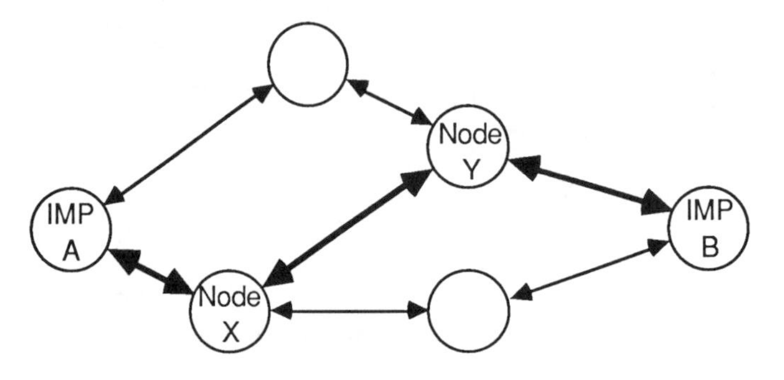

Figure 11.11.

11.3.2. Ambiguities in the ISO-OSI Model

In Tannenbaum [Tan81], Ethernet is a network layer protocol. In the January, 1988 issue of *MacWorld*, it is placed in the data link layer. This contradiction shows that the ISO-OSI model has not been an unqualified success in resolving ambiguities, especially regarding concepts like Ethernet which predate it.

The hosts on an Ethernet network are connected to a common cable, and operate like panelists with individual microphones tied to the same public address system. They are courteous enough to hear out the current speaker, but, as soon as he or she stops, they may all attempt to "grab the ether" simultaneously. Ethernet rules then resolve which one gets it.

Since these rules serve to sequence the transmission of packets through the network, they form a *network* protocol. However, in such a network, all nodes are adjacent, and the rules can also be viewed as means to ensure reliable transmissions between adjacent nodes, making Ethernet a *data link* protocol.

Such imperfections do not diminish the value of the ISO-OSI model. They only show that we should have limited expectations on the results of such a standardization effort if in the area of materials handling.

11.3.3. Similarities and Differences between the Models

When occurring across continents, data communication and transport of material transport are both constrained by the available infrastructure: the telephone network for data, and roads, railroads, and airlines for materials. On the other hand, within one site, the designer has the freedom to tailor the infrastructure to the needs, and this is true both for data and materials.

Some types of networks ensure that what is being sent is received in the same order, while others don't. This distinction is relevant both for data and materials. A *connection-oriented* service[1] connects two points as if by a conveyor belt, preserving the sequence in which data packets are sent. A *connectionless* service[2], on the other hand, sends data like parcels in trucks with destination addresses, and guarantees their arrival but not their order. Conveyor belts and gravity feeds preserve the order in which parts are loaded onto them, while hoppers, bowl feeders, and baskets return them randomly.

In data communication, messages are broken down into packets, which are themselves split into frames and eventually transmitted as bits. The difficulty associated with reception in a different order than shipment is the reconstruction of the original message.

Although large machines may have to be disassembled to be moved, it could be argued that the order in which identical screws are delivered is irrelevant. However, this point of view ignores the secondary function of the parts as carriers of quality information. An engineer can read the sequence of screws leaving a machine for clues on tool wear, but a material handling system that breaks this sequence also scrambles this message.

Many techniques of data communication rest on the possibility of resending packets of data destroyed in transit. If, in the Aloha system, two stations broadcast simultaneously, the messages jam each other, and the conflict is resolved by resending them at random later times. There is obviously no equivalent to this type of thinking in material handling, and the computer network literature devoted to it has no crossover value.

11.3.4. A Layered Model for Materials Handling

There is no layer-by-layer match of the ISO model to material handling, but the concept of a layered model itself is worth carrying over. Such a model could serve as a basis for industry standards and promote communication between system suppliers and users.

11.4. CONCLUSION

After discussing analysis methods for software and for manufacturing systems we have, in this chapter, examined similarities and differences between the objects they act on. The key idea was that the labels of "data" or "materials" do not refer to the physical nature of objects but to what we intend to do with them. We examined the consequences of this observation in two examples. In the first, we used a material flow model as a foundation on which to build a materials management system. In the second, we reviewed a standard model of data communications for insights into the construction of a similar model for materials handling.

---

[1] Formerly "virtual circuit service."

[2] Formerly "datagram service."

12

MANUFACTURING FUNDS FLOW

A factory is a system of funds, data, and material flows. Funds flow analysis must start from the purpose of a factory and take into account multiple resource uses, the passing of time, and risk. Net profit, relative profit, and cash measure various aspects of business performance.

Funds flow between a factory and its environment, but not between departments inside of it. Improvement projects must be assessed by their external impact. Costs are schedules of funds flows assigned to decisions or events. Manufacturing cost reductions and price erosion are modeled using inverse power "experience curves."

Direct use of accounting "unit costs" can lead to mistaken decisions. Project funds flow schedules can be reduced to payback periods, or transformed into immediate lump sum equivalents by discounted cash flow analyses based on rates modeling delayed gratification. Risk assessment is the weak point of these methods.

The movements and processing of materials and data inside the factory aim to reach goals set at the funds flow level. Data manipulations can be classified between quantity and quality control. Manufacturing problems are best solved by a holistic approach, combining actions taken at the materials, data, and funds flow levels.

12.1. A MISSING PERSPECTIVE

In the preceding chapters, we have seen that for manufacturing applications, structured analysis needed to be supplemented with modeling techniques for material flows. Unfortunately, this is still not sufficient. Even if a functional specification is based on a thorough understanding of a factory's material flows, is internally consistent, and describes useful outputs, it will not be approved without a prior assessment of its financial impact.

In his quality management training, J. M. Juran distinguishes the "language of things" spoken by the people of the shop floor from the "language of money" they must learn to get their proposals approved. Depending on perspective, a factory may look like a system that absorbs and produces funds, or one that takes in raw materials and ships finished goods. To be complete, a factory model must encompass all the relevant perspectives.

12.1.1. The Triple Flow Model

From every point of view, a factory is a flow system. The differences in perspective lie in the type of flow that is emphasized. The owners are motivated by financial results and see flows of funds. The managers, who issue instructions and follow through their execution see data flow. The workers and their supervisors see materials. The complete factory thus appears as the three-story structure of Figure 12.1.

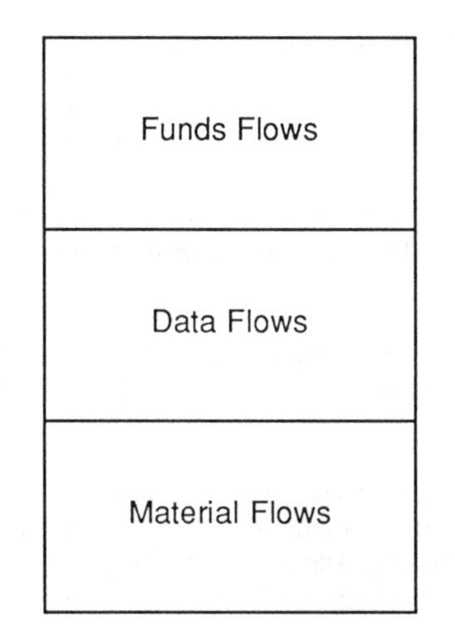

Figure 12.1. The Three-Level Model of Factories

Financial objectives are translated into directives affecting the internals of the manufacturing organization. The processing of these directives takes place in the data flow level and eventually affects material flows. Driving up the price of the company's shares in the stock market within the next quarter is a financial goal. To reach it, managers will be asked to operate without new equipment, reduce their staffs, and keep producing in the absence of demand. On the shop floor, the execution of these plans will change machine workloads and reliability — as well as inventory levels — in ways that may or may not have been anticipated.

Such textbooks on manufacturing systems as [Vol84] are written with only a cursory treatment of financial matters. The reader learns how to implement MRP but not how to estimate its economic benefits. At the opposite end of the spectrum are texts on operations research, like [Hil80], or statistical decision theory, like [Cox74], which provide methods to maximize economic utility, assuming such a quantity to be calculable. In practice, this assumption is rarely justified, and a manufacturing analyst can neither ignore financial modeling nor assume it already done.

Even though there is no universal criterion usable as a basis for optimization, an economic model is needed in each factory to decide how many parts to start next week or whether to implement a CIM system. There is no all-purpose model, but there are generic modeling approaches.

Data exchanges tell operators and equipment to process and move materials. These movements eventually provide the factory's response to external demands and create new sales opportunities. They also generate purchasing requirements and operating expenses. The combination of these factors determines the business performance of the factory.

12.1.2. Manufacturing Funds Flow Analysis Issues

The problems of manufacturing funds flow analysis are

- The purpose of a factory — As examples will show, saying that a factory exists "to make money" is not specific enough.

- Multiple resource uses — Buildings, power supplies, and even machines are shared resources. The economic effects of changes in allocation need to be modeled.

- The passing of time — The impact of many manufacturing decisions is spread over time, and the value of delayed expense and gratification must be taken into account.

- Risk — Finally, be it increasing production of one item by 10%, buying a new machine, or laying off employees, there is almost no risk-free decision in manufacturing. Projections frequently do not come true, and no analysis is complete, that does not take this into account.

An analyst is concerned with the effect of a project on a factory. The financial side of the analysis is intended to show decision makers that they should budget the project. The procurement of the financial resources by the decision makers, however, is outside the scope of our discussion.

12.2. THE BUSINESS OF MANUFACTURING

12.2.1. Business Performance in General

Following [Hel77] and [Riv75], we consider the funds flowing in a commercial transaction to be the amount that would be paid cash at the time of the transaction. The real transaction may involve credit, but all the various modes of payment are viewed as equivalent means of meeting the same commitment of *funds*.

Investments, such as equipment purchases, make durable changes in the factory system itself. Once an investment is made, the old factory model is obsolete. The factory after the investment presumably functions better than before.

By contrast, we use "operating funds" as part of the day-to-day operation of the factory, in transactions that do not change the factory itself. The boundary between investment and operation is fuzzy. For example, employee training programs can be viewed as operating expenses, but they may as well be thought of as investments.

Away from this gray area, operating outflows pay salaries, utility bills, and material purchases. Sales provide operating inflows. Decisions on product mix, production scheduling, or quality control, eventually made with the help of a CIM system, affect only the flows of operating funds.

Based on funds flow analysis, Helfert [Hel77] gives a number of ratios used to measure business performance: profit margin, turnover, return on assets, and others. The relevance of a ratio depends on whether the analyst observes the business from the point of view of its management, its owners, or its lenders.

Eli Goldratt proposes focusing on just three financial parameters:

- Net Profit — This number is the "net income" of operating statements. Among other things, it measures the scope of what the company can undertake. With $4 billion in net profit, there is little a company cannot do in product development or acquisitions.

- Relative Profit — The ratio of the "net income" to "net sales" entries from an operating statement measures the relative profitability of a business. In this sense, the production of $4 billion in profits on $100 billion of sales is a less impressive performance than profits of $1 million on sales of $10 million.

- Cash — The amount of cash and other liquid assets measures a business' ability to meet short term commitments and withstand fluctuations in the demand for its products. There is a limit below which this amount cannot drop without bringing about bankruptcy. The existence of this limit is known to management, but not its value, because it is a function of the lenders' dynamically changing willingness to extend credit.

There are many variations on this theme. For example, to better express the effectiveness of management, "net income" can be replaced by the net profit after taxes but before interest. The idea is that taxes are a normal expense of doing business, while interest is a compensation of a nature similar to dividends.

Relative profit, net profit, and cash are as relevant to retail and insurance as to manufacturing. In the following sections, we become gradually more focused. Net and relative profit generally cannot be maximized simultaneously. Figure 12.2 shows this in a simple manufacturing example with no claim to realism.

The total cost C(V) of producing V units in one period of time is assumed to be composed of a fixed component C(0), incurred even if nothing is produced, and a variable part, the convexity of which represents decreasing yields. As the production rate increases, the factory saturates and the cost of producing one more part becomes ever higher. On the other hand, the corresponding income from sales grows linearly as a function of the selling price p.

The relative profit is

$$V_R = \frac{[Vp-C(V)]}{C(V)}$$

and is maximum at the volume V_0 where

$$\frac{dC}{dV} = \frac{C(V)}{V}$$

In other words, V_0 is the volume at which the average unit cost matches the marginal unit cost. On the other hand, as long as the cost of making one more unit is lower than the price it can be sold for, it contributes to the net profit Vp - C(V), which is therefore maximized at the volume V_1, where the marginal cost dC/dV equals the selling price p. In general $V_0 \neq V_1$.

Figure 12.3 shows in the same simple model the type of tradeoffs that can be made between the two profit goals by setting production at the level maximizing net profit subject to the constraint that the relative profit will be at least r.

$$\frac{[Vp-C(V)]}{C(V)} \geq r$$

is equivalent to

$$C(V) \leq \frac{Vp}{1+r}.$$

Thus eligible volumes correspond to the part of the cost curve C(V) below the line

$$y = \frac{Vp}{1+r},$$

and the optimal volume is V_1.

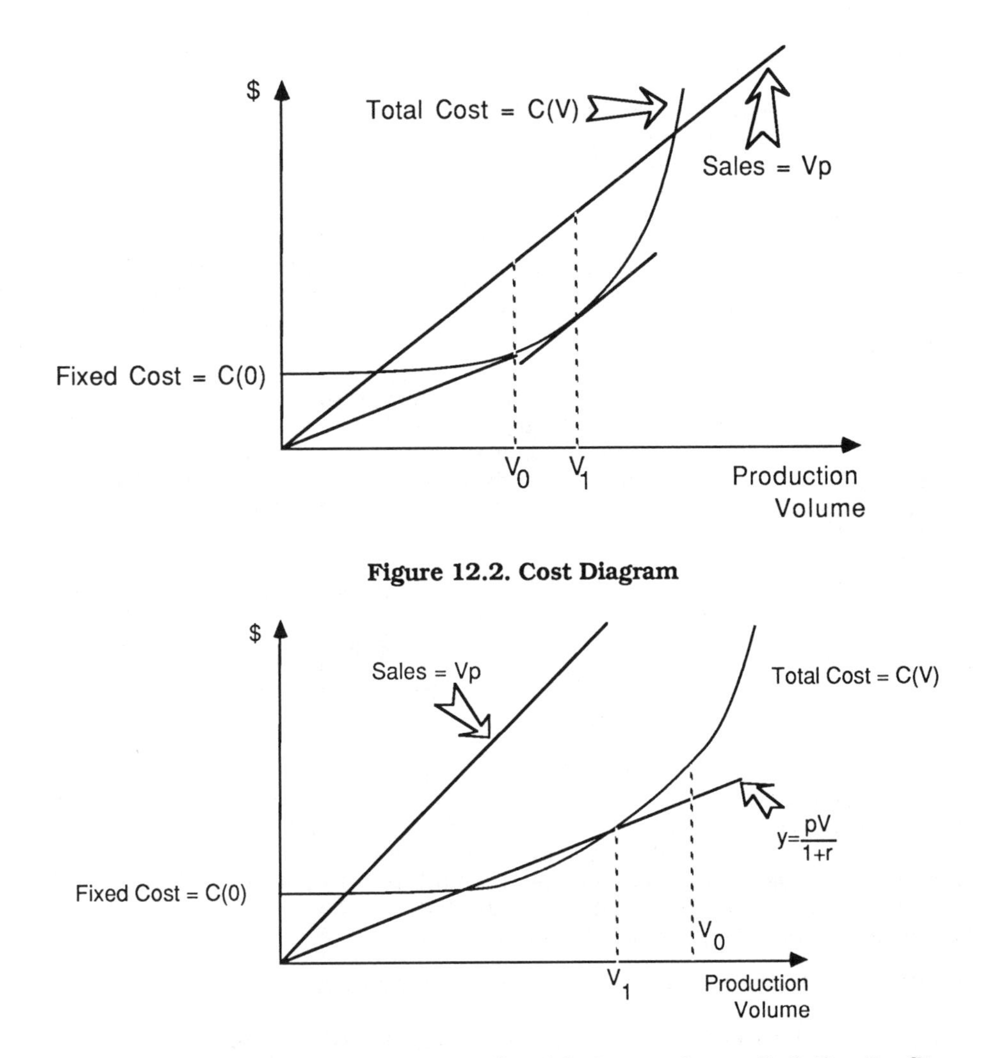

Figure 12.2. Cost Diagram

Figure 12.3. Maximum Net Profit with Constraint on Relative Profit

Except in periods of exceptionally high interest rates, businesses are better off with just enough cash to make sure they do not run out. The difficulty is in assessing the risk associated with a particular level, considering that the lower limit is dynamically determined by negotiations with lenders.

Financial analysis is weak in probabilistic modeling. The lenders' attitude is set by tradition and rules of thumb such as "the ratio of current assets to current liabilities should be 2:1." It is well known that, until the early 1980s, many Japanese companies prospered with debt levels that would have led to bankruptcy in the U.S. Rules of thumb are real constraints if lenders use them, but they may reflect an overestimation of risk.

12.2.2. The Paradox of the Pilot Line

Before proceeding, let us consider a common example in which the relevance of the preceding concepts is anything but obvious. It is a frequent pattern for manufactured products to go through (1) a prototype stage in a research laboratory, (2) a low volume, pilot production stage, and (3) high volume production.

Let us examine a facility in charge of the second stage. At first sight, this "pilot line" is like any other factory: it receives raw materials and turns them into finished goods. However, a second look reveals significant differences with a production line.

A pilot line does not and should not attempt to meet financial goals through the sale of finished goods. Its desired output is not a flow of goods for outside customers but a flow of manufacturing processes for production lines. A pilot line's production of finished goods is incidental. These goods may be shipped to customers as samples but they are not expected to make the pilot line financially self-sufficient. (See Figure 12.4)

Under normal conditions, a pilot line never produces goods reliably. As soon as it has perfected a reproducible process, it must be transferred to production, and the pilot line must start debugging a new one. Pilot lines usually do not sell processes to production lines, but are financed by a "tax" levied by the corporation on its operating divisions. Pilot lines must account for the use of their funds but cannot show a profit.

This situation is largely inevitable, but nonetheless a frequent source of conflict within companies. As soon as economic difficulties arise, the managers of operating divisions are prone to view a pilot line as a drain on their resources, for which they are not getting value. As a result, the pilot line may be pressed into service as a small production line. Equipment needed for experiments is routinely preempted by production, which makes it impossible to use for process development.

12.3. THE FACTORY AND ITS OPERATING FUNDS FLOWS

Funds flow between a manufacturing organization and its business environment. To raw material purchases correspond flows of funds to suppliers. Likewise, for each flow of goods there is a flow of funds from the customer. These transactions all involve real transfers of money, as shown in Figure 12.5.

On the other hand, funds flow neither between departments within a factory nor between several factories within the same organization. Actions taken on a machine inside a factory eventually affect the funds flows of the organization owning it, but not directly. Likewise, transfers of parts between factories within the same organization involve no real financial transaction.

At management's discretion, it is possible to model relations between departments by simulated "buy" and "sell" transactions, but the differences between such simulations and real funds flows are akin to those between Monopoly money and legal tender.

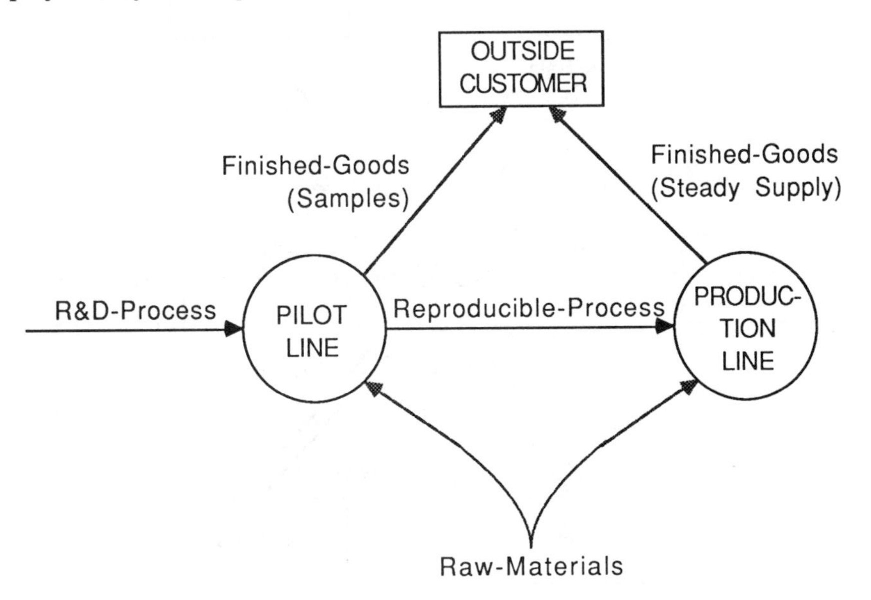

Figure 12.4. Material and Data Flow Diagram of Pilot Line

Hounshell [Hou84] describes the "inside contractor system" that was prevalent in the U.S. in the nineteenth century, and in which real funds flowed between departments inside a factory. An inside contractor hired, paid, and managed workers, and contracted with a company to make parts using its facilities and equipment.

This system was a feature of New England armory practice and was later used in such companies as Singer Sewing Machines. Today, it has all but disappeared. The significance of this evolution is worth pondering. The inside contractor system was simple and decentralized, with production activities regulated by genuine commercial exchanges. Yet it has not survived. This has to mean that the relations between departments are too intimate and too complex to be expressible purely in terms of trading.

To assess the economic effect of a project on a factory, the analyst has to relate it to funds flowing across the boundary between the factory and the outside world. Outflows contribute to costs and inflows to benefits. The comparison of the two lets managers decide whether the project should be carried out.

Used carelessly, cost figures produced by accountants can be misleading. Expressions like "unit cost" or "cost of goods" lead one to think that a cost number can be assigned to a part just like a weight or dimensions, and that the cost of a set of parts is proportional to its size.

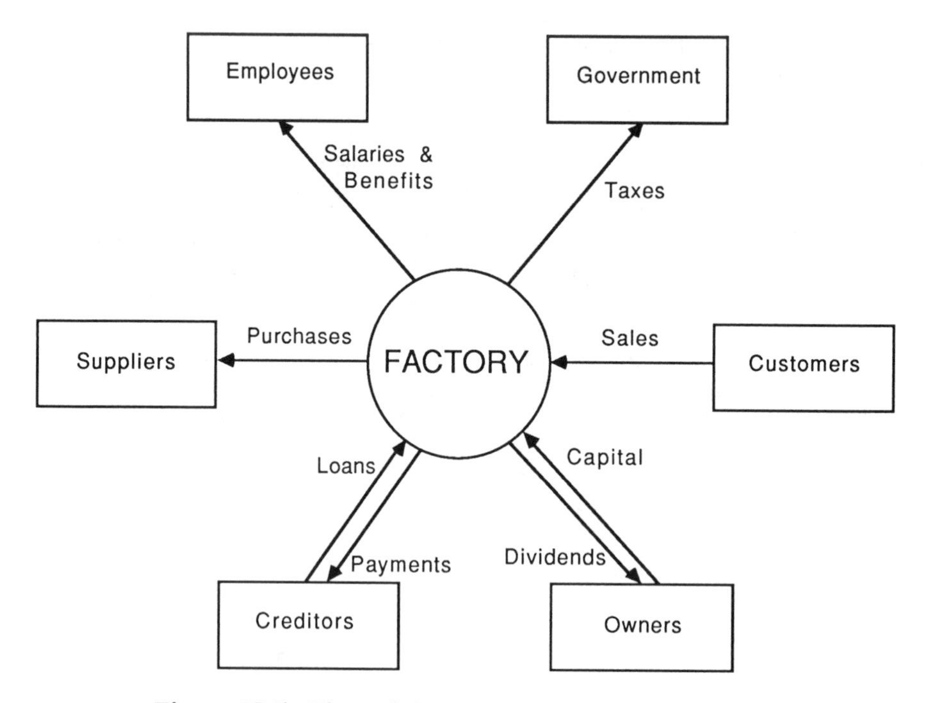

Figure 12.5. Financial External Context of a Factory

To get a better grasp on the concept of cost, we follow Riveline [Riv75] and say that, to a given observer, the cost of a decision or an event is the schedule of differences in his or her funds flows between (1) the situation in which the decision is made or the event occurs, and (2) a reference situation.

First, it is not goods that have costs, but decisions or events. Although everyone equates the cost of the decision to buy a newspaper with its price, there is no such straightforward equivalence in more complex cases. The cost of processing a part through a machine changes based on whether the *decision* is (1) to build a factory or (2) to increase production in an existing factory. An *event* such as an equipment failure has a cost that is well defined as the difference between what has to be spent because of the equipment failure that would not have to be spent otherwise.

Second, the cost of a decision or an event is not a single number but a *schedule* of numbers. Many decisions, such as renting a facility, cause funds to flow over time. One-time lump sum equivalents of schedules of outflows can be determined by present value analyses. However, those analyses rest on the choice of a particular discount rate, whereas the schedules themselves do not. Thus the present value equivalents of a schedule are varying interpretations of the same facts. By using the full schedule, we separate the cost measure from more subjective analyses.

Third, costs depend on the point of view of the observer. Riveline illustrates this by the example of a spare part stockout in a maintenance department. If the spare part is needed on a machine in order to ship products to customers, then, to the factory as a whole, the cost of being out of it is the income that would have been generated from sales if it had been available. On the other hand, from the point of view of the maintenance department, waiting for the next regular delivery of this part may be cheaper than making special efforts to procure it sooner.

Finally, costs are evaluated using a reference situation which does not have to be achievable, as long as it is definable. For example, a cost can be assigned to operating a line with a 5% scrap rate by comparison with a line producing no scrap at all even though no such line exists. Like perfect technology in essential systems analysis, this imaginary line serves as a yardstick against which real lines can be rated. A 5% and a 2% scrap rate can be compared to each other via the common reference.

In a chapter on the economics of process planning, Doyle et al. [Doy85] discuss the cost of machine tools to be selected by manufacturing engineers. Their discussion is focused on the choice between machine types, given that some machines will be purchased. The results would be different if the decision were on the need for any machine.

12.3.1. Effect of Throughput Time Reductions

This example, as shown in Figure 12.6, is an abstraction of the problem of failure detection delays, which is found in factories ranging from die casting to integrated circuits. The scrap cost considered here is only one of many associated with long throughput times. It is dominant in some cases and not in others.

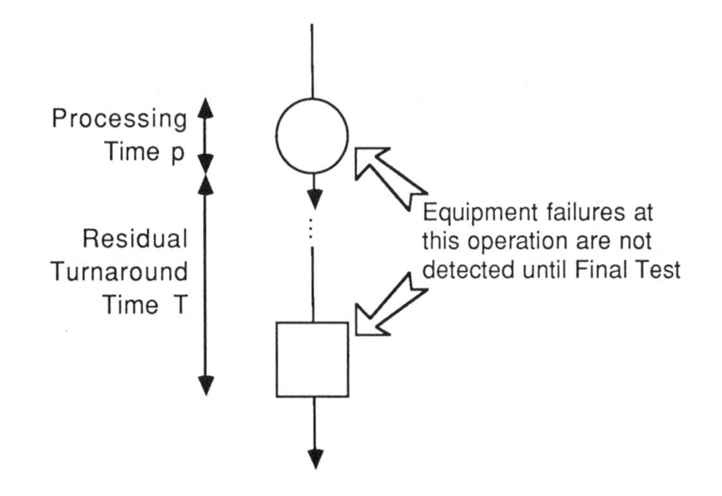

Figure 12.6. Fabrication Line Model

We consider a fabrication line making a single product at a constant rate, with equipment that breaks down in a manner that is not detected until the first part processed after the failure reaches a final test station at the end of the line.

We make the following further assumptions:

- The equipment processes one piece at a time.
- Its processing time per piece p is small with respect to the residual throughput time T from the operation to final test — that is $p << T$.
- The parts move along the line first-in-first-out.
- Once a failure has been detected, the machine can be promptly repaired.

These conditions enable us to view the parts as a signal continuously emitted by the machine, but perceived with a delay. The condition that $p<<T$ is satisfied in the two cases we mentioned. Shooting one casting in a die may take two minutes in a factory with a total throughput time of 12 hours. Implanting ions into a silicon wafer may likewise take 10 minutes while the integrated circuits on it will need another six weeks before testing. Long repair times can be another major contributor to cost. The purpose of assuming short repair times here is simply to restrict our discussion to detection delays.

Figure 12.7 shows the effect of one failure of the machine. It affects all the parts processed during the detection delay between the failure and the time the first defective reaches test. Clearly, the cost of a failure could be reduced to that of a single part by eliminating the detection delay.

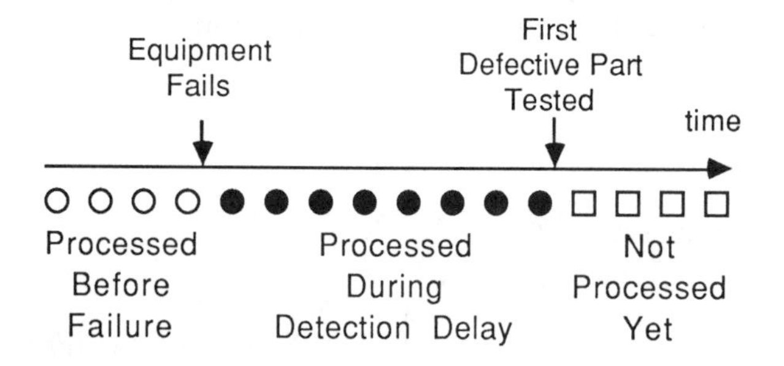

Figure 12.7. Equipment Production History

Once a failure has occurred, its cost in terms of defective parts is easy to establish, but of limited value. However, with a model of the equipment's reliability, we can answer a more difficult question: what does the factory lose *on the average* because of failure detection delays?

Let us use the most simple reliability model and treat equipment failures like a Poisson process: there is a constant rate μ such that, at any time t, the machine breaks down between t and t+dt with probability μdt, regardless of what has happened before.

As can be seen from Figure 12.7, a part reaching the machine at time t will be processed properly if and only if there has been no failure of the machine between t-T and t, because any failure occurring prior to t-T will have been detected by time t. Therefore the yield Y of the machine will equal its probability of not failing in an interval of length T. In the model we are using

$$Y=e^{-\mu T}$$

This model can be extended to several machines along the process with varying failure rates and residual throughput times, but doing so would be tedious and would not make the example more enlightening.

An objection to the model is that the residual throughput time is not a constant but a random variable. It would be difficult to model it that way, because the detection delay is not determined by the *average* of the throughput time distribution but by its extreme values — that is, by the first parts to reach testing after the failure has occurred.

In practical applications, we want to estimate the effect on Y of reducing T. Y and T are directly accessible, but μ is not, so that the preceding formula would have to be used in the form

$$\mu=\frac{-\ln Y}{T}$$

μ cancels out when we consider going from T_1 to T_2 , the yield Y_2 is obtained from the original yield Y_1 by

$$Y_2=Y_1^{\frac{T_2}{T_1}}$$

If the time detection time T is cut in half on a process with a yield Y_1= 90%, the yield is brought up to Y_2= 95% for that reason alone.

At this point, the analyst can estimate how much the yield will go up as a result of a throughput time reduction project, but one more step is needed to translate this into financial terms. This requires consideration of the business environment. If the product is in high demand, scraps translate to lost sales opportunities. On the other hand, if demand is fixed, the cost of scraps is found in excess raw materials and operating expenses.

12.3.2. Price Erosion and Experience Curves

The electronics industry offers the most easily perceptible examples of price erosion over the life of a product. That integrated circuits sold for $20 today will only fetch $2.50 a year from now has come to be expected by vendors. This phenomenon is particular to manufacturing, and not found in other types of businesses.

As a consequence, some means of anticipating the decrease in the prices of manufacturing goods is necessary to properly model income. *Learning curves* were first introduced during World War II to quantify increases in *labor* productivity. Later, in the 1960s, B. Henderson (cf. [Hen72]) extended that theory to encompass all manufacturing costs, and coined the term "experience curve" to designate the resulting relation between full "unit costs" and production volumes.

One apparent contradiction between our discussion of costs and experience curve theory is that the latter seems based on the assumption of the existence of a number that can legitimately be called "unit cost." In fact, all the evidence he presents is based on prices, because, as he admits, "accounting practices with respect to depreciation, allocated costs, and capitalization make comparative costs unreliable."

The more one has made of something, the better one knows how to make it, and therefore, the cheaper it becomes. Henderson's theory is that manufacturing costs drop by a constant percentage every time cumulative production volume doubles, both for whole industries and for individual participants.

For investment planning purposes, Henderson becomes more specific on what he means by cost. The decision he considers is whether to build an incremental volume dV of a product after a volume V has already been produced. If dC(V) is the cost of that decision, the unit cost Henderson plots on the Experience Curve is the expected value

$$c(V) = E\left(\frac{dC(V)}{dV}\right)$$

As investments are made, dC/dV will go up and down over time in ways that cannot be precisely anticipated when V=0. What Henderson's theory says is that the trend, or expected value c(V) of that ratio drops by a fixed percentage for each doubling of V. In Henderson's words, c(V) is "the smoothed rate of change in accumulated cash flow for any given level of accumulated experience."

dC/dV has the dimensions of a marginal cost, which, according to the law of decreasing yields, is supposed to *rise* with V. There is no contradiction with experience curve theory, because it is not the same "V" that is considered. The law of decreasing yields says that, *within a given time period*, the marginal cost increases with production volume because the factory saturates. Experience curve theory says that, *over time*, the average unit cost decreases as a function of the *cumulative* production volume.

The formulation of the relation as a percentage drop per doubling of volume is an intuitive way of describing an inverse power law. Experience curve theory says that c(V) will appear as a straight line in a bilogarithmic plot. This means that c is of the form

$$c(V) = a \times V^{-b}$$

with $a \geq 0$ and $b \geq 0$.

There is no mathematical justification for the use of inverse power laws in experience curves. To find empirical laws, engineers frequently take the approach of plotting data on various kinds of graph paper, in the hope of finding a straight line.

Given the task of modeling cost decline as a function of cumulative volume, it is to be expected that an engineer would think of plotting historical records on cartesian, semilogarithmic, and bilogarithmic paper. Although there is no a priori reason for the bilogarithmic plot to work, it is easy to see why none of the other two could.

If c had the form $c(V) = a - bV$, with $a \geq 0$ and $b \geq 0$, then it would become negative as soon as $V > a/b$. Therefore there is no way a linear cost decrease curve could fit data over an extended period of time and on a variety of products.

A straight line on a semilogarithmic plot would describe an exponential cost decay of the form

$$c(V) = a \times e^{-bV}$$

with a and b positive. The reason why such a model could not possibly fit is slightly more subtle than in the linear case. The total cost $C(V)$ of producing the first V units would be

$$C(V) = \int_0^V c(v)dv = \frac{a}{b} \times \left[1 - e^{-bV}\right]$$

and the total cost of producing an infinite volume would be $C(\infty) = a/b$, which is absurd: the best manufacturing technology cannot produce an infinite volume at a finite cost.

On the other hand, it is easy to see that an inverse power law

$$c(V) = a \times V^{-b}$$

with $0 < b < 1$ provides a cost decline pattern such that the cost of the first unit has a finite value, but an infinite volume cannot be produced at a finite cost.

$$C(V) = \int_0^V av^{-b}dv = \frac{a}{1-b}V^{1-b} = AV^{1-b}$$

The following table shows the evolution of the price of a Model T Ford from 1908 to 1916. Similar results can be obtained for integrated circuits, with the difference that the life cycle of a product is four years. Any semiconductor maker introducing a new product must anticipate cost decreases and therefore price erosion due to the competitive environment.

The unit price P fits

$$P(V) = 8345 \times V^{-.22}$$

from 1909 onwards. The large discrepancy in 1908 may represent a low introduction price aimed at conquering market share.

$P(2\times V) = P(V)\times 2^{-.22} = P(V)\times .87$, so that the price of a model T dropped by 13% in *current* dollars for each doubling of the cumulative volume. Henderson quotes rates in the 20 to 30% range in *constant* dollars and supplies supporting data from such industries as integrated circuits, gas ranges and facial tissues.

Ford Model T Data from Hounshell [Hou84]

Year	Cum. Volume V	Retail Price($)	$8345\times V^{-.22}$
1908	5,986	$850.00	$1,231.56
1909	19,826	$950.00	$946.31
1910	40,553	$780.00	$808.47
1911	94,041	$690.00	$671.89
1912	176,429	$600.00	$585.03
1913	365,517	$550.00	$498.41
1914	596,305	$490.00	$447.53
1915	991,093	$440.00	$400.20
1916	1,576,481	$360.00	$361.36

The validity of the experience curve concept is indisputable, but its application in practice is not straightforward. [Bau85] evaluates the effect of a CIM system by treating it as a way to increase the rate at which costs drop as a function of cumulative volume. In other words, the CIM system is given credit not for solving specific problems, but for helping engineers and managers solve all problems faster than previously.

12.3.3. The Example of a Memory Factory

In the following pages, we are going to examine how this theory could help us decide whether or not to build a factory with the material flow structure of Figure 12.8. DRAM stands for "dynamic random access memory" and designates one of the most common types of integrated circuits. DRAMs are commodities and a business in which success hinges on manufacturing ability to a greater extent than for custom products.

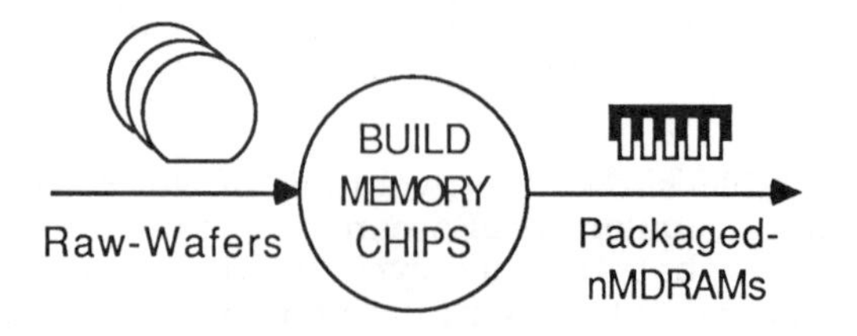

Figure 12.8. Material Flow Diagram

An "nMDRAM" is a DRAM with n Megabits. The memory industry quadruples the density of its products every two years, so that if 1MDRAMs are the standard product one year, 4MDRAMs will be two years later and 16MDRAMs four years later.

Large sums will be expressed in thousands (K$) and millions of dollars (M$), and volumes in millions of parts. The following assumptions are in line with the reality of the semiconductor industry:

- Before any part can be made, 100M$ must be spent on facilities and equipment.
- Raw wafers can be purchased at 20$ a piece.
- Labor comprises 1000 employees, of which
  400 direct labor employees receive salaries and benefits averaging 30K$/year and
  600 indirect labor employees receive salaries and benefits averaging 50K$/year.
- Indirect materials, inventory carrying costs, and utilities amount to 13M$/year.

The high ratio of indirect to direct labor reflects the industry's intense need for engineering support. On the factory's production, we assume the following:

- Volume will ramp up from 4000 to 24,000 wafer starts/month in the first 6 months and stay at the level of 24,000 wafer starts/month thereafter.
- The first 1000 wafers will yield 10,000 shipped devices.
- Each doubling in the cumulative volume of shipped devices will reduce by 20% the number of raw wafers needed per shipped device.

In a semiconductor factory, most of the costs are fixed. Changes in unit costs come in the form of yield variations. As experience accumulates, the same number of wafers produces more shipped devices. This corresponds to a reduction in the number of raw wafers needed for one device.

We predict the market behavior to be as follows:

- The ten thousandth shipped device will sell for $200.
- After a year of operation a shipped device will sell for $2.50.

Based on our assumptions, every year, the operating expenses will be as follows:

Yearly Operating Expenses

	Operating Expenses (M$/year)
Indirect Labor	30
Direct Labor	12
Indirect Materials, Inv. Carrying Costs, and Utilities	13
Total	55

Given the initial ramp-up, the factory will buy 228,000 raw wafers in its first year for 4.56M$, and 288,000 every year thereafter for 5.76M$. Our calculations are not precise to the point that ±0.5M$ will make a difference in the decision to build the factory. We will consider raw wafers to cost a flat

5M$/year, so that the total of operating expenses and direct materials amounts to 60M$/year.

There is an additional outflow that we do not have the means to calculate yet: taxes to be paid on reported profits. For the time being, we shall assume that we are building the factory in a tax haven, and later on, we will make the necessary adjustments.

In the tax haven, the cost of the factory is therefore the following schedule of funds flows:

Year	Outflows (M$)
0	100
1	60
2	60
...	...

The assignment of the initial 100M$ to "Year 0" indicates that these funds have to be spent *before* any device can be built. Now, to estimate what we will get in exchange for these outflows, we need to find out how many devices the factory will produce and the income their sales will generate, anticipating both steady increases in productivity and concurrent price erosion.

We have postulated the cumulative number W be of wafers started, and we need the corresponding number D(W) of devices shipped. If w is the number of wafer starts required to get *one* device out when the cumulative number of devices shipped is D, then our assumption of a 20% decrease in w for each doubling of D yields

$$w = A \times D^{-b}$$

where

$$b = \frac{-\ln(1-0.2)}{\ln 2} = 0.32$$

and A is a constant to be determined. Then

$$W = \int_0^D w(u)du = \frac{A}{1-b} \times D^{1-b}$$

and

$$D = \left(\frac{1-b}{A}\right)^{\frac{1}{1-b}} \times W^{\frac{1}{1-b}} = B \times W^{\frac{1}{1-b}} = B \times W^{1.475}$$

To find B, we use the assumption that the first 1000 wafers yield 10,000 devices. With W expressed in thousands of wafers, and D in millions of devices, we have D(1) = 0.01 = B, and finally

$$D = 0.01\times W^{1.475}$$

Now we know how many devices we will ship, but we still need to estimate how much income their sale will generate. Price erosion is assumed to follow also an inverse power law — that is, the average selling price p of one unit at the time when D devices have been shipped is of the form

$$p = M\times D^{-c}$$

and we need to find values for the parameters M and c.

Again, expressing D in millions of units, the assumption that the ten thousandth unit sells for $200 takes the form

$$p(0.01) = \$200$$

The number of millions of devices shipped in the first year is

$$D(228) = 0.01\times 228^{1.475} = 30.05$$

and since we assumed the selling price after a year to be $2.50, form

$$p(30.05) = \$2.50$$

From which we deduce

$$c = \frac{\ln(200)-\ln(2.5)}{\ln(30.05)-\ln(0.01)} = 0.547$$

and

$$M = p\times D^{c} = 200\times 0.01^{0.547} = 16.107$$

The cumulative sales income is then obtained by integrating the price p(D) between 0 and D:

$$S = \int_0^D p(u)du = \frac{M}{1-c}\times D^{1-c} = 35.56\times D^{0.453}$$

The corresponding percentage of price decline per doubling of the cumulative volume is

$$\beta = 1 - 2^{-c} = 31.55\%$$

Therefore, our assumptions imply that prices decline faster than yields increase, and we may wonder whether the factory can operate profitably. Now that we have all the necessary elements, we can calculate the results and see.

Income From Sales For Seven Years of Operation

Year	W (×1,000)	D (Millions) (M$)	S (Millions) (M$)	Yearly Sales (M$)
1	228	30.05	166.14	166
2	516	100.26	286.73	121
3	804	192.86	385.63	99
4	1,092	302.95	473.17	87
5	1,380	427.87	553.28	80
6	1,668	565.89	627.98	75
7	1,956	715.76	698.50	71

Therefore, in the tax haven, we can anticipate the following schedule of inflows and outflows over seven years:

Schedule of Inflows and Outflows in Tax Haven

Year	Outflows (M$)	Inflows (M$)
0	100	0
1	60	166
2	60	121
3	60	99
4	60	87
5	60	80
6	60	75
7	60	71

Since so far we have considered neither legal reporting requirements nor taxes, the analysis is independent of the form those take in the country the factory is built. We will base our adjustments to taxes on the following additional assumptions:

- The accounting laws of the factory's host country require the company to write off 50% of the cost of building it as an outflow of the first year of operation, and 10% of that cost for each of the following five years.
- The reported profit for each year is then the difference between the inflow and the total reported outflow.
- The reported profit is taxed at a flat 50% rate.

The following table compares the reported outflows with the real inflows to calculate the tax according to the system just described.

Tax Schedule in M$

Year	Reported Outflows	Real Inflows	Reported Profits	50% Tax
1	110	166	56	28
2	70	121	51	25.5
3	70	99	29	14.5
4	70	87	17	8.5
5	70	80	10	5
6	70	75	5	2.5
7	60	71	11	5.5

The schedule of inflows is unchanged, and the real schedule of outflows is obtained by addition of the tax.

After Tax Schedule in M$

Year	Outflows (M$)	Inflows (M$)
0	100	0
1	88	166
2	85.5	121
3	74.5	99
4	68.5	87
5	65	80
6	62.5	75
7	65.5	71

The preceding schedule is based on many arguable assumptions about equipment costs, operating expenses, productivity gains, and price erosion, but none about what a desirable investment is. Other tools — and further assumptions — will be needed for this purpose.

Given our model of the environment, the schedule shows the behavior, over time, of the factory as a financial black box. Each entry describes funds that would actually have to enter or leave it. Our derivation of these numbers shows that it is possible to keep a global view and not think in terms of unit costs.

But how robust is this analysis? Some — but not all — answers can be given to this question. "What if" calculations can be easily carried out to anticipate first-order effects, such as different rates of yield increase or price erosion. However, given correct rates, we have no tools with which to attack second-order effects such as fluctuations around these rates.

In the late 1970s, the prices of 16KDRAMs remained far over the experience curve trend for a time, due to orders from a single computer maker for more than 1/3 of the world's production. Order cancellations by this same customer then caused prices to fall *below* the trend.

Political risks are also present. The higher a company's market share, the faster its experience accumulates. For this reason, experience curve thinking leads manufacturers to seek market share by aggressive pricing. But this may lead to accusations of dumping by foreign governments. Dumping is defined as "selling below cost," and, although it may not have a clear meaning economically, it is a *political* reality.

The nMDRAM is the only product considered in the analysis, because it is the reason for the factory's construction. Later on, the factory may eventually be used to produce other products after the nMDRAM becomes obsolete. However, the nMDRAM market is the only factor in the current decision.

Once the factory exists and another product compatible with its technology becomes available, management will have to decide whether the company is better off sticking with the original plan of dedicating the factory to the nMDRAM or allocating part of its capacity to the new product. The new decision should be affected only by the schedule of inflows and outflows influenced by its outcome. The amount of resources invested to date in the nMDRAM will be irrelevant.

12.4. FINANCIAL ANALYSIS METHODS IN MANUFACTURING

12.4.1. Cost Accounting and Manufacturing

Review by the comptroller frequently is the most difficult hurdle for a productivity improvement project. The emergence of an adversary relationship between accountants and manufacturing analysts is understandable when one considers that the latter want to spend, while the former want to keep the company from overextending itself.

This conflict is healthy if it is resolved by sound economics, as it can be when there is an understanding of the "language of things" by the financial side and of the "language of money" by the manufacturing side. The conflict is harmful if it is managed by haggling and compromise in mutual ignorance.

Cost accounting, as its name says, is the art of accounting for the past uses of funds. The major goal of accounting is to provide *universal* means of reporting the financial side of business activity to outsiders. Whether a company provides health insurance, extracts oil, or manufactures integrated circuits, the line items in its balance sheet and its operating statement will be the same. However, the economic significance of these results varies case by case.

Accounting is driven by legal requirements. Each country sets rules that local companies must follow. Economics does not change from country to country, but accounting rules do, depending on the government's direction of the economy. The most economically significant use of the accounting model is the computation of tax liabilities.

In the accounting model, the unit cost of a product is most often broken down into materials, labor, and overhead. Direct materials become part of the finished product. Indirect materials do not, but they are used in the transformation of the direct materials. A fraction of the labor costs of the factory is conventionally allocated to each unit. Overhead includes depreciation and indirect labor costs, and is allocated to each unit in proportion to labor.

The materials cost of a product is truly variable, in that it is a function of the production volume. It can usually be assumed proportional to volume although, if the demand stretches the suppliers' capacity, a premium may have to be paid. As one would expect, material costs tend to be dominant in assembly.

They can amount to 95% of the costs in a factory that assembles computers by

1. Subcontracting the stuffing of integrated circuits into boards.
2. Inserting the boards into drawers.
3. Loading the drawers onto racks.
4. Running validation programs.

Semiconductor wafer fabrication is at the opposite end of the spectrum. A raw silicon wafer is bought for $20, while the accounting cost of a completed LSI wafer is on the order of $300, so that materials account for less than 7% of the total. Furthermore, expensive equipment and large engineering staffs combine to produce overhead factors of 1,000%. In other words, of the $280 not going into direct materials, more than $250 are overhead, making the breakdown by materials, labor, and overhead look like a classification scheme in which almost everything falls into the "other" category.

Since business organizations are legally required to report their activity according to the accounting model, these cost numbers are always available. The temptation is great to use them in manufacturing decisions without any refinement, but it should be resisted.

The unit costs could be used, for example, to rank the factory's products in a table (c_i, p_i), with c_i being the i-th highest unit cost and p_i being that product's selling price. Now all products with $c_i \geq p_i$ appear to be unprofitable and it would seem that the factory would be better off not making them.

Stopping production of these "unprofitable" products would result in spreading fixed costs over a smaller mix, thus increasing the overhead cost components and resulting in new costs $c'_i \geq c_i$, and in the appearance of new products to be dropped on the grounds that $c'_i \geq p_i$. This cycle could eventually lead to closing the factory.

The methods outlined in the preceding sections are meant to avoid this type of mistake. Rather than being used directly, accounting data should be refined into such schedules as we produced in the case of the nMDRAM factory. These schedules can then be evaluated using some of the methods discussed next.

12.4.2. Payback Period

Because of its simplicity and its intuitive interpretation as the time it takes to recoup an investment, payback period analysis is the most heavily used approach. Managers use rules of thumb such as: "If your system has a one-year payback, I can justify it. Two years is borderline. Beyond that I can't."

The payback period is the time it takes for cumulative inflows to equal the cumulative outflows. In its simplest and most commonly used version, no allowance is made for the time value of money. In other words, the analyst considers a dollar received a year from now equal in value to a dollar in hand today.

Returning to the nMDRAM factory, we see that, by the end of Year 2, the inflows total 287M$, while the outflows total 273.5M$. If the decision to build the factory is made at the beginning of Year 0, its payback period is just short of three years.

Simple payback period analysis can be performed from a schedule of funds flows without making any more assumptions about the future. As we shall see shortly, the more sophisticated methods require the analyst to set additional parameters.

Payback period analysis gives more weight to the near than to the distant future. Although this may be criticized as shortsighted, the reliability of forecasts drops rapidly the further into the future one goes. A "farsighted" analysis could show that an investment with a payback period of 30 years would become highly profitable in 35 years. However, the decision maker might doubt the analyst's knowledge of the economic environment three decades in advance.

Payback period analysis has been criticized on the grounds that investments are not made for the purpose of being recovered but with the hope of a profit. Therefore, it is what happens *after* the payback period that determines the desirability of the investment.

The limits of payback period analysis appear clearly when one tries to apply to operational decisions such as whether machine X should be allocated to parts of type A or B. If A's are almost finished and a week from the point of shipment whereas B's still require six weeks of processing through many other machines, then A's will always win because they have a shorter payback period.

However, if, for example, A's sell for $100 and B's for $1,000, then always allocating the machine to A's may mean preferring a $10 profit in a week to a $500 profit in 6 weeks. The simple payback period criterion considers neither the absolute amounts involved nor the effect of a delay on the value of money.

12.4.3. Discounted Cash Flows

Discounted cash flow analysis is actively promoted by American business schools, and is the core of Kaplan's [Kap85] recommendations for manufacturing. It is a more sophisticated method than payback period analysis but is flawed in that it still fails to provide for the quantification of risk. It is *not* commonly used in Japan, but that shortcoming does not seem to have held back local industry.

The theory of the time value of money establishes an equivalence between the values of sums of money possessed at various points in time, in the absence of inflation or risk. This theory assigns a price to delayed gratification: even if the purchasing power of a dollar remains constant and we are certain to receive it, one dollar a year from now is less valuable than one in hand today because we cannot spend it for a year.

Immediately, the restrictions of time value theory raise questions:

- *How much* less valuable is it as a function of time?
- How can the anticipated *inflation* be factored in?
- How can *risk* be assessed?

The time value of money is assumed to be stationary, in the sense that the relative value of delayed versus immediate receipt of a sum depends only on the delay. The percentage of value loss associated with a one year delay is the same in the year 2010 as it is in 1988. Formally, we can express this by a function $f(\Delta t)$ which, for all t, represents the value at time t of receiving a unit of currency at time $t+\Delta t$. A consequence of this stationarity is that

$$f(\Delta t_1 + \Delta t_2) = f(\Delta t_1) \times f(\Delta t_2)$$

The value at time t of receiving one dollar at time $t + \Delta t_1 + \Delta t_2$ is equal to the value at time $t + \Delta t_1$ of receiving $f(\Delta t_1)$ at time $t + \Delta t_1 + \Delta t_2$. Therefore $f(\Delta t)$ is of the form

$$f(\Delta t) = e^{-a\Delta t}$$

Since financial analysis views time as a sequence of periods, delays in time value formulas are view as integral numbers of periods. If the period is of length T, then f(nT) can be written in the form

$$f(nT) = \left(\frac{1}{1+i}\right)^n$$

where $\ln(1+i) = a$ and i is called the *discount rate.*

Once a discount rate is chosen, all the future outflows and inflows associated with a decision or an event can be converted into immediate equivalents called *discounted cash flows.*

In the analysis of the decision to build a factory, T= 1 year. However, if time value concepts are to be applied to operational decisions, then T = 1 week or T = 1 day may be meaningful. A unit of work in process on the shop floor can then be viewed as an investment opportunity. If C_R is the present value of all the outflows needed to get it shipped, and T_R is the time until it is sold for a price P, then C_R and $f(T_R)\times P$ are the corresponding present values of the outflow and the inflow.

Let us carry out a present value analysis of the nMDRAM factory with i = 10%. With

$$d = \frac{1}{1+i} = 0.909$$

the present value of the outflows over the first seven years of operation is

$$Out=100+88\times d+85.5\times d^2+74.5\times d^3+68.5\times d^4+65\times d^5+62.5\times d^6+65.5\times d^7=387M\$$$

On the other side, the present value of all the inflows is

$$In=166\times d+121\times d^2+99\times d^3+87\times d^4+80\times d^5+75\times d^6+71\times d^7= 513M\$$$

If all our assumptions hold, the construction of the factory is equivalent to a commercial transaction performed today which would immediately yield 513M$ for 387M$. One of the common measure of the desirability of an investment is its "Present Value Index," which is the ratio of the present values of the inflows and outflows:

$$\text{Present Value Index} = \frac{513}{387} = 1.32$$

A value of 1.00 means that the investment is equivalent to a loan with an interest rate of i. This perspective suggests finding the discount rate that would make the present values of inflows and outflows *equal*. This value is the *yield* of the investment. To be more profitable than the investment, a loan would need an interest rate exceeding the investment's yield.

In our example, with i = 47.5%, the present values of the in- and outflows are both 241M$. 47.5% is an interest rate that no one but a desperate borrower would agree to pay, and it makes the investment attractive. The big unanswered question, however, is how likely the projected funds flows are to materialize.

Payback period analyses can be done on *discounted* flows too, giving a *present value payback*. Considering that the payback period method emphasizes the near future and that the effects of discounting are felt mostly on the distant future, the differences between ordinary and present value payback periods will be frequently insignificant.

Kaplan [Kap85] recommends using a 5% to 8% discount rate, on the grounds that it is the range of historical, inflation-adjusted rates of return given to investors in the equity and fixed-income markets between 1926 and 1984. Helfert [Hel77] describes the discount rate as the "normal earning power of funds in the enterprise," which may be higher than the average offered by financial markets.

In Kaplan's view, a present value index of 1.0 means that, to an external investor, the investment equals what can be expected from stocks and bonds. In Helfert's view, it means that, to management, it is as good as the average of the previous investments of the company.

12.5. RELATION TO DATA FLOWS AND MATERIAL FLOWS

As soon as one moves towards the insides of the factory, away from the interface with its environment, the dollar, yen, or pound sign disappears from communications. This change is a signal that the decisions made by the initiators and the recipients of these messages no longer have a *direct* impact on funds flows.

As we have seen, the estimation of indirect financial impact is error-prone. Everything else being equal, if the purchasing department finds a raw material source that is 20% cheaper than before, the financial effect of switching suppliers is readily calculated.

On the other hand, the effect of reducing the number of labor hours needed per unit shipped by 20% varies depending on whether (1) overtime can be cut, (2) employees can be laid off, (3) employees can be transferred to other activities for which additional personnel would otherwise have been hired, or (4) none of these.

Work in process (WIP) inventory levels in numbers of units, turnover rates, manufacturing cycle times, and scrap and rework rates are examples of non-financial parameters used to evaluate the performance of managers inside a factory.

The juncture between the funds flow external context of the factory and the data flow model of its internals is the definition of policies and performance goals for departments, the pursuit of which presumably leads the factory towards its goals, financial or otherwise. It is conceivable, and frequently happens, that goals set inside the factory encourage behaviors that are detrimental to the factory as a whole. No general approach exists today with which to avoid such problems, although it is the subject of active research in business schools.

The projections on the future of the nMDRAM factory were predicated on a 20% yield increase per doubling of the number of devices shipped. The history of technology suggests that it is a reasonable expectation for such a yield increase to occur within the industry. Those factories in which it happens may survive; the others will get out of the race. Whether or not it happens in our example factory depends on what its managers, engineers, and workers do.

Decisions taken at the data flow level fall into two categories. The first one could be called "quantity control" and is the subject of the rest of this book. The second one is quality control.

Quantity control has to do with where parts should be, the orchestration of their movements, and the allocation of processing resources. The data manipulations associated with quantity control comprise the generation of instructions, their communication to the material flow level, and the comparison of actual, quantitative production results with the instructions.

Quality control deals with the physical properties of parts, means of detecting discrepancies from expectations, and actions that can be taken to avoid them. These activities include the definition of products, processes, and critical parameters on the one hand, and the analysis of measurements on the other.

At first sight, these activities appear unrelated. It seems natural to divide the work of controlling quality and quantity between separate departments with minimal communications. However, our analysis of throughput times with unreliable equipment showed that, even though getting parts through a line faster is a matter of quantity control, the main benefit is to scrap fewer parts, which is a quality control goal.

Better results are obtained by a *holistic* approach to manufacturing, in which no territorial limit is placed on the problem solvers' actions. The road to quality may go through throughput time reduction. To achieve this may in turn require changes in machine layouts, so that the funds, data, and material flow levels of the factory are all involved and affected.

Part Two: Application to Production Scheduling

13

THE COMPONENTS OF A PRODUCTION SCHEDULING SYSTEM

To provide a common framework, terminology is introduced for the description of times, part aggregates, activity types, resources, and inventories. For each of these components of the whole production scheduling system, analysis tools and potential problems are reviewed. Because only some of these problems are present in any real factory, there can be no universal solution. In the following chapters, we are going to examine in this light the scope of applicability of several systems in actual use.

Once a demand is placed on a factory in the form of customer orders or a market demand, the tasks remain of (1) setting the times and quantities of raw material releases to the shop floor, (2) determining quantities to be moved through each operation for each type of work in process over time, (3) comparing what actually happens on the floor with what was anticipated, and (4) forecasting the output of finished goods.

Several approaches are available. Some take the form of software products, which may or may not be part of CIM systems. Scheduling any but the most trivial factories is a mathematically daunting problem, but one that is practically solved somehow everyday in every operating factory. The schedules that are used are rarely optimal in any sense, and the financial stakes in improvements are high.

Unlike quality control, production scheduling is a discipline in which there is no universally accepted approach, dissatisfaction with current systems is widespread, and the search is on for better ones. To derive some general principles, we shall review three approaches:

- Manufacturing Resource Planning, also known as MRP-II, as opposed to "Materials Requirements Planning," which was the original meaning of the acronym.
- Optimized Production Technology, or OPT.
- Just-In-Time, or JIT.

There are no usable standards in the terminology of production scheduling. The American Production and Inventory Control Society (APICS) has published a dictionary, but it describes inconsistent usage rather than prescribes consistency. For the "cycle time" of an operation, the APICS dictionary gives the two meanings of standard time to go through it as in MRP-II, and of average time between two subsequent part completions as in Just-In-Time.

To compare MRP-II, OPT, and Just-In-Time, we need a common framework, and this is what we will construct in this chapter. However, the reader should be aware that internal consistency within this book is obtained at the cost of deviation from other authors.

We will also examine the properties of each component of the factory for factors of complexity in production scheduling, knowing that only a small number of these factors is present in any one given factory, and that no real scheduling system could possibly take all of them into consideration. This will help us highlight, in later chapters, the tradeoffs made in MRP-II, OPT, and Just-In-Time.

13.1. TIME

Scheduling production involves deciding when activities should take place, and therefore timing is of the essence. Whenever possible, we will qualify the word "time," because of otherwise troublesome ambiguities: in "start time" it refers to the *instant* at which something starts, but in "activity time," it means the *duration* of the activity.

Figure 13.1. shows what we mean in this book by the *lead time* and the *throughput time* of an order. After issuing an order, a customer cares most of all when it will be shipped, and may rate suppliers by their lead times. Within the manufacturing organization, the order is integrated into the production schedule. Its "release to the floor" is the instant at which the production department is made aware of its existence and assumes responsibility for it until shipment. The throughput time of the order is therefore the length of the interval during which it is in the hands of production.

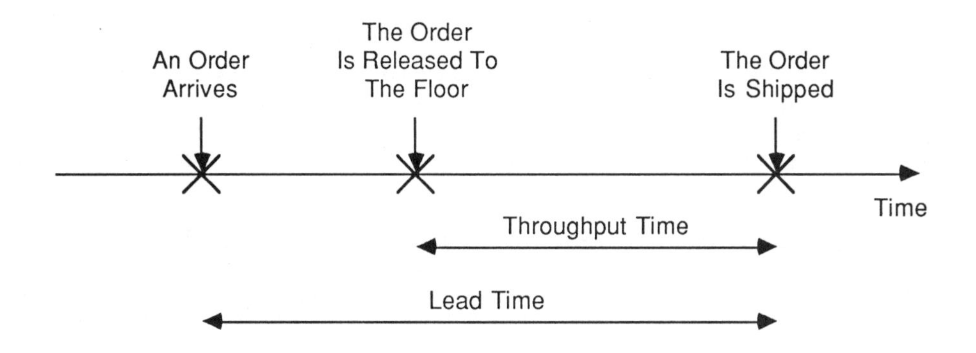

Figure 13.1. Lead Time and Throughput Time

Not all orders have the same lead or throughput time. At order receipt, it is usually impossible to predict them exactly. If the factory's operations are in some sense stationary, then the flow of orders can be assumed to have *probability distributions* for these times.

Some customers may accept a statement that an order may take "six to eight weeks" to fulfil. Those who are in a position to do so, however, usually demand a commitment to deliver by a certain time, called *due date.* Strictly speaking this expression is misleading because it carries the implicit assumption that no interval shorter than a day is considered, and *due time*" would be more precise.

Figure 13.2. shows the result of setting due times by a fixed offset from order receipt with a given lead time distribution. The area under the tail end of the distribution gives the percentage of missed due times. Lead times are sometimes also called *turnaround times.*

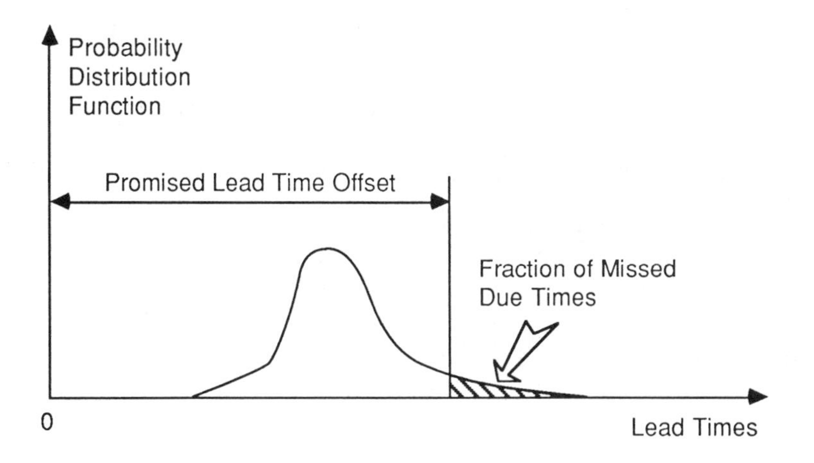

Figure 13.2. Due Times and Lead Time Distributions

Lead times are visible both internally, within the factory and externally, to customers. As a consequence, the lead time of an order accrues by one second for every passing second. The throughput time, on the other hand, looks different, depending on whether it is viewed from the outside or the inside of the shop floor. From the external perspective, it accrues like the lead time, but, as seen from the shop floor, it only accrues while the factory is working.

From 8:00 a.m. one day to 8:00 a.m. the next, the external throughput time increases by 24 hours, even if the factory only works one eight-hour shift. It is no concern to an outsider whether the factory works all the time or only eight hours a day. On the other hand, all employees below the level at which these decisions are made will be concerned only with the allocation of their work time. The external throughput time is said to be measured in *calendar time*, as opposed to its value in *work time* (see Figure 13.3).

If a machine processes parts one by one and takes exactly the same amount of time for each, then that time is defined to be the *processing time* of the part through the machine. The machine has a capital of time available, from which every part going through it withdraws an amount equal to its processing time. The extension of this concept to machines processing several parts at a time or to groups of machines is not straightforward, and is discussed in the section on resources.

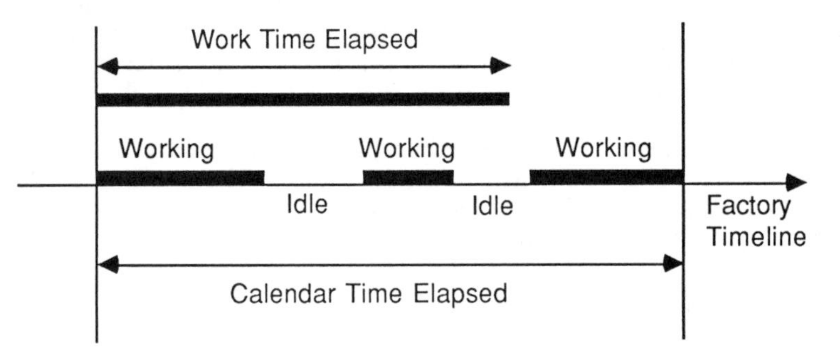

Figure 13.3. Calendar Time versus Work Time

By *queue time* we will mean the time spent by parts in front of an unavailable resource. The use of *waiting time* will be reserved for parts held up by the unavailability of other parts with which to be assembled.

Following industrial engineering usage, we will call *cycle time* the time between two successive completions of an activity, and *throughput*" its inverse. These terms usually require qualifiers. In particular, it is necessary to specify whether we are talking about the minimum cycle time a resource is capable of, the average of a variable value, or a fixed, deterministic one.

It is a common occurrence that parts be placed "on hold" until a responsible engineer disposes of them. The *hold time* contributes to the calendar throughput time, but it is often excluded from the work throughput time, on the grounds that it is time during which the production organization is not responsible for the parts.

13.2. THE PRODUCTION NETWORK

We would like to base the production scheduling models of factories on a "perfect *process* technology" assumption, which would state that every machine in the plant will faithfully carry out any legitimate order to process parts through an operation, and that the resulting parts will have all the desired properties.

It would be a valid approach if we could lump under the label of "quality control" all the activities that must be carried out to make this assumption appear true when viewing the factory from a sufficiently high level. With all the problems of imperfect processes out of the way, scheduling a factory to respond to a given demand would become a purely deterministic problem.

Unfortunately, as we have already seen in the previous chapter, scheduling and quality control issues can rarely be decoupled. Quality problems disrupt production. Quality control activities, such as inspections, must be scheduled. In turn, the scheduling methods of a factory impact the ease or difficulty of diagnosing quality problems. As will be shown in later chapters, the treatment of this relationship is one of the major differences between MRP-II and OPT on one side, and Just-In-Time on the other.

If, temporarily, we do make the perfect process technology assumption, it will allow us to identify problems that are *purely* of a scheduling nature in the sense that they would be present even in the most reliable, fully debugged factory. For example, the scheduling model would have to include those activities with requirements in time or resources such that their omission would have an externally perceptible impact on shipments. It does not matter whether these activities transform parts.

13.3. PART AGGREGATES

In semiconductors or pharmaceuticals, there are engineering reasons to want a set of parts to undergo a complete processing sequence without being separated. If a customer is poisoned by medicine bought in a supermarket, the supplier must be able to trace all the capsules made with the tainted one to diagnose the origin of the poison and protect other consumers. This type of concern leads to the definition of the first type of part aggregates relevant to scheduling.

13.3.1. Lot

A *lot* is a set of parts required to undergo a manufacturing process together. Maintenance of lot integrity is a constraint on production scheduling, no matter what approach is used.

13.3.2. Load

A *load* is a set of parts processed simultaneously by a machine. Some machines process parts one at a time. Others take several simultaneously. All these parts begin and end processing at the same time. If the machine is stopped halfway through an operation, all the parts in it are processed to the same degree. This type of aggregate we shall call a *load*.

13.3.3. Process Batch

A *process batch* is a set of parts processed on a machine between two setups. Once a machine is set up for an operation, it processes a number of loads before it is changed over to another operation. In calling this set of parts a *process batch*, we are following OPT's terminology. A process batch can also be called a *production run*.

13.3.4. Transfer Batch

A *transfer batch* is the smallest set of parts allowed to move from one operation to the next. To keep materials flowing, it is often desirable for parts having completed an operation to move on without waiting for the end of the process batch. Transfer batch is also an OPT's term.

13.3.5. Kit

All the preceding aggregates are sets of *identical* parts. In factories assembling a large number of products in low volume, it is common to withdraw all the components of a unit at one time from a single inventory location and place them in a container to travel together along an assembly route. Such a set of parts is called a *kit*.

A kit is a set of parts to be assembled together in a series of operations but placed in a single container beforehand. One effect of kitting is to make assembly look like fabrication. Instead of distinct material flows converging to each assembly operation, kits in various stages of completion flow along a sequence of operations.

This simplification is obtained at the cost of pegging components to finished goods earlier than would be necessary otherwise, which restricts the production manager's options. Maintenance of complete kits on a shop floor is also a problem wherever this system is used. When a part in a kit is defective, it is more expedient for workers to replace it from another kit than from the distant stock room.

Figure 13.4 shows the relationships between various types of part aggregates.

13.4. ACTIVITY TYPES

In Chapter 10, we introduced the ASME conventions to represent manufacturing activities. Here we take a second look at each of these categories from the point of view of its input and output characteristics in time and quantities.

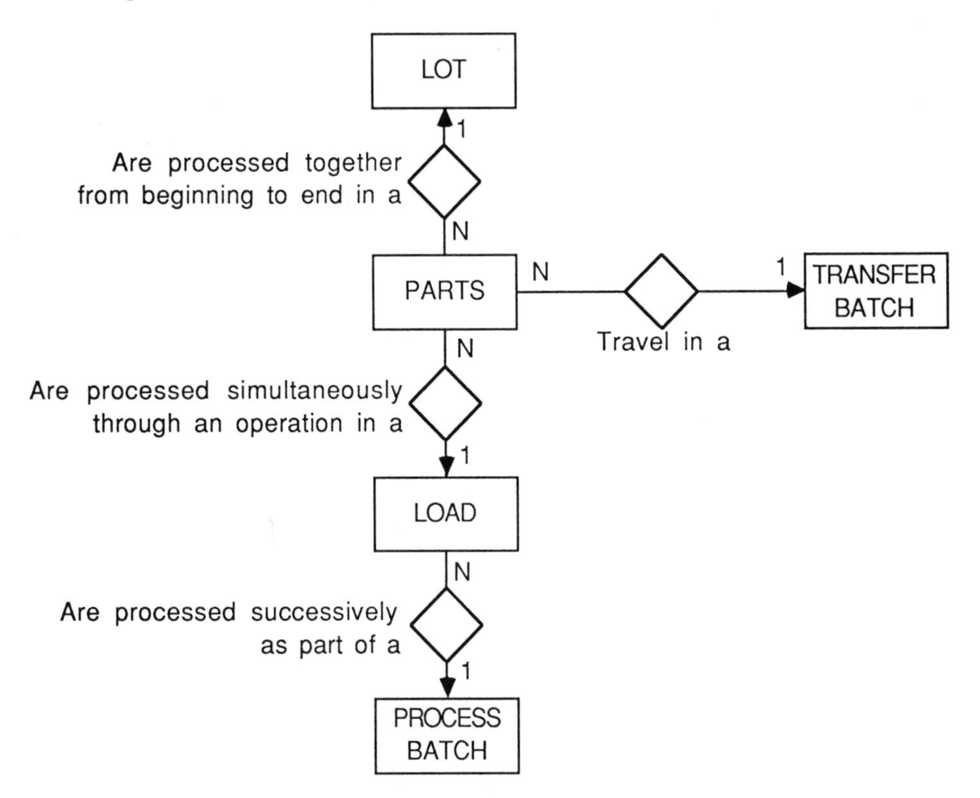

Figure 13.4. Relationships between Part Aggregates

13.4.1. Fabrication

A fabrication has a single input flow and a single *desired* output flow. There may be undesired output flows to scrap and rework (see Figure 13.5). One input part yields at most one output part. Fabrication activities are strung into routes in which the output of an operation is the input to the next one.

Scheduling problems in fabrication are most commonly caused by equipment, either in the form of insufficient capacity or of inadequate process capability. The former may result in missed shipments if capacity has been overestimated. The latter curtails output by causing parts to be scrapped. It also increases and randomizes processing times due to reworks.

The effect of reworks and scraps on processing time through a fabrication operation is analyzed in [Bau85-2] using the control theory model shown in Figure 13.6. Parts flow like a "signal" transformed by the operation, and the rework loop provides a destabilizing *positive* feedback.

If an "impulse" of n parts is fed to the operation at time $t=0$, processing completes at the rate $n \times h(t)$, where $h(t)$ is the impulse response of the operation. $h(t)$ is also the probability density function of the processing time of the operation without rework: a part started at $t=0$ is completed between t and $t+dt$ with probability $h(t)dt$. The impulse response $r(t)$ of the "undo" step performed prior to repeating the operation is defined the same way.

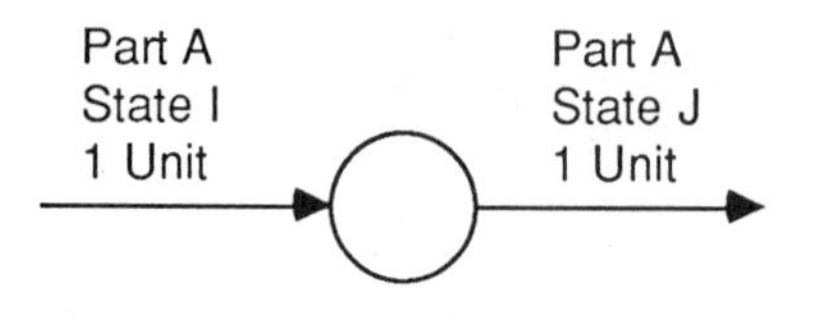

a. Perfect Fabrication

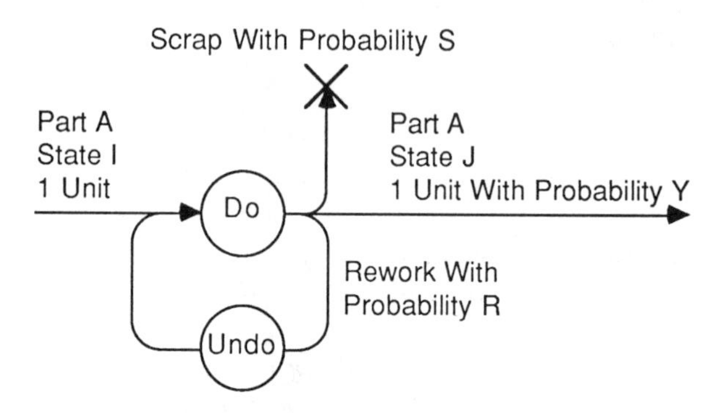

b. Fabrication With Problems

Figure 13.5. Fabrication Models

In control theory, the Laplace transforms of $H(s)$ and $U(s)$ of $h(t)$ and $u(t)$ are called *transfer functions*. In the probability perspective, they are the moment generating functions of the processing times. The transfer function $W(s)$ of the system with rework is

$$W(s) = \frac{Y \times H(s)}{1 - R \times H(s)U(s)}$$

And, from this formula, we deduce various results. The final yield of the operation with rework is the probability Y_f that an input part is *eventually* processed successfully, and

$$Y_f = W(0) = \frac{Y}{1 - R}$$

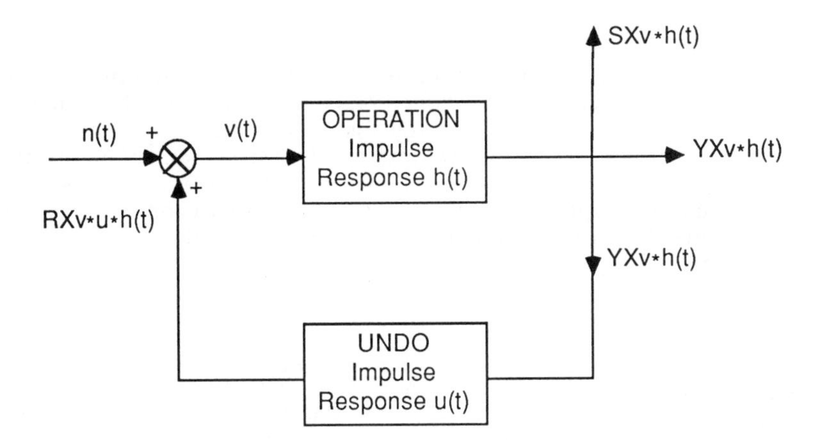

Figure 13.6. Control Theory Model of Fabrication with Rework

The expected value and variance of the processing time T through the operation with rework are given by

$$E(T) = -W'(0)$$

and

$$Var(T) = W''(0) - W'(0)^2$$

These formulas enable us to appraise the effects of rework both on the average processing time and on its variability around the average. In the most common and most simple case, the processing time through the operation *without* rework is a deterministic T_0. If the UNDO step likewise always takes exactly T_1, we have

$$W(s) = \frac{Y \times e^{-T_0 s}}{1 - R \times e^{-(T_0 + T_1)s}}$$

and therefore

$$E(T) = T_0 + (T_0 + T_1) \times \frac{R}{Y}$$

$$Var(T) = (T_0 + T_1)^2 \times \frac{R}{Y^2}$$

Although these simple formulas are usable, the method by which they were derived is not easily extended to more complex cases.

13.4.2. Assembly

An assembly activity has one or more input flows for one output flow. With only one input flow, it can still be called "assembly" if several input units are needed for one unit of output. The key issue in assembly scheduling is the availability of all *input* parts. The greatest single reason for delays in assembly is a shortage of a small number of components keeping "almost finished" goods from being completed and shipped. The operations in Figure 13.7 cannot be started until at least n, m, and p units of the respective components arrive.

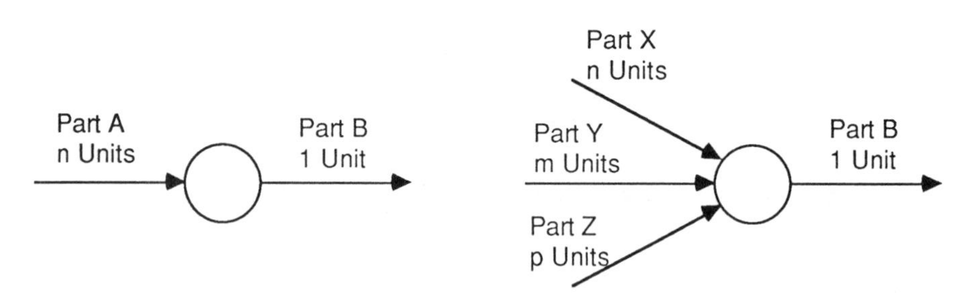

Figure 13.7. Scheduling Assembly Operations

Queueing theory provides usable network models for fabrication shops in which parts can move as soon as processing resources are ready, but none so far applicable to the synchronization problems of assembly, the avoidance of which is the motivation for kitting.

13.4.3. Disassembly

A disassembly activity has one input material flow and one or more outputs. If it has only one output flow, it is disassembly only if one unit of input produces several units of output — for example, sawing a finished silicon wafer into individual integrated circuits (see Figure 13.8).

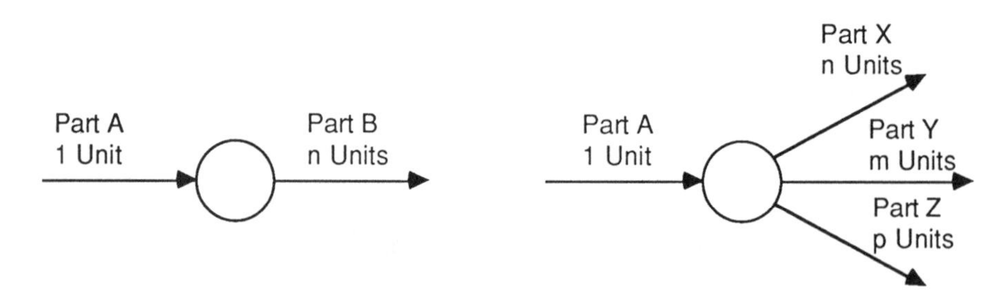

Figure 13.8. Scheduling Disassembly

Disassembly with multiple output flows constrains the product mix. In a slaughterhouse, a pig yields two hams, four feet, and one liver, regardless of the relative market demand for these products. The same pattern is found with composite molds in metal foundry work and with multiproduct wafers in semiconductors.

The demand for food products can be modified by changing prices. Because of this elasticity, the demand for the various cuts of meat can be adjusted to the supply. On the other hand, the demand for different engine parts cast together in a composite mold depends on that of the cars they go into, and price manipulations can only affect the secondary market for spare parts.

13.4.4. Inspection, Sorting, and Binning

In this type of activity, a single input flow feeds multiple outputs, but, in contrast with disassembly, each part follows one of the output flows instead of being split between them. These activities are branching points, where attributes are assigned to parts, which determine their subsequent paths (see Figure 13.9).

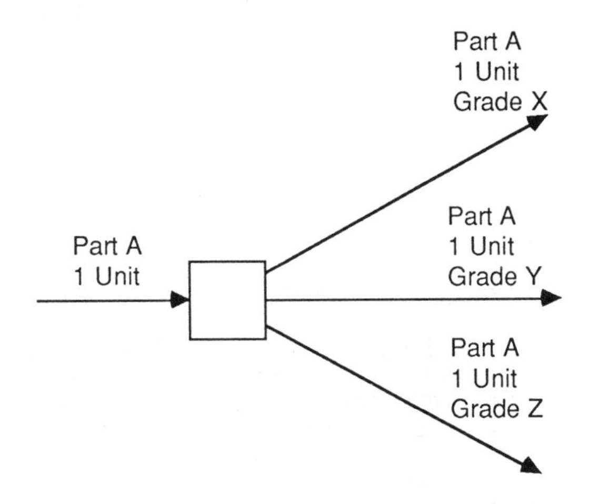

Figure 13.9. Scheduling Inspections, Sorting, and Binning

From a scheduling standpoint, the volumes of output parts have two undesirable properties:

1. They are random.
2. They are independent of market demand, and determined instead by the process capability of the factory.

After integrated circuits are tested, they are put into bins corresponding to the specifications of various customers. The number of chips falling into each bin varies randomly from lot to lot, which makes it difficult to decide how many parts should be produced to fill any given order. The ratio of military to commercial parts after testing is independent of the demand for each type.

If the inspection consists of making a pass or fail decision on the basis of a single measurement, then the processing time of a part can be modeled as a constant. However, many inspections are comprised of several tests, sequenced by decreasing values of the following function:

$$\text{Figure of Merit} = \frac{\text{Fraction Rejected}}{\text{Test Time per Part}}$$

In other words, the tests that find most rejects fastest are done first, and parts go through the tests until they are rejected or complete the whole sequence. The inspection time of a part is therefore a random variable with a distribution depending on the quality of the population of parts.

13.4.5. Transport

Transport activities move parts from one location to another. As a scheduling problem, the internal transportation system of a factory is no different from a bus or railroad system. One key scheduling decision affecting the transport system is the selection of a transfer batch size (see Figure 13.10).

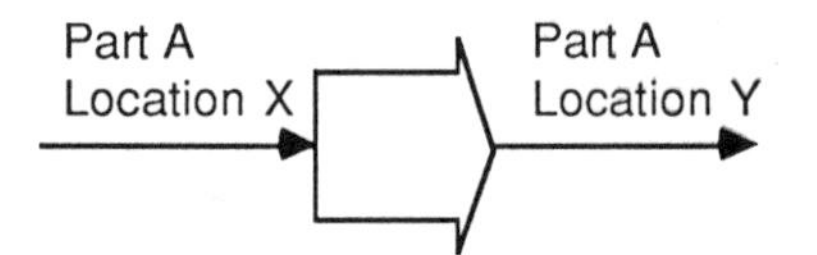

Figure 13.10. Scheduling Transport

If, for example, forklifts will not move anything less than two full crates, then the average queue time of a part in the transport system will be twice as long as if they were allowed to move only one crate. The latter situation may be achieved by using a larger fleet of smaller forklifts.

13.4.6. Storage and Retrieval

Wherever parts are allowed to accumulate, there is such an activity, and it may have any number of input and output flows. From a scheduling standpoint, the key issues are the numbers of parts stored and the times they spend in storage. We shall discuss these matters in the section on inventory.

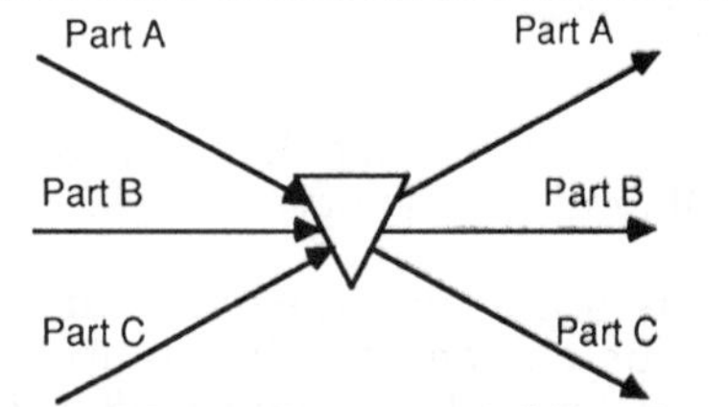

Figure 13.11. Storage and Retrieval

13.5. RESOURCES

A resource, in a scheduling system, has a name and is required for a period of time in order to carry out an activity.

13.5.1. Hierarchical Model

Depending on the scheduling problem under consideration, the same entity may or may not be modeled as a resource. Company wide production planning in a vertically integrated organization like General Motors is a scheduling problem in the sense that it involves setting sequences of tasks over time. However, in this context, a resource may be an entire division. Within a division, each factory is then treated as a resource. Within each factory, each department becomes a resource, and this hierarchical breakdown goes on all the way to individual machines.

The challenge of resource modeling is the development of quantitative characteristics measurable at *all* levels and giving consistent results with every "change of magnification." The characteristics of a resource must be calculable from those of its lower level components, and errors in resource modeling at one level of the hierarchy propagate upwards to all the higher levels.

E. Lazowska et al. [Laz84] use "flow equivalent service centers" or "FESCs" to model resource aggregates. In [Laz84], this approach is applied to performance evaluation in computer systems, not to production scheduling. Its relevance to manufacturing resources is examined next.

13.5.2. Work Centers versus Work Cells

First-level groupings of machines are critical to production scheduling and the model should reflect the shop floor layout. Two patterns are found throughout industry. In a *work center*, machines that can perform the same activities are arranged in parallel. In a *work cell*, machines feeding one another are laid out in sequence.

The work center organization is characteristic of the *job-shop* layout, and the work center concept of sets of interchangeable machines is central to the MRP-II approach. Work cells comprise *flow lines*s (see Figure 13.12).

13.5.3. Resource Characterization

Some resources, such as computer programs, can be used *simultaneously* in an unlimited number of activities; they are said to have "infinite capacity." Other resources, such as cutting tools, can be used at one time in a finite number of activities, and are said to have "finite capacity."

Most production scheduling systems assign to each resource standard queue and processing times *per part*. In reality, this characterization is simplistic except for machines processing one part at a time at a constant rate. In every other case, waiting and processing time depend on the workload of the machine, even in the absence of randomness.

If a furnace needs one hour to process a load of 10 parts, then its standard processing time is 0.1 hours per part. However, if there is not enough work to keep the machine full and it is run with loads of only five parts, then the effective time per part rises to 0.2 hours. To take an extreme example, if a single part is moved in an otherwise empty factory, then the processing time of the furnace for that part is one hour.

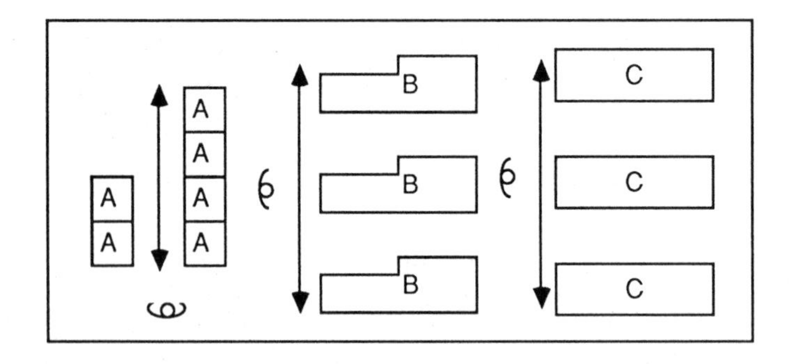

a. Job- Shop With Work Centers

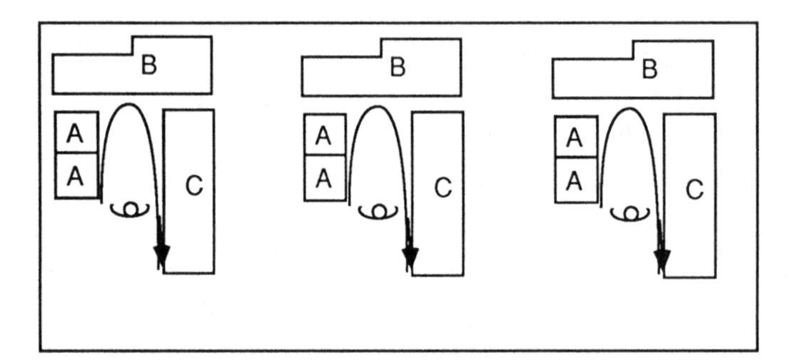

b. A Flow Line With Work Cells

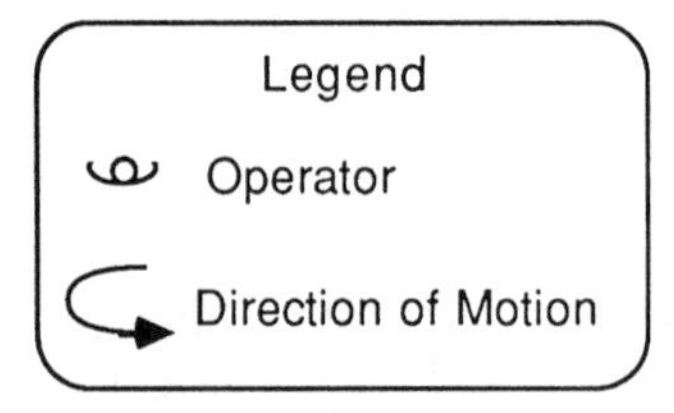

Figure 13.12. Job-Shops and Flow Lines

The load dependence of process and queue times can be modeled simply for resources carrying out only one operation, by plotting its throughput as a function of the total number of parts in it and queued in front of it.

Let us assume a work center is composed of n machines, each of which processes parts one by one with a cycle time of T, and the total number of parts queued or in process is L. When L=1, one and only one machine is working, so that the throughput is

$$\lambda(1)=\frac{1}{T}$$

The throughput increases linearly as more and more parts are allowed into the system, as long as L<n. As soon as N≥n, all machines are busy, and

$$\lambda(L)=\frac{n}{T}.$$

In one single formula, the throughput λ as a function of L is

$$\lambda(L)=\begin{cases}\frac{L}{T} \text{ if } L<n, \text{ and} \\ \frac{n}{T} \text{ otherwise}\end{cases}$$

λ as a function of L is shown in Figure 13.13.

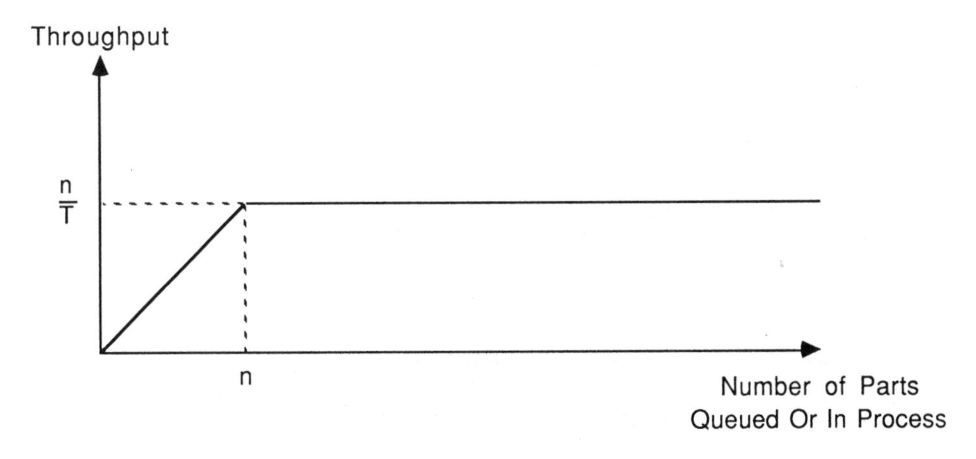

Figure 13.13. Work Center Instantaneous Throughput

The throughput λ given by Figure 13.13 is an instantaneous value, in the sense that it is what the throughput would be if the number of parts at the work center could be set exactly at a level L. In reality, even in a stable factory, L will fluctuate, and the throughput, as a function of that average, will take the form shown in Figure 13.14.

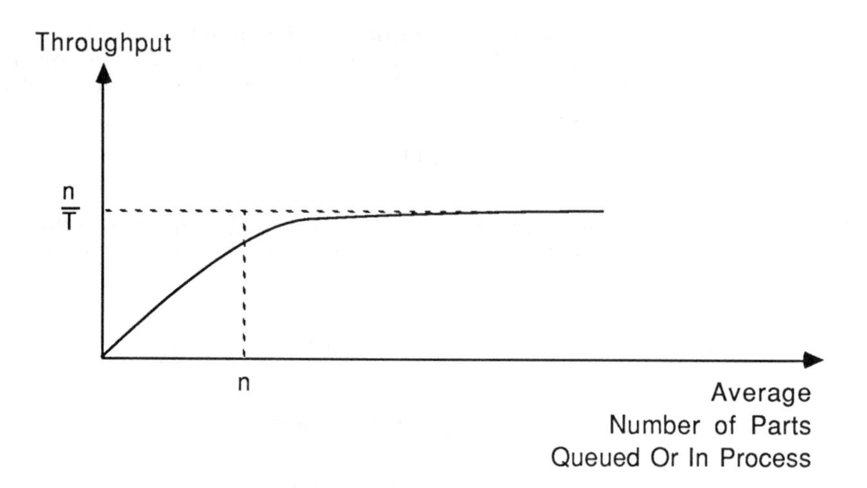

Figure 13.14. Work Center Throughput

13.5.4. Little's Law

The example in the preceding discussion showed that it is in general not possible to assign to a resource a single *processing* time per part. The dependence of *queue* times on workload is obtained by means of a simple but powerful result known as Little's law.

It says that, if the throughput of the work center is stationary λ, the number of parts queued or in process is L, and the average throughput time is W, then

$$L = \lambda \times W$$

In other words, the number of parts at the preceding work center is the product of its throughput by its throughput time. Little's law applies more generally to any queueing system with a stationary workload, and, without exaggeration, can be said to be to this type of system what Ohm's law is to electronics.

If we draw a boundary around a single machine processing parts one at a time, then the machine utilization ρ is the average number of parts in it, and the processing time p is the throughput time. In this case, Little's law says that

$$\rho = \lambda \times p$$

If, instead, we look at a whole manufacturing organization, then Little's law relates the number of pending orders, their arrival rate, and the turnaround time. Little's law also applies to the example of engineering change orders given in Chapter 5, by relating the number of change orders in the approval cycle to the rate at which they are submitted by engineers and the average time it takes for management to reach a decision.

Little's law is proven by looking at two alternative ways of calculating the area between the two curves in Figure 13.15. We consider a queueing system, and give it N discrete inputs over time. Let T be the completion time of the last arrival. In Figure 13.15, we plot the cumulative numbers of arrivals and completions over time between 0 and T.

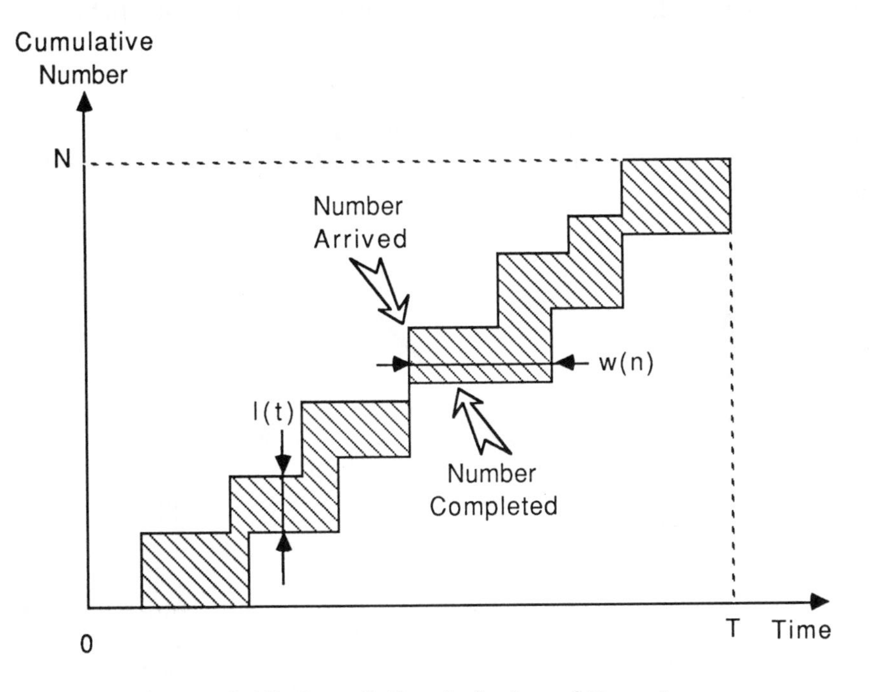

Figure 13.15. Cumulative Arrivals and Departures

Between 0 and T, the throughput was

$$\lambda_T = \frac{N}{T}$$

At any time t, the number of units in the system is the difference l(t) between cumulative arrivals and completions, and the average of that number between 0 and T is

$$L_T = \frac{1}{T}\int_0^T l(t)dt$$

But the gray area in Figure 13.15 can be viewed either as $\int_0^T l(t)dt$ or as the sum of the waiting times of the N requests. If the n-th request is satisfied in time w(n), it contributes 1×w(n) to the gray area. Therefore, if W_N is the average throughput time of the N requests,

$$L_T = \frac{1}{T}\int_0^T l(t)dt = \frac{1}{T}\sum_{n=1}^{N} w(n) = \lambda_T \times \frac{1}{N}\sum_{n=1}^{N} w(n) = \lambda_T \times W_N$$

From there, it is apparent that, if the workload of the resource fluctuates around a constant long term average, letting T grow to infinity yields

$$L = \lambda \times W$$

In Figure 13.15, we plotted λ as a function of L. Using Little's law, we can now plot W as a function of L, as in Figure 13.16, and finally W as a function of λ, as in Figure 13.17.

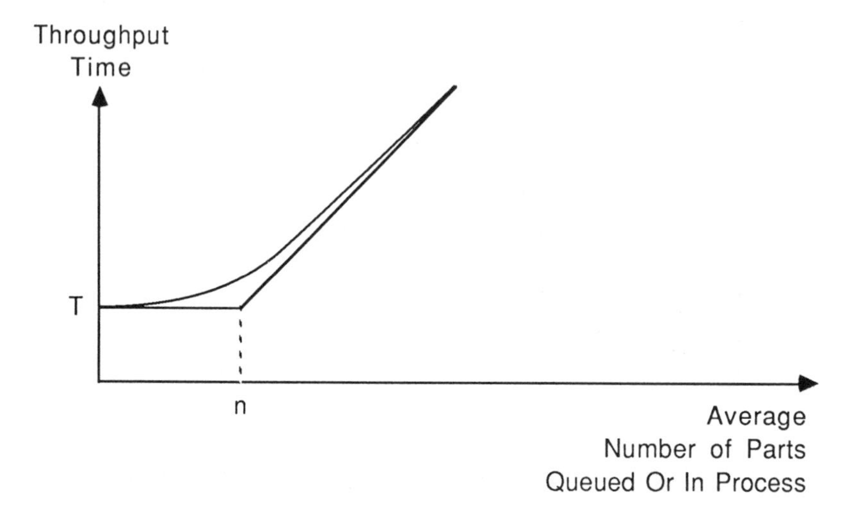

Figure 13.16. Throughput Time versus Work in Process

Figures 13.15, 13.16, and 13.17 together show that, when the work center is lightly loaded, its throughput is proportional to the work in process and its throughput time is constant. On the other hand, when it is heavily loaded, its throughput is constant but the throughput time is proportional to the work in process. Throughput and throughput time cannot be simultaneously constant unless the work in process is, too.

13.5.5. Infinitely Divisible Resources

Most resources share their time between different activities, and the software tools available for capacity requirements planning usually allocate a resource as shown in Figure 13.18. They decide whether a workload is feasible simply by adding up time requirements and checking whether there is any idle time left. In this model, all resources are implicitly assumed to have the property we will shortly define as *infinite divisibility*. We make this assumption explicit in order to tell the circumstances in which it is applicable from those in which it isn't.

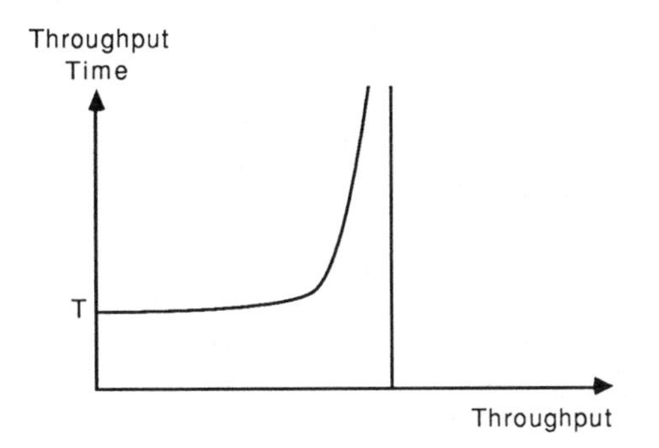

Figure 13.17. Throughput Time versus Throughput

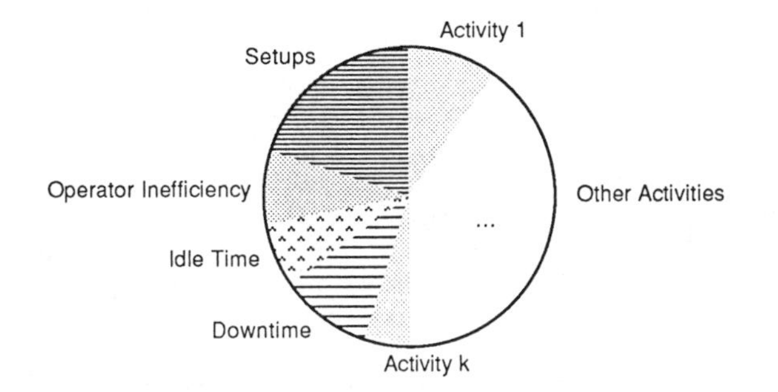

Figure 13.18. Time Usage of a Resource

We will say that a resource is infinitely divisible if, and only if, it can simultaneously perform any number of tasks, continuously allocating to each task a fraction of itself.

Infinitely divisible resources are idealizations of resources which can perform a large but finite number of tasks simultaneously. If a work center is composed of 100 identical machines and loaded as follows:

Activity	Work Center Time
A	15%
B	10%
C	75%

This workload can be handled by allocating 15 machines to A, 10 to B, and 75 to C. The set of activities and the numbers of machines needed for each could vary, while maintaining a situation in which a fraction of the work center is continuously working on each activity.

If, on the other hand, a work center is composed of only one machine, working 100 times faster than the components of the divisible work center would be an indivisible resource: with it, activity A must have the whole work center 15% of the time instead of 15% of the work center all the time.

Scheduling an infinitely divisible work center is much simpler than an indivisible one, because the sequencing problem disappears. However, the dependence of throughput times on workload does not.

Real work centers are *never* infinitely divisible, and a question that must be answered is the degree to which this model is an acceptable approximation. Setups present a major difficulty, which we discuss in a separate section. For the time being, let us assume that we are dealing with an indivisible resource that can switch instantly from one activity to another.

Figure 13.19 shows that, over time, the cumulative output of an indivisible resource becomes progressively closer to that of an infinitely divisible resource of equal capacity and subjected to the same workload. The fit of the approximation is good if, as in Figure 13.19b, the resource time taken by one process batch is small with respect to the time period considered. In the case of Figure 13.19c, a time period taken at random will show a great variability around the output of Figure 13.19a.

It is generally possible to determine how long a manufacturing task would require on an infinitely divisible resource that is temporarily dedicated to it. For example, if all the machines in a work center have a load size of 1, the processing time P of a batch is proportional to its part count and inversely proportional to the number of machines.

If several activities are carried out concurrently, then the preceding activity is allocated a fraction f of the resource. The processing time of the batch then becomes

$$P_s = \frac{P}{f}$$

If two batches each get half the resource, then each will take twice as long to get through as it would if it had the resource to itself. This formula is applicable to an indivisible resource with process batches small enough for the divisible resource approximation to hold.

13.5.6. Composite Resources

A manufacturing activity usually requires not one but a combination of resources. An *operator* may be needed to attach a *tool* to a *jig* before affixing it to a *machine,* and to dismantle the setup afterwards. The italics in the previous sentence all identify resources used in conjunction.

If all but one of the resources are abundant, then production schedulers only worry about the one that isn't. They do not even need to be aware of the others. But if several resources are in short supply, the scheduling problem rapidly becomes inextricable.

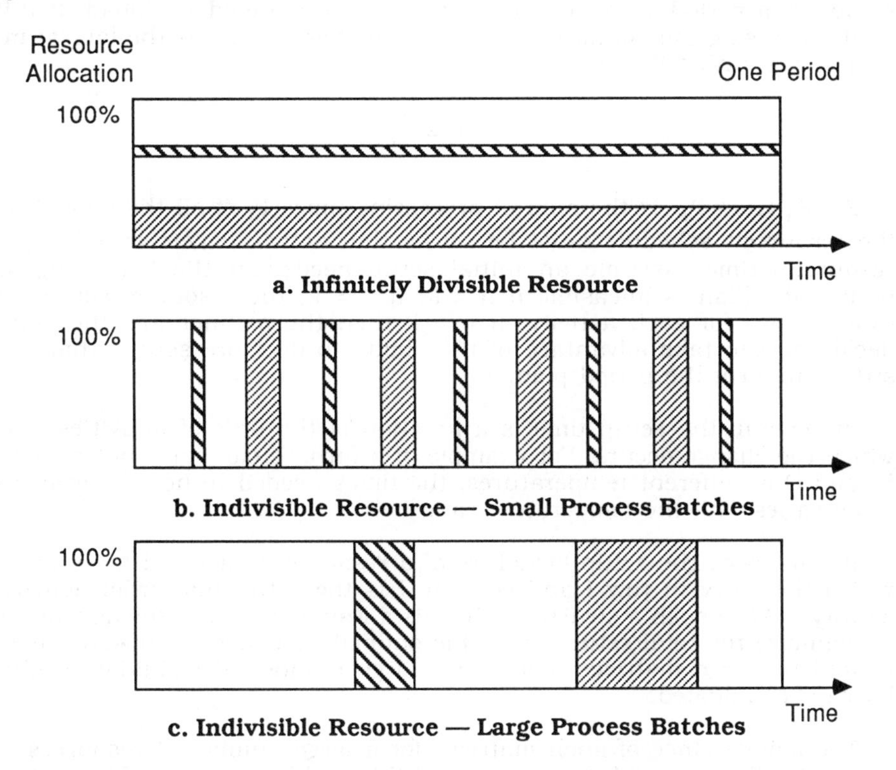

Figure 13.19. Three Hypothetical Resources with the Same Workload

The operator, in the preceding example, may be needed only for setup before the activity starts and changeover after it ends, but not while it is in progress. On the other hand, the tool and the jig will be tied up during that whole sequence. Therefore, it is generally not sufficient to consider the time load on each resource. Their availability must also be synchronized.

13.5.7. Setup Times and Setup Matrices

When an indivisible resource is changed over from one activity to another, it is frequently necessary for settings to be adjusted, and jigs and fixtures to be replaced. In the absence of such time consuming *setup* activities, there would be no reason not to set

$$1 \text{ Process Batch} = 1 \text{ Load}$$

This would make the resource's output as close as possible to that of an infinitely divisible resource. Thus large setup times not only restrict the resource's capacity by tying it up for a fraction of the time but also force process batches to be larger than they could otherwise be.

If all setups take the same amount of time S, the maximum number n of setups in a period T can be deduced from the planned workload. If P if the total processing time of all the planned activities, then n is the largest integer such that $P + nS \leq T$ — that is,

$$n = \left\lfloor \frac{T-P}{S} \right\rfloor$$

$P = P_1 + \cdots + P_k$ is the sum of processing times P_i of all the activities, and the resource's schedule must be set so that activity i gets exactly P_i of the resource's time. Assume an initial setup needed at the beginning of the period, the plan is infeasible if $n < k$. If $n = k$, the resource will be set up exactly once for each activity. If $n > k$ then the planner has the option to decide how to take advantage of the ability to do more setups than needed strictly to meet the period plan.

In general, the setup time is a function of the *pair* of activities between which the change occurs. For example, if a furnace can carry out activities A, B, and C at different temperatures, the times needed to heat it up or cool it down varies with the temperature range.

In this case, the setup structure of a resource is modeled by a matrix, in which the entry in row I and column J is the setup time from activity I to activity J. With such a structure, there is a way to sequence the activities so as to minimize the total setup time. This optimal sequence is, however, not easy to find for large matrices, because a large number of candidate sequences have to be evaluated.

The maintenance of such matrices for a large number of resources would be practically impossible. A way around this problem, used in OPT, is to enter resources into "setup groups" sharing setup matrices.

13.5.8. Interruptions of Service

Even within the constraints of the work calendar, resources are not constantly available. Service interruptions by a resources are either scheduled or accidental. Scheduled interruptions include breaks for employees and preventive maintenance for machines. Accidental interruptions are failures.

Preventive maintenance is easiest to schedule production around when it is performed at fixed intervals, and when the maintenance activity itself is a fixed sequence of tasks. Cleaning a filter once a week would be one such example. If the activity includes inspections which may or may not uncover problems, then its duration becomes a random variable.

Usage-based maintenance requires more caution. If photolithography masks are to be cleaned every 50 uses and there are 10 masks, then production will be least disrupted if the mask usage counts are kept evenly spread between 0 and 50. But this is not an easy discipline to maintain.

Given the choice, quality-conscious operators will always want to use the cleanest mask on hand — that is, the one with the lowest usage count. This would tend to keep usage counts equal. All masks would cross the 50 limit and become unavailable almost simultaneously.

As discussed in Chapter 1, equipment failures generate the need for large safety stocks. In most factories, equipment reliability is only characterized by first-order statistics such as a percentage of downtime. This makes no distinction between machines subjected to failures that are frequent but brief or rare but long. Reliability models characterize failure processes by a *mean time between failures* (MTBF) and a *mean time to repair* (MTTR) but this data is usually unavailable.

If it were, elaborate tradeoffs could be made to find optimal preventive maintenance schedules, but the improvement of equipment reliability is a preferable endeavor. To this end, maintenance can be integrated with production activities into *total plant maintenance* (TPM).

13.6. INVENTORY

"Queue," "buffer," "stock," "inventory," and "bank" are so many words used in industry to designate accumulations of parts. These words differ not in meaning but in connotation. *Queue* evokes in the listener's mind the unpleasant experience of standing in line. If a pile of crates is referred to as a queue, it will not be perceived as a desirable possession. *Bank*, on the other hand, conjures up the image of "money in the bank," vaults, and treasure troves. The word makes the pile of crates a valuable asset, even though the analogy between it and what one keeps in a bank is tenuous.

More than any of the other scheduling system components we have discussed so far, inventories are *observable* and *controllable*. Industrial engineering studies are needed to characterize resources, but all it takes to find out how much inventory there is, is to look. Managers cannot decree throughput times, but they can set and enforce caps on inventory which affect throughput times via Little's law.

Manufacturing inventories are usually broken down into the three categories of raw materials, work in process, and finished goods. Bulk supplies of indirect materials may be counted separately from direct raw materials. Work in process is comprised of all materials that are waiting, queued, or being processed in the factory. Finished goods are ready for shipment to the factory's customer, which may be other factories to which they are raw materials.

The material flow network of the factory contains points at which parts are allowed to accumulate into inventory buffers for two reasons:

1. to compensate for the differences in process batch sizes between two subsequent operations.
2. to protect the flow of production against random variability in the factory.

Like geological profiles, the size and content of inventory buffers contain historical information. A series of snapshots cannot unequivocally point to the reason for its existence. However, if we know that reason, we can usually deduce the structure it should have, and build a catalog which an analyst can use to confirm or refute assumptions about the factory.

13.6.1. Schedule-Based Buffer Analysis

One approach, used by OPT, is to plot as a function of future time t, the fraction of the material scheduled to be in a buffer at time t that is already there at time 0. An example is given in Figure 13.20, showing 100% of the material due by the beginning of the next shift already in the buffer, as well as 30% of what is due four shifts from now. If the buffer is in front of a resource processing many different types of parts, each part on the vertical axis is scaled by it processing time — that is, a part requiring three times more of the resource time than another contributes three times more to the bar graph.

If too much material is processed ahead of time, the early bars will be too high. Bars dropping off too fast may show a risk of shortage. However, the problem with these plots is that they require *explicit schedules* both to draw and interpret them. Next, we will examine in more detail a tool with no such prerequisites.

13.6.2. Age Distribution Analysis

Figure 13.21 is a similar plot, but based on the age of the material in the buffer. It can be drawn from the arrival times into the buffer and the processing time requirements of the following resource if there is a need to scale the histogram bars. In Figure 13.21, the vertical axis is normalized by the height of the bar representing the latest arrivals, for reasons that will become clear shortly.

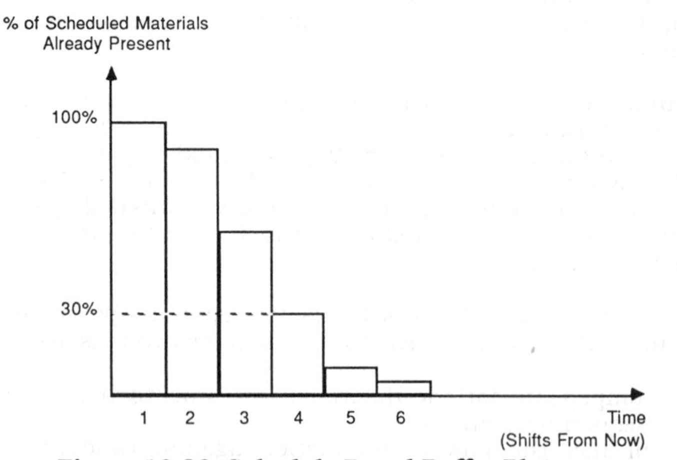

Figure 13.20. Schedule-Based Buffer Plot

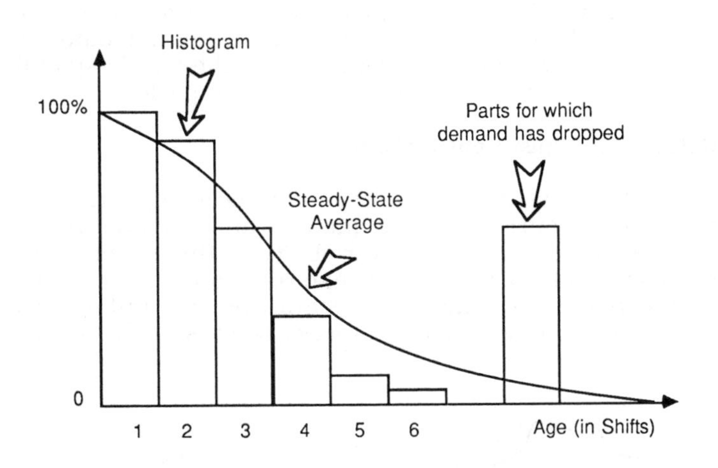

Figure 13.21. Age-Based Buffer Plot

Saying that a part in the buffer is of age t is equivalent to saying that its waiting time in the buffer will be at least t. If the factory is in some form of a steady state, the queue time and the age distribution of parts in a buffer can be inferred from one another.

More specifically, parts in the buffer reach age t+dt if and only if they reach age t and their waiting time is not between t and t+dt. Therefore, if

n(t)dt is the expected number of parts in the buffer between the ages of t and t+dt, and

w(t)dt is the probability that the queue time of a part is between t and t+dt,

then n(0) is the arrival rate into the buffer and

$$n(t+dt) = n(t) - n(0)w(t)dt$$

If we normalize, as in Figure 13.21, by using

$$p(t) = \frac{n(t)}{n(0)}$$

we have

$$\frac{dp}{dt} = - w(t)$$

and

$$p(t) = 1 - \int_0^t w(s)ds = 1 - W(t)$$

In other words, the cumulative distribution W(t) of the waiting time in the buffer is obtained from the curve in Figure 13.21 by a symmetry around the horizontal axis at y = 50%. p(t) cannot have a peak at any time t > 0. If, as in Figure 13.21, such a peak is present in the actual data, it points to a departure from steady state. One would expect the steady-state distribution to emerge when the results of several snapshots are pooled.

13.6.3. Batch Size Adjustment Buffers

Figure 13.22 shows an inventory buffer receiving various types of parts from upstream resources in the factory and feeding them to a single downstream resource. There is no variability in this factory, and the purpose of the buffer is to adjust the flow of materials to differences in process batch sizes between resources. The steady-state age distribution of material in a buffer that serves strictly to adjust process batch sizes between two resources is convex.

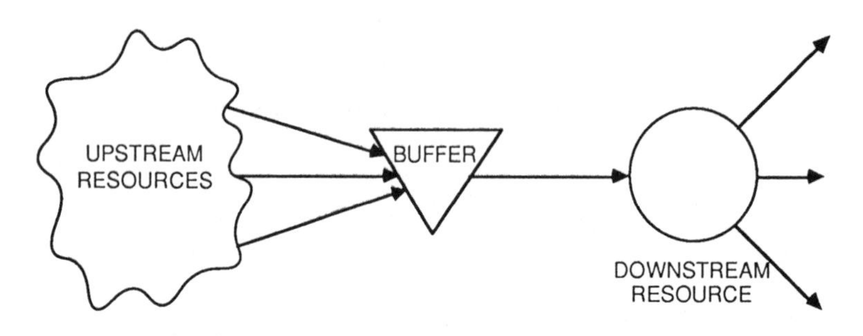

Figure 13.22. A Batch Size Adjustment Buffer

The process batch size of the resource fed by the buffer may be larger or smaller than that of any of the feeding resources. We model the first case with the approximation that the set of upstream resources collectively behaves like one single, infinitely divisible resource. Figure 13.23 shows the resulting timelines.

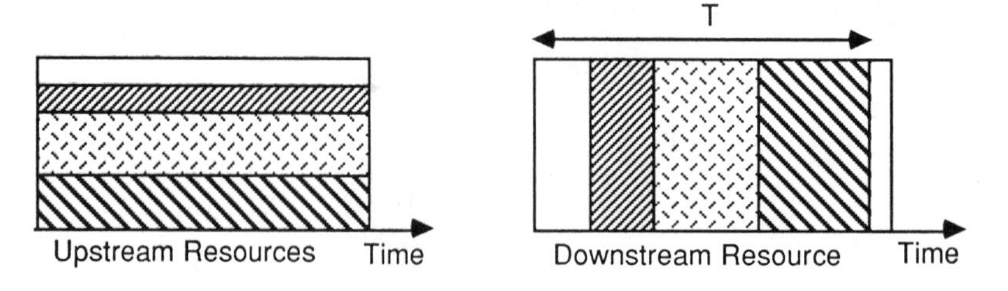

Figure 13.23. Case 1 Resource Timelines

For the downstream resource, we let T be the time between two process batch starts for a particular activity. For that activity, time is broken down into periods of length T, during which inventory accumulates in the buffer, and at the end of which the buffer is emptied by the transfer of the process batch to the resource.

The buffer content in input parts for that activity is shown in Figure 13.24. It rises slowly up to the size of one process batch, and drops abruptly back to 0 when that batch moves into the downstream resource. Over time the average buffer size is half that of a process batch.

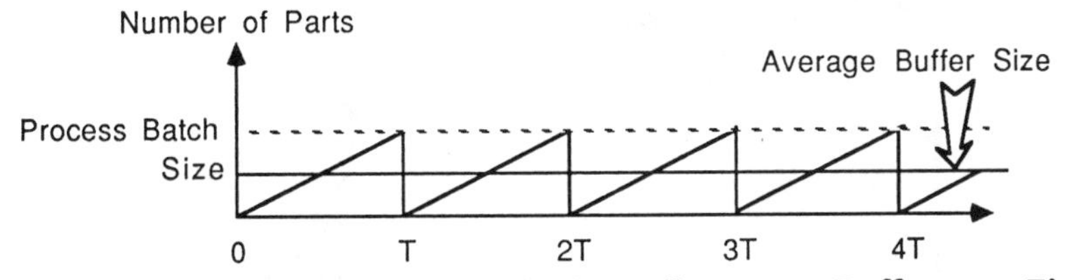

Figure 13.24. Number of Parts in Batch Size Adjustment Buffer over Time

In this buffer, the waiting time of a part is equally likely to take any value between 0 and T, and therefore follows the uniform distribution on that interval. The resulting age distribution is wedge-shaped, and derived as shown in Figure 13.25.

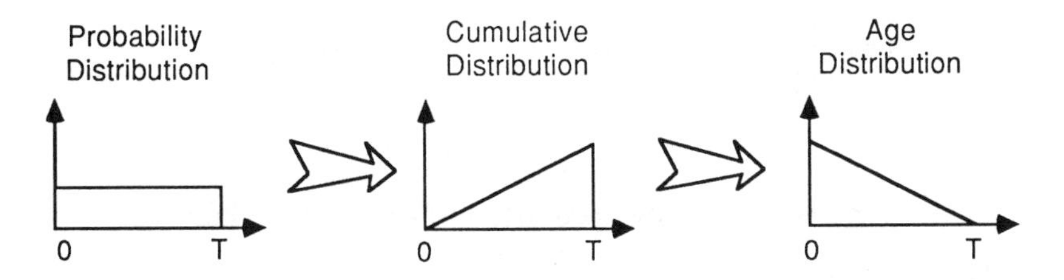

Figure 13.25. Age Distribution in Batch Size Adjustment Buffer

The opposite type of batch size adjustment buffer, in which large batches delivered by upstream resources are fed to an infinitely divisible resource, behaves as shown in Figure 13.26. The preceding discussion of age distribution applies here verbatim, so that a batch size adjustment buffer for one type of part has a wedge shape, no matter what direction the adjustment is in.

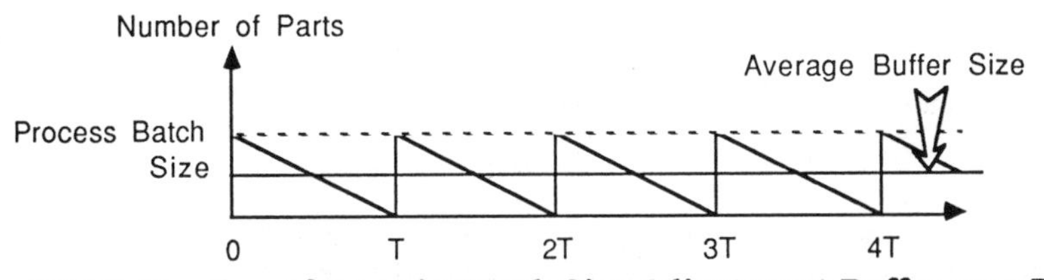

Figure 13.26. Number of Parts in Batch Size Adjustment Buffer over Time

If a batch size adjustment buffer contains parts of different types, the accumulation of wedges of different sizes results in the convex age distribution shown in Figure 13.27.

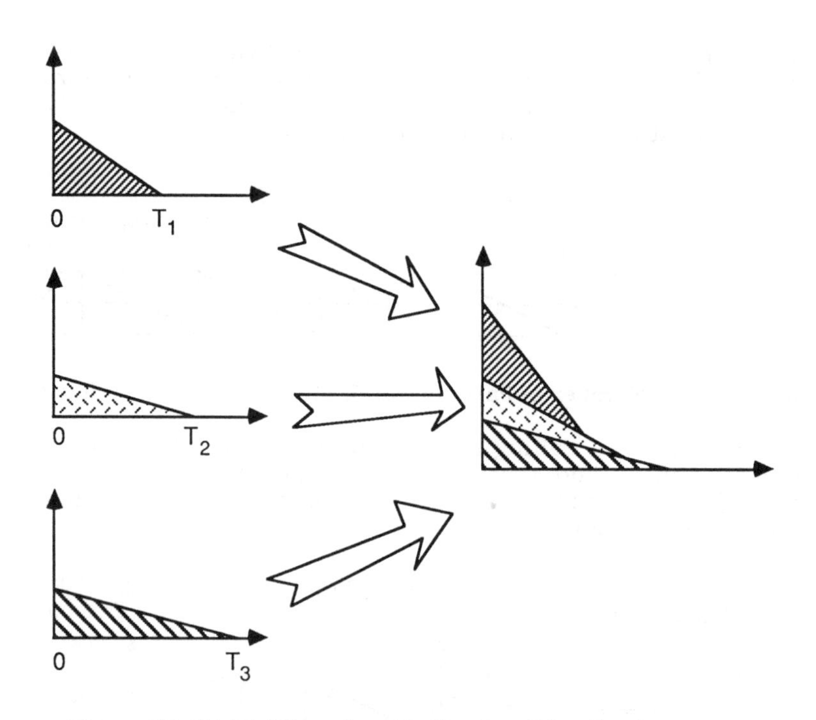

Figure 13.27. Multiproduct Buffer Age Distribution

13.6.4. Safety Stocks

With perfect process technology, safety stocks would be needed to protect the factory against interruptions in supplies of raw materials. With imperfect technology, they are also needed for work in process and finished goods to protect shipments against disruptions to the flow of production.

Figure 13.28 shows two possible age distributions in a safety stock. The most commonly used model in queueing theory has an exponential distribution

$$w(t) = \alpha e^{-\alpha t}$$

for queue times, resulting in the age distribution

$$A(t) = e^{-\alpha t}$$

Since this function is convex, it might be difficult to distinguish it from that generated by batch size adjustments. One possible method is a semi-logarithmic plot, on which the exponential distribution generates a straight line.

The other curve in Figure 13.28 corresponds to queue time probability distribution functions with a maximum, or mode. This pattern is clearly distinguishable from others we have discussed.

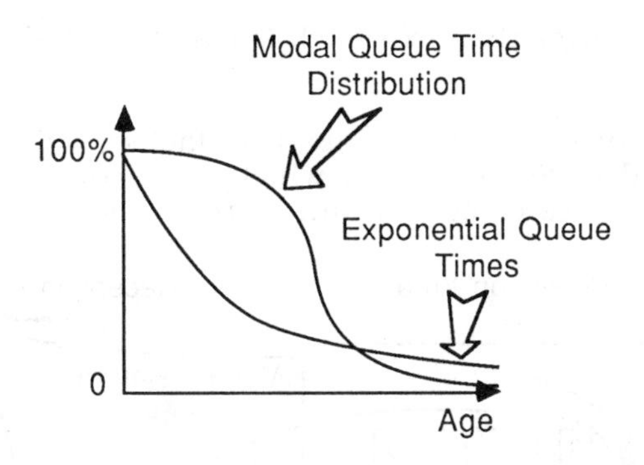

Figure 13.28. Age Distribution In Safety Stocks

13.6.5. Queue Organization

So far, we have looked at means of finding out how the factory behaves from the observation of its inventory. In later chapters, we will review in some detail the effect of inventory policies on production schedules. To give a foretaste, we can examine the effect of various ways of organizing queues.

As human beings, we are much more sensitive to the queues we have to wait in than to the queues of inanimate objects on a shop floor, but the concepts we will describe, of feeding several resources from a single queue and of using tokens, apply to parts as well as to us.

In a supermarket, each check stand has its own queue. If a queue moves faster than the others, it only benefits those customers who are in it. Opening or closing a check stand requires an explicit reallocation of customers, and the order in which they leave the store does not necessary reflect that in which they arrived at the line of check stands.

At American banks, customers are in a single, first-in-first-out queue, and the next customer to be served is helped by the first available teller. Most of the statements made about the supermarket case no longer apply. In particular, it is easy to adjust the number of available tellers. Differences in service time to the customers currently being helped offer the only opportunity to break their arrival sequence.

Japanese banks offer yet another type of queue structure. Customers submit the paperwork of a transaction in a basket at a window and receive a numbered token. Until the number is called, the customer can sit down and read a newspaper. The token is an abstraction of the customer, waiting in his or her place. This system maintains customer sequence rigidly, but it does not cope well with errors in the paperwork or other exceptional situations.

The following example is a real case of combined use of generic queues and tokens to process vendor calls on the purchasing department of a large assembly factory. In most business organizations, the recipient of the visit reserves a conference room for a time exceeding the anticipated length of the meeting. If the meeting is short or starts late, conference room time is wasted; if it lasts longer than expected, it may be cut short by a conflict with another reservation.

An employee suggestion in a Japanese factory led to the adoption of a method illustrated in Figure 13.29, which shifts the administrative work to the visitor and allocates resources with more flexibility.

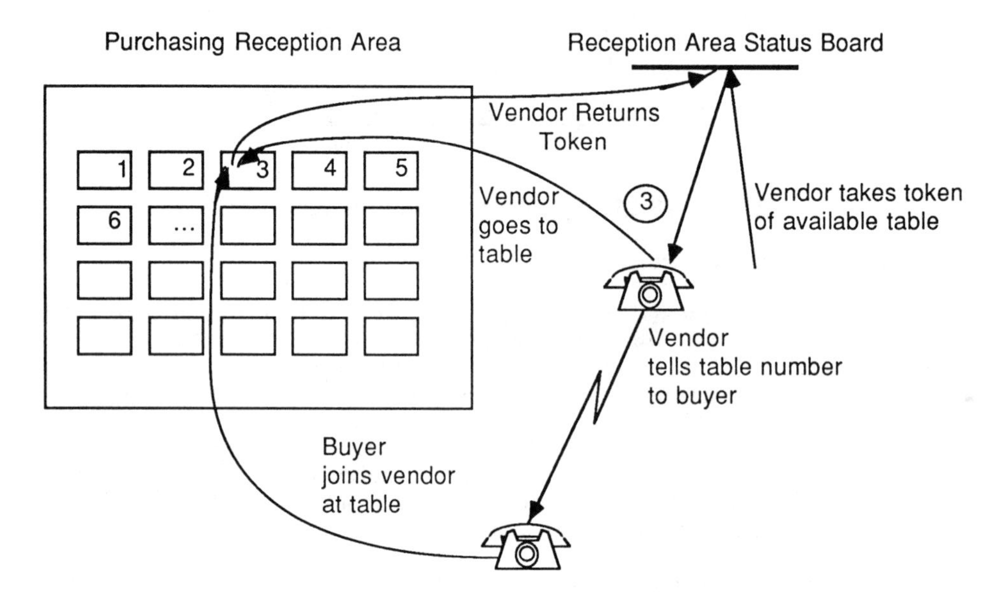

Figure 13.29. Physical Setup of Purchasing Area

The purchasing department has a pool of conference tables in a large reception area, the status of which is displayed on a board near the entrance. The board is a map of the reception area, in which the image of each free table is covered by a plastic token bearing its number.

Each arriving vendor draws a token from the board, or waits for one to be available if all tables are occupied. The vendor calls the buyer, tells him or her the number of the token, and goes to sit and wait at the table. When the meeting ends, the vendor returns the token to the board and thus makes the table available for the next party.

Figure 13.30 shows the data flows associated both with the traditional method and the new, and highlights the transfer of responsibility to the vendor. If any waiting in line has to be done, the vendor is the only one affected. Given that the purpose of the visits is to sell, vendors willingly cooperate in the operation of this system. It could not be expected to work if the visitors were customers instead.

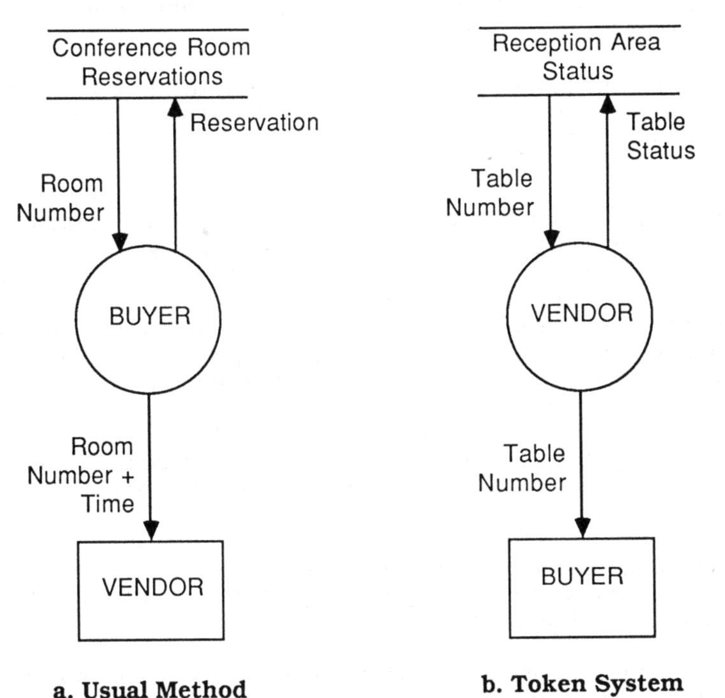

a. Usual Method **b. Token System**

Figure 13.30. Data Flow Comparison

The Kanban method, discussed in Chapter 22, uses tokens as a regulating mechanism on a much larger scale. In all the preceding examples, parts, or visitors, were processed in the order in which they arrived — that is, first-in-first-out. Production schedulers can also affect the flow of parts in a factory by introducing alternative queue disciplines. These will be discussed in chapter 16.

13.7. PUTTING IT ALL TOGETHER

In Chapter 10, we built a factory model in several stages:

1. As a network of *activities* connected by *material flows.*
2. As a network of *resources* connected by material flows.
3. As a network of resources connected by material flows with intermediate *inventory buffers.*

Eli Goldratt has proposed a classification of *activity* networks into three elementary types, called V-, A-, and T-plant, combinations of which can be found in actual factories.

A V-plant is characterized by gradual product differentiation from a small number of raw materials. Integrated circuits started on identical silicon wafers follow gradually divergent routes though film growth, patterning, and doping operations.

In an A-plant, many raw materials converge through assembly operations into a small number of finished products. This pattern is found, for example, in factories making automobile components, such as roller bearings or shock absorbers.

In a T-plant, a large number of products are assembled from a small number of subassemblies or purchased components drawn from a common stocking level, and which may themselves be the outputs of fabrication processes. An A-plant can be turned into a T-plant by the addition of options and variants to the products. Factories building mainframe computer systems are T-plants because of the large number of possible combinations of CPUs, memory boards, storage systems, and other peripherals.

Goldratt's purpose in this classification is to help the analyst find *bottlenecks*, a concept discussed at length in later chapters. Goldratt's claim is that critical resources are identified by different signs in V-, A-, and T-plants:

- By the largest input inventory buffers in a V-plant.
- By chronic shortages of the parts they process in an A-plant.
- By their contributions to late shipments in a T-plant.

The resource network is derived from the activity network by merging into one all the nodes using the same resource. Strictly speaking, when activities require composite resources, there usually is only one immobile resource, the other ones being mobile fixtures, tools, or people. The immobile resource is used as an "anchor" for the activity in the resource network. The resource network may differ dramatically from the activity network. In semiconductor manufacturing, a V-structure is collapsed into a hub and spoke structure around photolithography.

Whether or not Goldratt's classification is sufficient, it should be obvious that there are many different types of factories, in each of which *some* of the difficulties described in this chapter will be present. If a production scheduling system attempted to take *all* these factors in consideration, it would be hopelessly complex. There can be no single solution analysts can recommend for all factories. In the following chapters, we are going to examine the scope of applicability of several.

14

THE BACKWARD SCHEDULING LOGIC OF MRP-II

MRP is the most common software approach to production scheduling, but few implementations are successful. Based on order due times, MRP schedules part movements and evaluates resource work loads, but produces no resource timelines. Final resource allocation decisions are left for shop floor personnel to make, with the help of dispatch lists.

Time-phased part requirements are deduced from a master production schedule using a bill of materials and routings with standard throughput times and yields. Processing requirements are determined by netting part requirements against available inventory. Each resource has a capital of processing time, and capacity requirements planning highlights shortfalls. After adjustments to make the schedule compatible with initial inventory and capacity constraints, a forecast is produced as the best response the factory can offer to the master schedule.

14.1. MRP'S IMPLEMENTATION RECORD

The use of MRP systems is so widespread in the U.S. that many managers have come to think of them as a necessary part of manufacturing, along with buildings, machines, and workers. Their diffusion is such that they are purchased on the grounds that "everybody else has one."

MRP methods are computer-based, and many software companies offer products variously labeled "MRP," "Closed-Loop MRP," or "MRP-II," and the implementation of MRP is itself a thriving business for consultants. Large accounting firms have teams of employees on loan to their manufacturing clients to implement MRP.

Originally, MRP stood for "Materials Requirements Planning" and its goal, as stated in a frequently quoted phrase by Oliver Wight, was to ensure the availability of "the right parts, at the right time, in the right quantities." This statement is not as specific as it first sounds in the absence of an explanation of what "right" means. One explanation for this early emphasis on materials was a large overcapacity in American industry in the aftermath of World War II: when production was held up, it was usually by shortages of components, not capacity. Today, after many refinements, MRP stands for "Manufacturing Resource Planning" and many systems include a Capacity Requirements Planning (CRP) module.

MRP systems are widespread, but even the MRP literature concedes that there are few "Class A" users — that is, users who do everything the system says and reap its benefits in full. MRP is supposed to yield lean work in process inventories and good due time performance, but, today, American manufacturing is not renowned for either. In Japan, users of Just-In-Time methods do not rely on MRP-II for resource allocation on the shop floor, but they *do* use MRP in the original sense of materials requirements planning, particularly through a network of component suppliers.

Based on MRP-II's results to date, its applicability as a method of shop floor scheduling is not a foregone conclusion. Before buying it, managers should evaluate not only several MRP-II products, but alternatives to the approach as well.

We focus here on scheduling logic, to the exclusion of parameter editing and other administrative functions, including the management of multiple versions of pro forma schedules. Also excluded are auxiliary functions such as order entry support which may be supplied with MRP systems. The decisions to make are not just on how flexible or user-friendly a particular MRP system is, but whether its fundamental logic is applicable.

Why do factories who "absolutely needed" one three years ago still prosper today without one? Why are some factories with MRP-II systems not better off than others without one? Why are so many purchased systems not used or underutilized? These are questions that must be answered.

14.2. THE PHILOSOPHY OF MRP-II

MRP aims for due time performance, as opposed to such goals as minimum throughput times. It schedules where material should be over time rather than what resources should do. It is then necessary to find out which parts are behind schedule and somehow expedite them.

This often gives rise to a complicated array of categories of hot lots. On the other hand, systems aiming for minimum throughput times try to treat all parts equally and make them all fast. Short throughput times make due time performance a moot point because all customers get prompt delivery.

Although it specifies future events, MRP-II does not attempt to produce a timeline for every resource in the factory. None of the alternative systems we will look at do either: OPT does it for a few critical resources, and Just-In-Time, via the Kanban system, sequences only one step ahead.

MRP-II leaves the scheduling decisions within each work center to the supervisor in charge. MRP-II (1) lets each supervisor know how many units of each product and operation should be moved out of a work center in each time period, and (2) schedules the parts to arrive at the work center early enough so that the supervisor should be able to meet the demand. It is then up to the supervisor to find how to do it.

To assist supervisors, dispatch lists are produced, assigning priorities to all the parts in process by work center, but the supervisors have the authority to process parts in a different order. Besides component shortages, the most common reason for supervisors to disregard their dispatch lists is conflict with their performance measures. Strict compliance with the dispatch lists does not guarantee that the schedule will be met.

The algorithms of MRP are simple: anyone who can balance a checkbook knows enough mathematics to understand them. This simplicity makes MRP principles accessible to a broad audience, but it also restricts the class of scheduling problems it can solve.

Given that throughput times, yields, and resource capacities can be modeled each by a single number that is independent of the level of activity within the factory, the scheduling problem is completely constrained. MRP has removed all the mathematical difficulties from the model. Unfortunately, it does not have the power to remove those difficulties from reality.

14.3. WIGHT'S DISCUSSION OF MRP VERSUS REORDER POINT

The reorder point method is older than MRP and consists of issuing orders for parts whenever stocks fall below a threshold. The orders are for quantities just large enough to raise the stock back to a specified upper limit. They are issued to suppliers for raw materials and to upstream work centers for components and subassemblies. The idea is to keep parts flowing fast enough to meet the average demand, with a safety buffer for protection against fluctuations.

Wight [Wig74] criticizes this method for maintaining stocks of unneeded parts while failing to guarantee availability of all the components needed for assembly. Reorder Point replenishes stocks in response to past activity, regardless of future demand. If safety buffers on 10 parts guarantee availability 95% of the time for each part, all 10 parts are available *simultaneously* only 60% of the time, which is unacceptable if they must be assembled together.

Wight sees MRP solving these two problems by distinguishing between *dependent* and *independent* demand. Independent demand originates outside the factory and is for finished goods or salable spare parts. The demand for all other parts is dependent, and deduced from the independent demand by a time-phased bill of materials explosion (see the following). Thus all the needed parts are produced in a timely fashion, while none are made for which there is no demand.

Writing in 1974, Wight blamed the academic establishment for publishing research reports on the mathematical theory of Reorder Point, while staying aloof from down-to-earth, pedestrian MRP which is much more applicable in practice. Twelve years later, MRP receives extensive coverage in college textbooks on operations management, but the method that is currently getting the best results in the field is not MRP but Just-In-Time, which is a philosophical descendent of Reorder Point. The Kanban system may be viewed as Reorder Point with equal upper and lower limits and an order quantity of one.

Wight's assessment that the academic operations research establishment has failed to make a significant contribution to the practice of production scheduling remains valid in the late 1980s. The blame for this unfortunate situation does not lie entirely with the academics, who receive a cold welcome in factories. Many American production managers combine anti-intellectualism with a low tolerance for abstraction and the perception that theirs is the only "real world." Academics who try to help are faced with a wall of skepticism that discourages all but the most determined.

14.4. THE EXTERNAL CONTEXT OF MRP-II

Figure 14.1 shows the interactions between an MRP system and the factory it serves. The diagram describes as coming from engineering the material flow model needed for scheduling. The routings describe sequences of fabrication activities, and the bill of materials the component structure of shipped products.

MRP's view of the factory tends to match that of its main user, the production control department. The division of work between production control and manufacturing management varies from one factory to another, but, as a general rule, production control is responsible for

1. Releasing raw materials to the shop floor.
2. Keeping track of work in process and expediting as needed.
3. Promising deliveries.

The outputs of MRP shown in Figure 14.1 are tailored to these functions. The Start-Schedule times releases of materials to the shop floor. The move schedule, combined with the production reports from the parts tracking system gives the means to find out what is running late, and the Dispatch-Lists sent to the shop floor do the expediting. Finally, the Production-Forecast is the basis for promising deliveries.

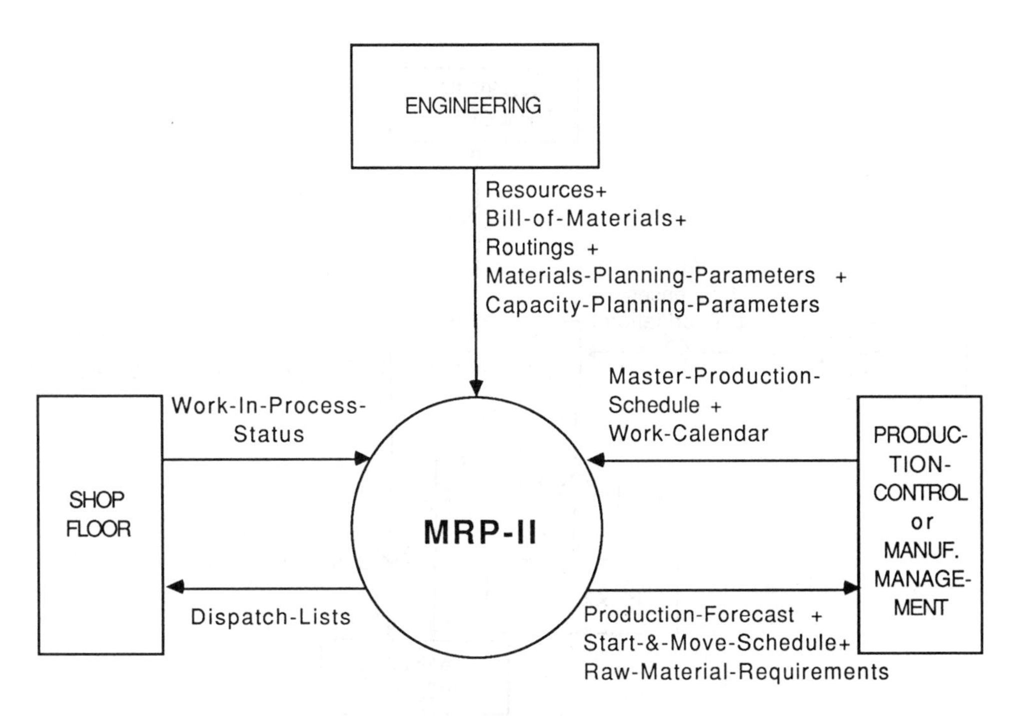

Figure 14.1. The Logical External Context of MRP-II

In general, production control does not make scheduling decisions at the machine level. The final arbiter of what parts go into an available machine is the manufacturing supervisor. Production control can coax, beg, or threaten, but the supervisor has the last word. Likewise, the Dispatch-List recommends a processing sequence, but does not enforce it.

We shall next examine in detail the inputs which production control must supply in order to get the outputs. Resource descriptions, with setup times and processing times per part, are engineering parameters. However, the average processing times per part used by MRP-II in capacity planning include allocations for setups, which depend on process batch sizes, and are also influenced by whether machines are allowed to process partial loads. Those are scheduling decisions, and this is why those parameters are not shown as coming from engineering. The factory model of MRP-II is shown in Figure 14.2.

14.5. MODULAR BREAKDOWN

Figure 14.3 shows the planning functions of MRP-II. Each is discussed in detail next.

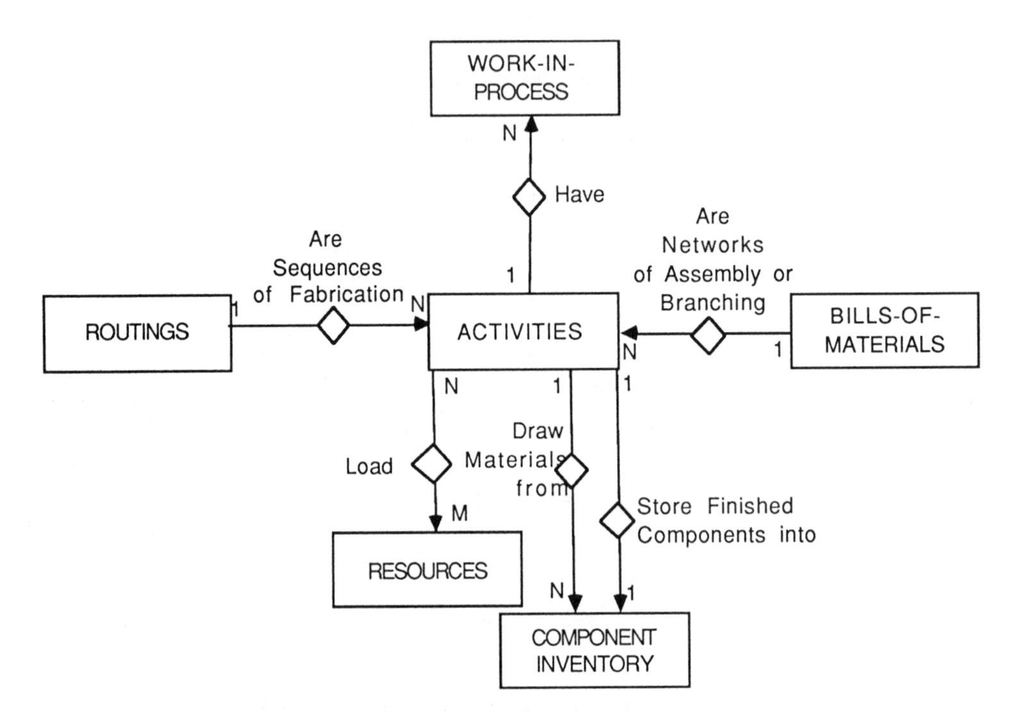

Figure 14.2. Factory and Inventory Data Model

14.5.1. The Master Production Schedule (MPS)

The master production schedule (MPS) usually is a mixture of firm orders for the near future and forecasts for the more distant future. Its structure for one product is shown in Figure 14.4, in a case where planning is run at intervals that are shorter than the average throughput time.

Only that part of the MPS which is between now and the next update actually determines what is to be shipped. Subsequent runs will do that for later periods. For times between the next run and the standard throughput time, the MPS serves to determine work in process. MPS entries between the Standard Cycle Time and the Planning Horizon do not result in any material being started or moved, but can be used to anticipate capacity problems.

The relative length of order lead times and standard throughput times determines how much the MPS depends on orders versus forecasts. The diagram illustrates a number of forecasting methods in use, with seasonal variations around a long term trend inferred from history.

Plugs are peaks in the MPS that are due solely to upper management's demand of a sufficient level of sales in the business plan. With a total prescribed from above, planners have to plug in numbers that are acceptable to management but devoid of a basis in reality. They are much less conspicuous when forecasts are presented as tables of numbers rather than in graphic form.

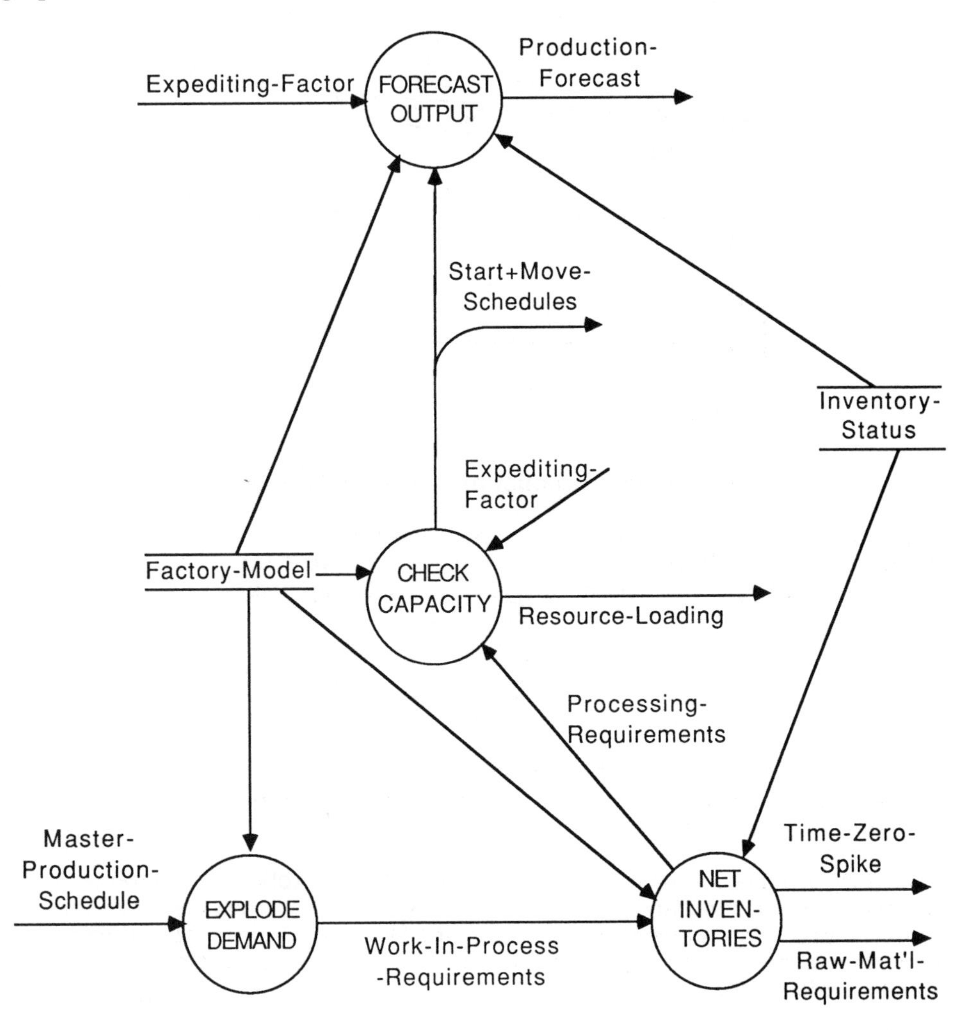

Figure 14.3. MRP-II Data Flows

Example: In a factory, orders received and shipped within the same accounting month were referred to as "turns." One month, the turns forecast for the following month was $487,000. Since turns were comprised of many small, independent orders the following model could be used for the turns volume U(p) of month p:

$$U(p) = a \times p + b + E(p)$$

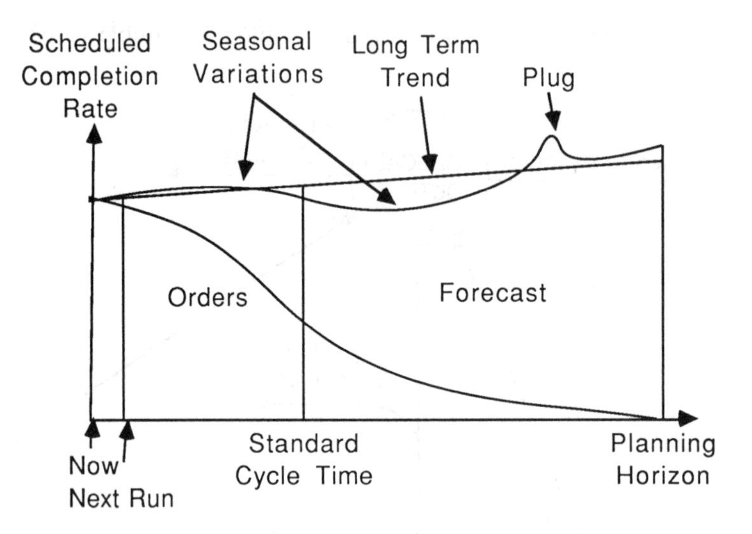

Figure 14.4. Master Production Schedule for one Product

where the E(p) are independent and identically distributed Gaussian estimation errors. Parameters a and b were estimated by least squares over the results of the past 18 months. The variance of E(p) was estimated from

$$\sigma^2 = \sum_{p=1}^{18} \frac{\left[U(p) - a \times p + b\right]^2}{17}$$

For the following month, this model's forecast was 19a+b $\pm 3\sigma$, or \$350,000 ± 50,000. The actual turns of the following month amounted to \$360,000. The \$487,000 figure had been a plug.

Starting from a forecast is a way to bypass the problem of generating it. However, a production scheduler must be concerned by the quality of such a critical input. The preceding example showed the danger of subjectivity. There are many others, and the forecaster needs solid statistical analysis skills to avoid them.

If lead times exceed throughput times, then the factory strictly makes to order. If it has to deliver in less time than it takes to make the product, then it must make to stock, based on a forecast. When an MPS overloads some resources, it must be revised. Whether material is pegged to a firm order or not is one of the revision criteria.

The minimum data definition for an MPS is

Master-Production-Schedule = {Due-Date + Product + Quantity}

Auxiliary information, such as who the customer is, may be included for use in dispatching.

14.5.2. Backward Explosion of Demand

Backward explosion of demand derives time-phased work in process requirements from orders. This is feasible with fabrication and assembly, but not with disassembly or binning because the output mix of such activities is not controlled by the planner. Disassembly and binning would lend themselves instead to *forward explosion of supply.*

With only fabrication and assembly, the bill of materials and the routings combine into an activity network where multiple output arrows indicate branching points — that is, the output of the activity follows one of the paths but is not split between the various paths. The activity network, and the parameters required for backward explosion of demand are shown in Figure 14.5.

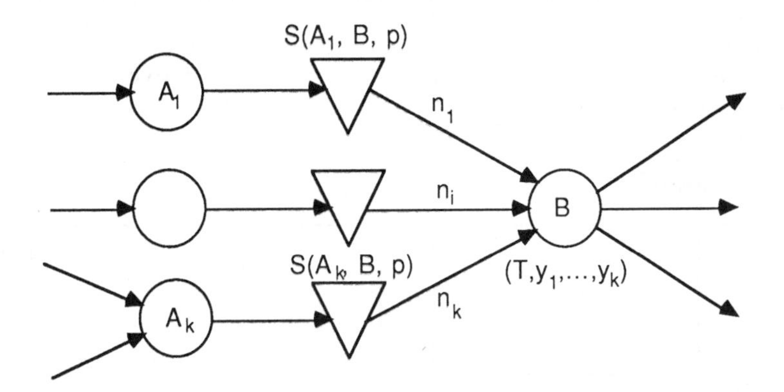

Figure 14.5. A Section from the Activity Network

Each unit of B shipped as a product or as a component of another product contains n_i units of the output of A_i. The activity producing B's is modeled as a delay with a length T that is its standard throughput time.

In the course of production, components are occasionally damaged and must be replaced. For A_i, let y_i be the fraction of the units that are successfully assembled into B's on the average. y_i is also the survival probability of each unit of A_i output fed to assembly of B's. If n units of B are needed at time t, backward explosion of demand says that $n \times (n_i/y_i)$ units of A_i must be available in front of B by t-T.

This logic clearly determines time-phased material requirements before every node of the network by working backwards in time and from successor to predecessor activities, starting from the master production schedule and ending with raw material purchases.

The yields y_i can be embedded in the bill of materials by replacing the n_i with

$$n_i^* = \frac{n_i}{y_i}$$

The disadvantage of this approach is that it mixes in one number the product design parameter n_i with the manufacturing performance parameter y_i, which are subject to independent changes. Over time, the fraction of screws damaged in assembly should drop, but the number required to hold together a television set only changes by designer decision.

The direct application of the delay formula would produce, for each activity, a list of times and part counts required to be available by the assigned times. Each entry in the master production schedule would generate time-phased withdrawal orders for all buffers containing needed components.

MRP does not produce such lists, but instead lumps the withdrawal orders by time periods, or *buckets*. Parts needed at any time during period p are taken to be needed at the end of the period. This reduction of time into a discrete sequence of periods is not done only for convenience: unless the period is long with respect to the time it takes to process a batch, the capacity planning logic discussed next falls apart.

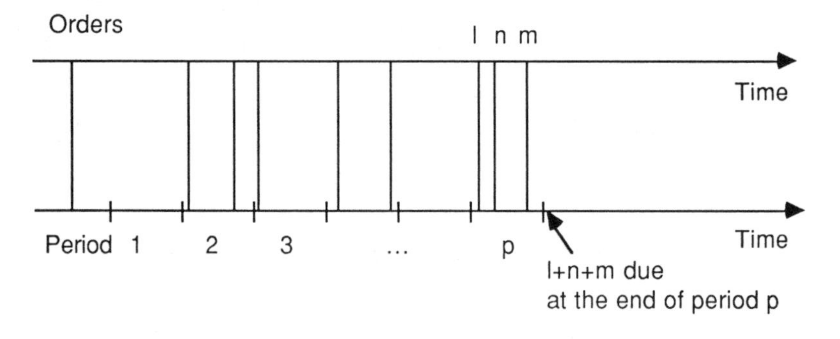

Figure 14.6. The Effect of Bucketing

If H and D are respectively the planning horizon and the period length, then the end result of backward explosion of demand is a table of numbers S(A, B, p), for p=0,...,H/D and for all pairs of activities (A, B) linked by an arrow (and therefore with work in process in between). S(A, B, p) is the number of units out of A that must be fed to B by the end of period p.

14.5.3. Net Inventories

Let

- L(A, B, p) be the content of the buffer between activities A and B by the end of period p.
- Q(A, p) be the output planned for activity A during period p.

If, for all buffers (A,B), $L(A,B,0) \geq S(A,B,0)$, there is enough material *initially* in the line to meet the schedule, and the calculations are simple because we are not considering capacity limitations yet.

For $0 \leq p \leq \frac{H}{D} - 1$, we have

$$L(A, B, p+1) = L(A, B, p) - S(A, B, p) + Q(A, p+1)$$

by conservation of flow. Since we only produce what is needed to meet the master production schedule, we have

$$Q(A, p+1) = \text{Max}\{ S(A, B, p+1) - [L(A, B, p) - S(A, B, p)], 0 \}$$

In other words, we are only going to process through A in period p just enough parts to avoid a shortage when S(A, B, p+1) parts are withdrawn at the end of period p+1. By induction, we can calculate Q(A, p) for all activities and all periods, without ever needing to reference the parameters of backward explosion of demand. Inventory netting is carried out using only the calculated material requirements and the initial work in process status.

14.5.4. The Time-Zero Spike

It is a different matter when there are buffers (A, B) such that $S(A,B,0) > L(A,B,0)$. L(A,B,0) is the initial condition of work in process at the time the planning software is run, and, as such, is the factory's response to earlier planning runs. If there is a time zero spike and the standard throughput times are followed, then the schedule cannot be met, even though we have yet to introduce capacity constraints.

When orders received for a product exceed the forecasts on which earlier runs were based, then there will be buffers where the amount S(A,B,0) to be withdrawn at time $t = 0$ exceeds the available supply. The same effect frequently results from disruptions in the line, but variations in demand would cause it even in a perfect line.

To respond to this situation, let us introduce two new concepts. Let us call

- S*(A, B, p) the number of parts planned to be withdrawn from the (A,B) buffer, which may be less than S(A, B, p).
- G(A, B, p) the requirements backlog for withdrawals from the (A,B) buffer at the end of period p.

The backlog G(A, B, p) is given by

$$G(A, B, 0) = S(A, B, 0) - L(A, B, 0)$$

and

$$\forall p \geq 1, G(A, B, p) = S(A, B, p) - S^*(A, B, p) + G(A, B, p-1)$$

The simplest case is that of fabrication, where

$$S^*(A, B, p) = \mathrm{Min}\{L(A, B, p), S(A, B, p) + G(A, B, p-1)\}$$

and

$$Q(B,p+1) = y(B) \times S^*(A, B, p)$$

In the case of assembly, if $A_1, \ldots, A_k$ supply components for the assembly activity B, and if some of the A_i are short at the end of period p, then $S^*(A_i,B,p)$ is determined by the component in shortest supply. Formally, if there is an i, $1 \le i \le k$, such that

$$L(A_i, B, p) < S(A_i, B, p) + G(A_i, B, p-1)$$

then, if n_i and y_i are as in Figure 14.5, and

$$m = \underset{i=1,\ldots,k}{\mathrm{Min}} \left\{ L(A_i, B, p) \times \frac{y_i}{n_i} \right\}$$

the number of parts to withdraw from the (A_i, B) buffer at the end of period p is

$$S^*(Ai, B, p) = m \times \frac{n_i}{y_i}$$

and

$$Q(B, p+1) = m.$$

The problem is further complicated if B is a branching activity — that is, if each part produced by B follows one of a number of material flows. If there is a shortage in any input buffer and there are requirements on several output flows, then rules are needed to allocate the feasible production to the various requirements.

One refinement of this method consists of using "expedited" throughput times for the backlog. By definition the positive G(A, B, p) are overdue. Therefore it makes sense to give priority to material pegged to the backlog and thus make it move faster through the line.

But how much faster will it actually move? And how much slower will the rest of the material move if it is given a lower priority? MRP does not answer these questions. One approach consists of letting the planner set an acceleration factor q between 0 and 1, and use qT as the expedited throughput time of an activity with a standard throughput time of T.

14.6. CAPACITY CALCULATIONS

14.6.1. Capacity Modeling in MRP-II

Up to this point, we have made no use of capacity constraints. Capacity Requirements Planning maps activities to the resources they need and calculates time-phased resource work loads. Each resource has a capital of work time, from which activities withdraw until it is exhausted (see Figure 14.7). The resources are considered infinitely divisible, in that average time requirements per part are multiplied by part counts for each activity and time period, and summed up over all the activities using the resource.

However, unlike the infinitely divisible resource model (see Chapter 13), MRP and CRP do not consider the throughput time of an activity to be related to the fraction of the resource allocated to it. The standard throughput times are expected to hold whether a resource is loaded to 10% or to 90% of capacity. Only loads above 100% of capacity are flagged as infeasible.

A CRP resource is a set of machines or tools qualified to do the same work. If machine X and machine Y can both perform activity A, then they are considered part of the same resource. It is assumed that, if X and Y can both do A, they are also interchangeable for any other activity. That is, anything X can do, Y can do, too. There is no activity for which X is qualified but Y is not. An activity may require several resources, such as a machine and a jig, or a setup tool. CRP loads all the resources needed but does not consider the fact that they may be needed simultaneously.

The resource time needed is proportional to the number of parts to process, and the processing time is assumed to be the same for all individual machines within the resource. The possibility of putting to use an old, slow machine when the new, fast ones are overloaded is not considered. Setups and batching are not considered either.

14.6.2. Capacity Requirements Planning

CRP calculations lend themselves to modeling as a simple matrix multiplication. If $\mathbf{M} = (t_{AX})$ is the time per part required by activity A on resource X, and $\mathbf{Q}(p) = (Q(A,p))$ is the row vector of output requirements of all activities in period p, then

$$\mathbf{R}(p) = \mathbf{Q}(p) \times \mathbf{M}$$

is the row vector of all resource time requirements of period p. If $\mathbf{V}(p)$ is the row vector of resource times available in period p, then the negative coordinates of $\mathbf{V}(p) - \mathbf{R}(p)$ indicate capacity shortfalls.

A resource may be globally overloaded — that is, it does not have enough work time in the planning horizon to cover the total demand. In this case, the master production schedule is infeasible, and the question arises again as to what should be dropped from it. To find out the effect on the master schedule of removing a particular batch from the work load of a resource, we need each batch to be pegged to an order.

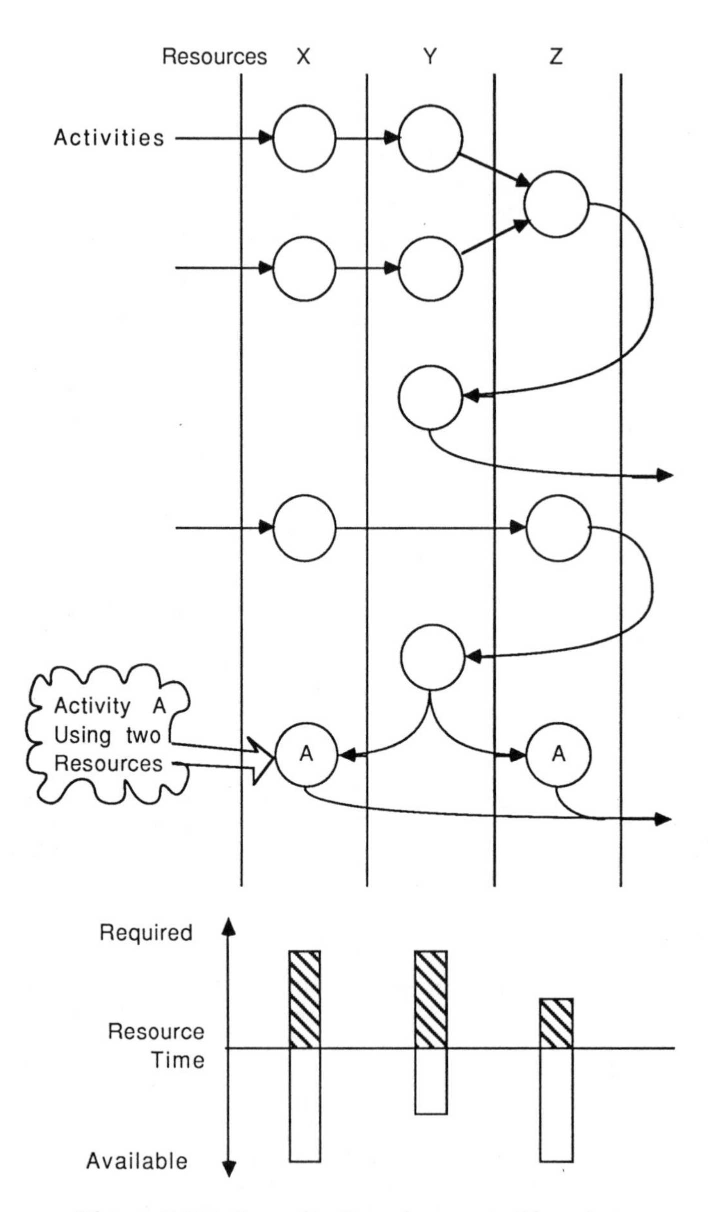

Figure 14.7. Capacity Requirements Planning

A resource may have sufficient capacity globally, but be overloaded in a given period p. The problem is then to reschedule the excess work to periods in which there is spare capacity. If this is done by rescheduling to an earlier period q, then the work in process requirements upstream from the activity must be modified.

Let N parts be rescheduled from period p to period q for activity A. A backward explosion of demand for N out of A at the end of p determines quantities to be

1. deducted from current work in process requirements.
2. shifted forward in time by p-q periods.
3. added back, after shifting, to work in process requirements.

If it works, this procedure has the advantage of making the master schedule feasible. However, 2 may create a new time-zero spike, and 3 may create new resource overloads.

The search for spare capacity later rather than earlier eliminates the risk of a new time-zero spike. It may also create other resource overloads in later periods, which can be solved likewise. This approach, however, does not result in meeting the master schedule, unless the delayed material is sufficiently expedited. The problems with the use of expedited throughput times are similar in this context as in the resolution of time-zero spikes.

14.7. FORECAST PRODUCTION

As a result of adjustments made to match the material requirements with the initial conditions of the work in process and the processing requirements with the available resources, we now have a plan which is the closest the MRP logic can take us to the master production schedule.

The production forecasting module goes forward in time from period 1 to period H/D and applies the equation

$$L(A, B, p+1) = L(A, B, p) - \sum_{B} S^*(A, B, p) + Q(A, p+1)$$

to determine a planned output schedule.

15

DISPATCHING IN MRP-II

Dispatching sorts the work in process by static and dynamic priorities, but batching, setups, and resource contention are not considered. The effect of a dispatching rule is difficult to anticipate except in trivial cases. Throughput times are primarily determined by queue sizes, and dispatching only has a secondary effect.

Dispatch lists are produced in subscription mode, on-line, or in real time, the latter being required for automation. The most common sequencing criteria are First-In-First-Out (FIFO), Earliest Due Date (EDD), Shortest Processing Time (SPT), anticipated lateness (Slack), and anticipated lateness ratio (Critical Ratio).

In the deterministic case of n jobs on 1 machine, EDD minimizes the maximum lateness, SPT minimizes the average throughput time, and Slack maximizes the minimum lateness. If jobs arrive as a Poisson process in front of one machine and have a processing time distribution with finite first- and second-order moments, then the queue discipline affects the average throughput time only if it takes processing times into consideration.

15.1. FROM PLANNING TO DISPATCHING

Dispatching is the most popular component of MRP. Even in factories where materials and capacity requirements planning modules are not used, production supervisors welcome the dispatch lists telling them what to work on next.

The benefits to be expected from a dispatch system should not be overestimated. Manufacturing throughput times are determined essentially by how large queues are; what sequence the queued items are processed in only has a second-order effect on performance. The Just-In-Time system (see Chapters 20 to 22) focuses on the amount of work in process and makes no attempt to use sophisticated dispatching rules.

Trying to anticipate the effect of a dispatching rule on a nontrivial factory is grasping at straws. As we shall see, the theoretical results even on very commonly used rules are scant, and sometimes do not show the rules to be particularly desirable. In Appendix A we show in a special case that the Slack algorithm maximizes the minimum lateness of the jobs in a queue.

Appendix B shows how the average waiting time in a queue with random arrivals (Poisson) in front of an indivisible resource is unaffected by the dispatching rule unless the priority assignments depend on the processing time through the resource. The most commonly used rules do not, or do so only weakly, and their local effect is to make some jobs faster at the expense of others in a way that balances out.

The only general method available to estimate the global effect of a dispatching rule is discrete event simulation, but it requires so much programming time and computing power that an answer is frequently obtained faster and cheaper by trying the approach on a real factory.

Most dispatch systems simply produce lists of the batches queued in the input buffer of each work center, sorted by a priority function. The supervisors are not obligated to follow the priority list exactly. The worst that may happen if they don't is that they may have to justify their choices to superiors.

Production supervisors are subject to pressure from expediters with parts they think are hotter than everything else and engineers who want to run experiments. Dispatch lists enable supervisors to fend off all these people. Anyone trying to influence their decisions can be referred to "the system" as the arbiter of urgency.

The priority function used to rank batches usually depends on the output of the planning system, but there is neither theoretical nor experimental proof that following the dispatch lists to the letter would lead to meeting the master production schedule.

15.2. REAL-TIME, ON-LINE, AND OFF-LINE DISPATCH

A dispatch system is not completely described by its sequencing algorithm and the data content of the dispatch lists. The manner and frequency in which they are produced is just as critical a factor in their usefulness.

The subscription mode, where lists are periodically printed and handed out to "subscribers," is the easiest approach because it does not require the end user to have a terminal, and the computer time needed by the list generation is not constrained by someone waiting in front of a terminal (see Figure 15.1).

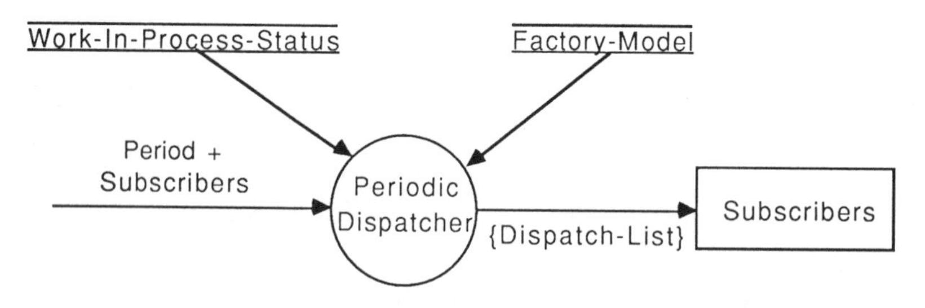

Figure 15.1. Dispatch System in Subscriber Mode

The downside of this approach is that the list of a work center becomes gradually obsolete as new material arrives and dispatched material leaves. To slow list obsolescence, some systems anticipate material movements, for example, by including the high priority material of a work center in the dispatch list of the next downstream work center.

Trying to compensate for the restrictions of the subscription mode complicates data processing, rapidly reaching a point where the on-line use of a simpler method becomes a desirable alternative. In this mode, shown in Figure 15.2, dispatch lists are generated on demand using the work in process status maintained by the parts tracking system. Users must have access to a terminal. The ability to generate up-to-date lists at any time removes part of the need to anticipate arrivals.

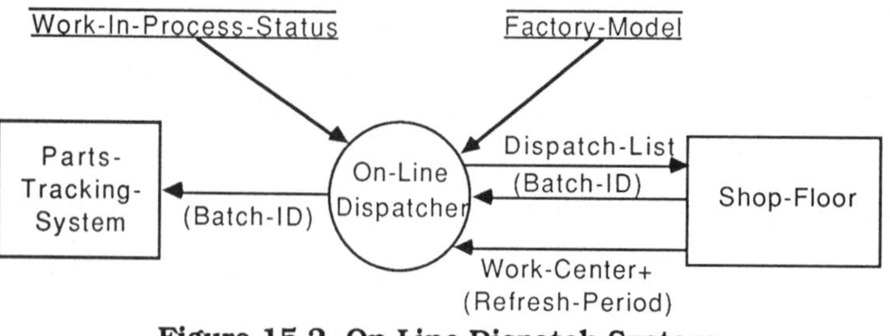

Figure 15.2. On-Line Dispatch System

Some on-line systems allow operators to initiate tracking transactions directly from a dispatch screen. The limitation of the on-line method is that the action of the operator will not be reflected in a display of the same list viewed by the supervisor at the same time unless he or she decides to regenerate it.

If current displays are refreshed automatically at a sufficiently high frequency, such as every 30 seconds, then no user will notice a difference between this mode of operation and true real time, where updates to active displays are triggered by part movements.

Real-time dispatching is required for automation. When an empty automated guided vehicle reaches an unattended loading station, the control system must know which batch(es) to load onto the vehicle and notify other affected stations. Real-time dispatching is technically feasible with most common algorithms, with the exception of critical ratio, as discussed soon.

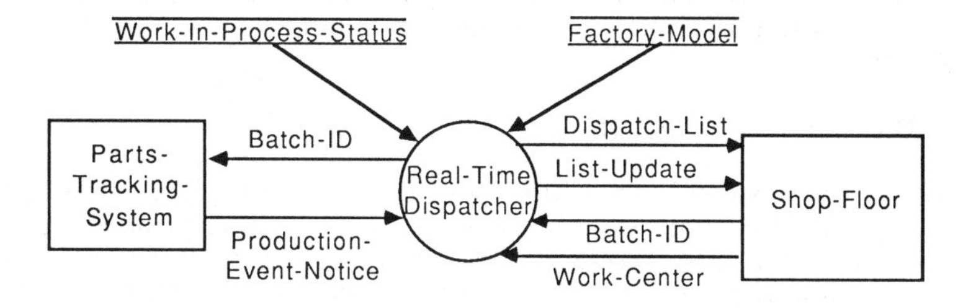

Figure 15.3. A Real-Time Dispatch System

15.3. DISPATCHING ALGORITHMS

15.3.1. First-In-First-Out (FIFO)

In its simplest form, this method assigns the highest priority to the material that has been waiting for the longest time in every queue. It has the following advantages:

1. It preserves the order in which parts are processed, which is essential for quality analysis (see Chapter 12).
2. It gains acceptance on the shop floor easily, because people perceive it as "fair."
3. The only data it requires is the history of arrival times into each queue.
4. Performance characteristics of queueing systems are rarely calculable with any other queue discipline.

The perception of fairness has nothing to do with scheduling and everything to do with psychology. In queues of human beings, such as airline passengers checking in their luggage or spectators buying theater tickets, any queue discipline other than FIFO is a fail-safe way to start a riot. The only exceptions tolerated by people are cases of greater need: patients with minor ailments will usually willingly yield their turn at a hospital to a woman in labor.

The fairness perception about FIFO sequencing of work in process is a projection by the operators of their own feelings onto inanimate objects. But if

FIFO scheduling is such a natural way to operate, what good can explicit FIFO dispatch lists do? Why is it useful to generate them if people spontaneously obey the FIFO discipline?

In the absence of any explicit priority sequence, the physical layout of inventory buffers determines how operators choose what to work on next. If carts line up in front of a resource and gradually roll towards it, then FIFO sequencing is followed and the dispatch list is unneeded. On the other hand, if boxes are stacked on a table, the most recent arrivals are on top and get processed first, thus following last-in-first-out discipline.

A scheduling reason to line up airline passengers in order of arrival times is that the ground staff does not know anything else about them. The nurses giving priority to the woman in labor have obvious visual information on the urgency of her case. On a shop floor, it is possible, at a cost, to maintain more data about the queued work in process and use more refined methods of sequencing it.

15.3.2. Earliest Due Date (EDD)

For work in process to have a due date, it must be pegged to an order. This can be done when raw materials are started, and redone periodically to make the due dates match the master production schedule. It seems reasonable that, if a component goes into an assembly that is due tomorrow, it should take precedence over material that is not due out for a month even if that material has been waiting longer.

There is, however, an enormous distance between such an intuitive justification and an understanding of how such a scheduling system behaves. Let us restrict ourselves to a one-machine factory and to the task of sequencing one morning the jobs of the day, with the understanding that the sequence will not be altered as a result of orders arriving during the day. In this context, French [Fre82] proves that EDD sequencing minimizes the maximum lateness of all the jobs (see Appendix A).

15.3.3. Shortest Processing Time (SPT)

If we know how much time a batch of parts will require to be processed through a resource, we can use this data as a selection criterion and choose to process next those parts with the shortest processing time (SPT). (This method is also known as Shortest Job Next, or SJN.)

SPT has the advantage over EDD of not requiring pegging, and of using a parameter that is intrinsic to the material flow system. Unlike a due date, the processing time of parts through a resource is determined by technology and is independent of scheduling. This desirable property is, however, true only with resources whose capacity is independent of their work load.

The most important result in the mathematical theory of scheduling is that SPT minimizes the average throughput time of jobs under the following conditions:

1. One single, indivisible resource.
2. When a job arrives with a shorter processing time than the current job in process, it preempts the resource — that is, the current job is interrupted and the resource is allocated to the new, shorter job.
3. There is no restriction on job arrival patterns.

Although we do not give a proof of this result, we give two analyses of SPT in special cases. Appendix A contains French's proof that, when scheduling n jobs on one machine, SPT minimizes the average throughput time of the jobs. In this model, no job arriving during the execution of the schedule will be inserted in it and there is no preemption. On the other hand, if SPT is used in real time, then jobs are inserted in the queue as they arrive, behind the shorter jobs and before the longer ones. The mathematical analysis of this model is shown in Appendix B, also excluding preemption.

The intuitive justification for SPT is that the throughput time penalty paid for letting a long job wait for a short one is less than that paid for letting the short one wait for the long one. This method is appropriate when the jobs are equally important to customers, because the number of satisfied customers increases fastest by going SPT.

Example: A programmer is given the task of coding six custom reports requested by users of a CIM system. These users track progress on all six projects at the same frequency, regardless of obvious differences in complexity. SPT sequencing is the natural way.

The problem with SPT is that the waiting times of long jobs can become unbearably long and that pure SPT at each resource does not necessary give good global performance. If the parts produced by the long job are to be assembled at a later stage with those produced by a short one, quick turnaround on the short job does not shorten the overall throughput time.

15.3.4. Slack

The slack of a batch of parts is the difference between the work time T_A available until it is due out of the factory and the work time T_R required to get it there. The slack method sorts batches by increasing slack, and the rationale for using it is as follows:

- If $T_A \geq T_R$, then there is more work time available than is required, and the material will be completed ahead of schedule if it follows standard throughput times from now on. It is therefore not urgent to process it.

- If $T_A < T_R$, then the material will be late unless it moves faster than standard. The larger the difference between T_A and T_R, the higher its priority should be.

To calculate the slack of a batch of parts, we need a due date, and therefore it must be pegged to an entry in the master production schedule. The "time required" T_R is the sum of the standard throughput times of the operations remaining to be performed until the parts reach finished goods. Thus this method requires considerably more data than FIFO or EDD.

There is no general theory of how sequencing by increasing slack affects the global performance of a factory. If we restrict the scope of investigation, as in the case of EDD, to n jobs on one machine, we show in Appendix A that this method maximizes the minimum lateness of the jobs, which hardly seems a worthwhile endeavor.

Slack can be viewed as the anticipated "earliness" of a batch: if, from now on, the batch proceeds according to standard, then its current slack will be its earliness when it reaches finished goods. Let L_1 and L_2 be two batches with the same negative slack and queued in front of the same resource, but with L_1 at the beginning of its route, while L_2 is one operation away from shipment. Although it would make sense to favor L_2 over L_1, slack does not do it. We shall see that Critical Ratio does, and that it is not always desirable.

Slack has another property distinguishing it from Critical Ratio: if L_1 and L_2 are two batches with different slack values and neither batch is moved for a time t, then the slack S_i of L_i becomes S_i+ t and the relative values of S_1 and S_2 are unchanged. If L_1 had a higher priority before, it still does. We shall see that, with Critical Ratio, the passing of time may reverse the relative priority of two batches.

15.3.5. Critical Ratio

This method uses the same data as Slack. The Critical Ratio of a batch of parts is T_A/T_R and is a measure of relative lateness. The lower it is the more urgent it is to process the lot. What if $T_R = 0$? Material requiring no time for processing presumably requires no processing and therefore should be excluded from dispatch lists.

Figure 15.4 shows two batches with the same Slack but different Critical Ratios. Batch 1 is still at the beginning of its processing sequence, so that T_A and T_R are both large and their ratio is near one. Near the end of the process, where Batch 2 is, the same difference yields a much lower ratio.

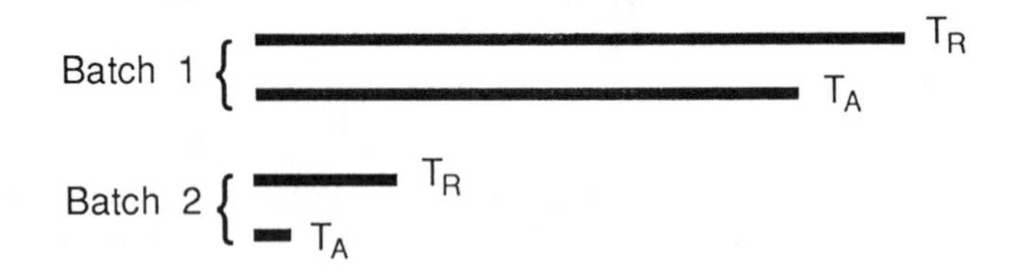

Figure 15.4. Materials at Beginning versus End of Process

Material that is near the end of the process can be sold sooner than material near the beginning, and therefore it appears desirable to give it priority. Figure 15.5 shows a factory with a hub and spoke structure similar to but much simpler than that of semiconductor wafer fabrication facilities. Wafer processing involves the use of the same photolithography resource 10 or 15 times, and the hub of Figure 15.5 is such a resource.

After being processed through the hub a first time, parts go to spoke S_1 and return to the hub, then go to spoke S_2 and return to the hub for a final operation. Dispatching parts through the hub involves choosing between parts at different operations in the process.

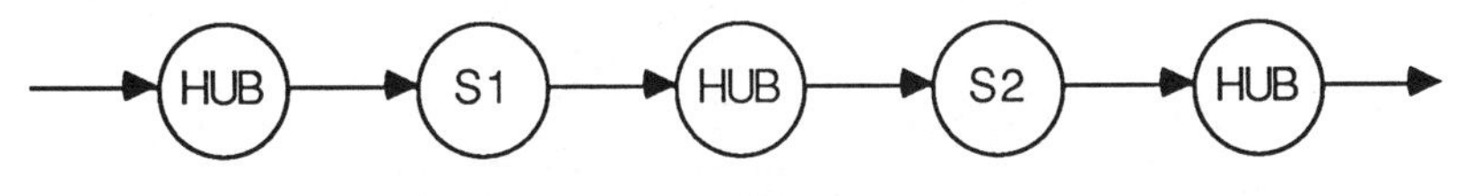

Figure 15.5. a. Hub and Spoke Logical Flow

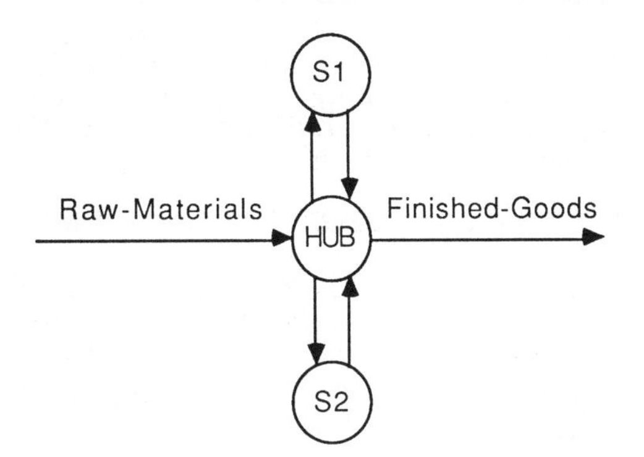

Figure 15.5.b. Hub and Spoke Physical Flow

In general, discrete event simulation seems an appropriate tool for the analysis of this system. When Critical Ratio dispatching was implemented in a real semiconductor factory with high inventories and in which almost all the material was late, the results were not what management expected.

The systematic preference given to the back end of the process and regular releases of new raw materials to the front end combined to further bloat the work in process inventories at the hub while starving the front-end spokes. The dispatching rule was then changed to slack. It is not clear what would have happened if the use of Critical Ratio had been continued, but it is conceivable that a cyclic pattern would have emerged in which the hub would have been dedicated to back-end material until it ran out, then to front-end material until it came back from the spokes.

If L_1 and L_2 are two batches with required times $T_{R(i)}$ and $T_{R(1)} > T_{R(2)}$ and neither batch is moved for a time t, the critical ratio of L_i decreases by $t/T_{R(i)}$, so that the critical ratio of L_1 decreases faster than that of L_2, this may result in the type of relative priority reversal shown Figure 15.6.

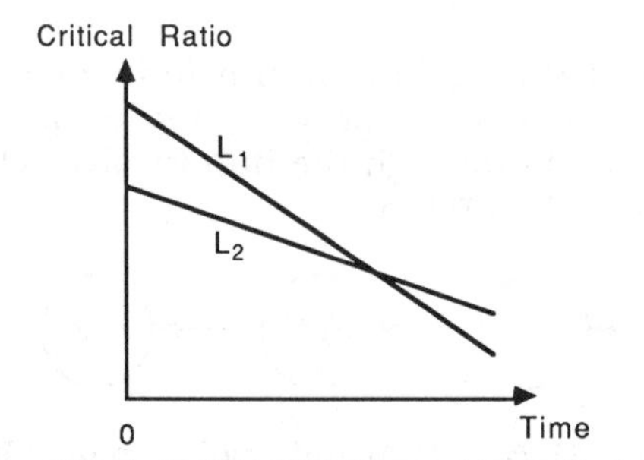

Figure 15.6. Critical Ratios of Two Batches

An event-driven, real-time dispatching system would normally not recalculate priorities unless something happened. But if the critical ratio method is used, then the priority ranking of a list of batches may change without anything happening to any of them.

Appendix A. EDD, SPT, AND SLACK FOR N JOBS ON 1 MACHINE

The proofs in this appendix are from French [Fre82]. They apply to various priority rules used to sequence n jobs for processing through one indivisible resource. Everything about the jobs is known beforehand. The n jobs to sequence are called J_α , $\alpha = 1, \ldots, n$, their processing times p_α, and their due dates d_α. Jobs which might arrive during the execution of the schedule are not considered and would not be inserted into the sequence.

A schedule is a permutation $S = [J_{i(1)}, \ldots, J_{i(n)}]$ of the jobs. The method used to prove that a sequencing rule maximizes or minimizes a performance measure is to start from a schedule violating it and show that its performance is improved whenever a violation is removed (see Figure 15.7).

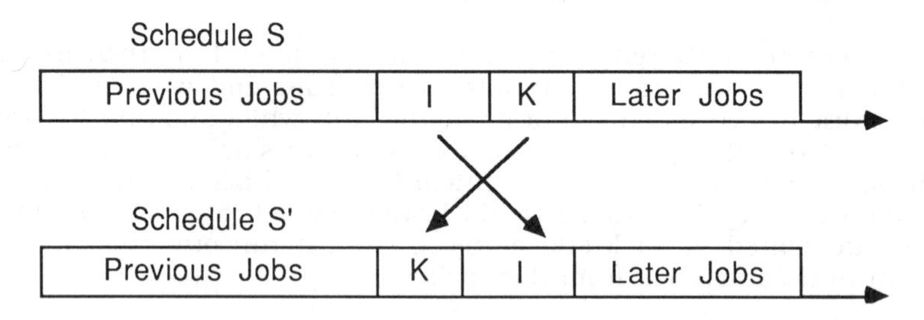

Figure 15.7. Interchanging Two Jobs

15.A.1. EDD Minimizes the Maximum Lateness

The lateness of a job is the difference between its due date and its completion time. If S is a schedule that does not follow the Earliest Due Date (EDD) rule, then it must include two *consecutive* jobs J_I and J_K such that $d_I \geq d_K$. The proof consists of showing that, when these two jobs are interchanged, the maximum lateness can only be decreased.

Let L_α be the lateness of job J_α. Then, if L is the maximum lateness of all jobs other than J_I and J_K, the maximum lateness of schedule S is Max{L,L_I,L_K}. If a is the sum of the processing times of all the jobs prior to J_I, then

$$L_I = a + p_I - d_I$$

and

$$L_K = a + p_I + p_K - d_K$$

The maximum lateness of schedule S' is Max{L,,L'_I,L'_K} where

$$L'_I = a + p_I + p_K - d_I$$

and

$$L'_K = a + p_K - d_K$$

But $d_I \geq d_K \Rightarrow L'_I = a + p_I + p_K - d_I \leq a + p_I + p_K - d_K = L_K \leq Max\{L, L_I, L_K\}$
and $p_I \geq 0 \Rightarrow L'_K = a + p_K - d_K \leq a + p_I + p_K - d_K = L_K \leq Max\{L, LI, LK\}$

and therefore Max{L,,L'_I,L'_K} ≤ Max{L,L_I,L_K}, q.e.d.

15.A.2. SPT Minimizes the Average Throughput Time

If S is a schedule that does not order jobs by increasing processing times, then it must include two consecutive jobs J_I and J_K such that $p_I \geq p_K$. The proof consists of showing that, when those two jobs are interchanged, the average throughput time of the whole sequence can only be decreased.

Let F_α be the throughput time of job J_α, and

$$\bar{F} = \frac{1}{n}\sum_{\alpha=1}^{n} F_\alpha$$

be the average throughput time of all the jobs.

If b is the throughput time of the last job prior to J_I , then the throughput times of J_I and J_K are $F_I = b + p_I$ and $F_K = b + p_I + p_K$. In the schedule S' deduced from S by interchanging J_I and J_K, the corresponding throughput times are

$$F'_I = b + p_K + p_I = F_K$$

and

$$F'_K = b + p_K \leq b + p_I = F_I$$

Therefore

$$n\bar{F}' = \sum_{\alpha \notin \{I,K\}} F_\alpha + F'_I + F_K' \leq \sum_{\alpha \notin \{I,K\}} F_\alpha + F_K + F_I = n\bar{F}$$

15.A.3. Slack Maximizes the Minimum Lateness

If S is a schedule that does not sort jobs by increasing slack, then it must include two consecutive jobs J_I and J_K such that Slack(I) ≥ Slack(K). The proof consists of showing that, when those two jobs are interchanged, the minimum lateness can only be increased.

Let us set the origin of time at the beginning of the schedule and call a the sum of the processing times of all jobs preceding J_I and J_K in schedule S. In S, at time a, the slacks of J_I and J_K are respectively $U_I = d_I - a - p_I$ and $U_K = d_K - a - p_K$. As in the discussion of EDD, we find that the latenesses of J_I and J_K are

$$L_I = a + p_I - d_I = -U_I$$

and

$$L_K = a + p_I + p_K - d_K = p_I - U_K$$

In the schedule S' obtained by interchanging J_I and J_K,

$$L'_I = a + p_I + p_K - d_I = p_K - U_I \geq - U_I = L_I$$

and

$$L'_K = a + p_K - d_K = -U_K \geq -U_I = L_I$$

The minimum lateness L of all other jobs is the same in S and S'. Therefore we have

$$\text{Min}\{L, L'_I, L'_K\} \geq \text{Min}\{L, L_I\} \geq \text{Min}\{L, L_I, L_K\}.$$

and sorting by increasing slack maximizes the minimum lateness!

Appendix B. WAITING TIMES IN QUEUES WITH POISSON ARRIVALS

In contrast with Appendix A, we are considering queue disciplines applied in real time — that is, a batch arriving in front of the resource is inserted immediately into the queue at a rank determined by the discipline and is subject to being passed over by later arrivals.

As before, the discussion is restricted to the case of an indivisible resource. It is processing one batch at a time and is not preempted until the batch is completed.

We make the additional following assumptions:

1. The processing times of the batches are independent and identically distributed, with

$$F(dt) = \text{Prob(Processing time is between } t \text{ and } t+dt)$$

   such that

$$\int_0^{\infty} t \times F(dt) < +\infty$$

   and

$$\int_0^{\infty} t^2 \times F(dt) < +\infty$$

2. Batch arrivals into the queue form a Poisson process of rate λ — that is,

$$\text{Prob(a job arrives between } t \text{ and } t+dt) = \lambda dt$$

   and the numbers of arrivals in two disjoint intervals are independent.

We are going to show that, if the queue discipline is independent of the processing time, the average waiting time W_q satisfies

$$W_q = \frac{W_0}{1-\rho}$$

where

$$W_0 = \frac{\lambda}{2}\int_0^{\infty} t^2 \times F(dt)$$

is the average time needed to complete the batch in process at arrival time and

$$\rho = \lambda \int_0^{\infty} t \times F(dt)$$

is the utilization of the resource.

To prove this, we decompose the total queue time W_q of a batch into the average time W_0 needed to complete the batch in process at the time the new batch arrives and the total processing time W_1 of the queued batches that take priority over the new one. W_1 includes the processing times of the higher priority batches already present at arrival time and arrived during the waiting time itself.

Obviously, $W_q = W_0 + W_1$. Let us concentrate first on W_0. Figure 15.8 shows a possible timeline for the resource, showing it to be sometimes idle and the rest of the time processing batches of various lengths.

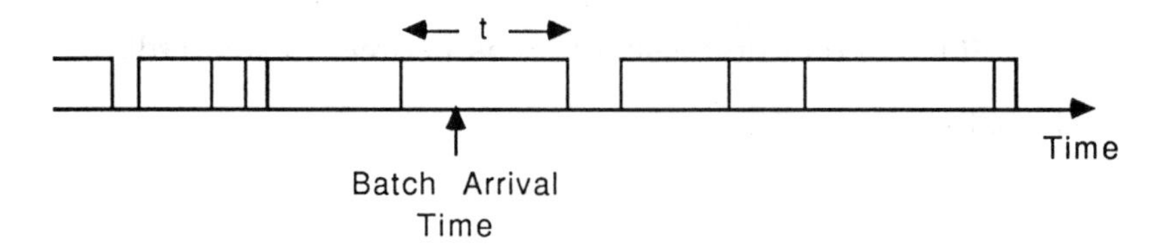

Figure 15.8. Batch in Process at Arrival Time

The Utilization law, a special case of Little's law, says that the steady-state probability ρ that the resource is busy at any time is the product of the arrival rate λ by the average batch processing time. Therefore

$$\rho = \lambda \int_0^\infty t \times F(dt)$$

More specifically, we can use the same law to find the probability that the resource is busy with a batch of processing time between t and t+dt when the new batch arrives. The arrival rate for such batches is λF(dt) and since their processing time is between t and t+dt, the utilization of the resource due to such batches is λF(dt)×t.

Knowing that the new batch arrives in the queue when a batch of processing time t is in progress, it is equally likely to arrive at any point within the processing cycle, so that the average time it has to wait for the current batch to complete is t/2. Therefore

$$W_0 = \int_0^\infty \frac{t}{2} \times t \times \lambda F(dt) = \frac{\lambda}{2} \int_0^\infty t^2 \times F(dt)$$

To calculate W_1, we must characterize the batches that our batch must wait for before it can begin processing. Figure 15.9 shows cumulative arrivals and departures so that, at any time, the number of batches in the queue is the difference between the two curves. Flow balance requires the arrivals and departure to occur at the same rate λ.

When our batch arrives, it finds a fraction of the queue having a higher priority. While it is waiting, all the departing batches have higher priority. The number of those varies due to new arrivals and to reversals of relative priorities due to the passing of time, as we have seen in the discussion of Critical Ratio.

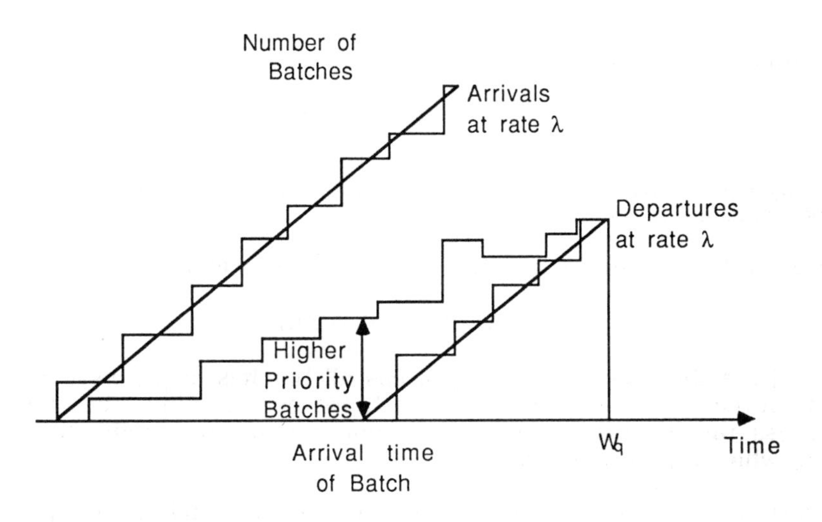

Figure 15.9. Evolution of the Queue

Under the condition that our batch waits a time T_q, the expected number of batches leaving the queue during that time is $\lambda \times T_q$. If we take the average with respect to T_q, we find that the average number of batches our batch has to wait for is $\lambda \times W_q$.

Since the priority rule is independent of the batch processing times, the average time to process a higher priority batch is the same as the overall processing time average, therefore

$$W_1 = \lambda\, W_q \int_0^{\infty} t \times F(dt) = \rho\, W_q$$

and

$$W_q = W_0 + \rho \times W_q \Rightarrow W_q = \frac{W_0}{1-\rho}$$

This result is a special case of Kleinrock's Conservation Law. It applies strictly to FIFO, LIFO, and EDD. Slack and Critical Ratio do consider the processing time at the current station as one of many parameters, but one that has an important influence only for parts that are near the end of their processes. For these rules, this result is an approximation.

SPT is the only rule we have looked at to which this result does not apply. In the case of SPT, the average waiting time $W_q(t)$ of a batch of processing time t is given by Phipps' formula:

$$W_q(t) = \frac{W_0}{\left[1-\rho(t)\right]^2}$$

where W_0 is as above and

$$\rho(t) = \lambda\int_0^t u \times F(du)$$

which, by the Utilization Law, is the utilization of the resource due to batches of processing times $\leq t$.

To prove Phipps' formula, we decompose $W_q(t)$ as follows:

$$W_q(t) = U(t) + A(t)$$

where U(t) is the time needed to process all batches of processing time $\leq t$ already in the queue when the new batch arrives, and A(t) is the time needed to process the batches of processing time $\leq t$ arriving while the batch is waiting, which pass it in the queue.

Little's law tells us that the average number of batches with processing times between t and t+dt in the queue is $\lambda F(dt) \times W_q(t)$. Multiplying by the processing time t and integrating between 0 and t, we get

$$U(t) = W_0 + \lambda\int_0^t u \times W_q(u) F(du)$$

The number of batches of processing times between u and u+du arriving in an interval of length W(t) is $\lambda F(du) \times W_q(t)$, and their contribution to $W_q(t)$ is $\lambda F(du) \times u \times W_q(t)$. Therefore

$$A(t) = \lambda W_q(t)\int_0^t u \times F(du) = \rho(t) W_q(t)$$

and

$$W_q(t) = W_0 + \lambda\int_0^t u W_q(u) F(du) + \rho(t) W_q(t)$$

Differentiating both sides, we get $dW_q(t) = 2W_q(t)d\rho(t) + \rho(t)\, dW_q(t)$, which implies

$$\frac{dW_q(t)}{W_q(t)} = \frac{2\, d\rho(t)}{1-\rho(t)}$$

and finally

$$W_q(t) = \frac{W_0}{[1-\rho(t)]^2}$$

16

APPRAISING MRP-II

Capacity requirements planning is rarely implemented due to the difficulty of gathering the requisite data and the ease of making rough calculations with electronic spreadsheets. MRP's standard throughput times resemble outside suppliers' lead times more than estimates of queue time.

The value of calculations based on average yields and throughput times is limited by the randomness of actual results. MRP sees time as a sequence of periods which must be long enough for parameter averages to be meaningful, and short enough to provide guidance. Planners have no direct control of throughput times, so that pro-forma schedules generated with changes in the standards only represent management wishes.

16.1. WHY CRP IS RARELY IMPLEMENTED

Of the functions of MRP, Capacity Requirements Planning (CRP) is the least frequently used. MRP is most often used as if the factory had no capacity constraints. More so than other modules of MRP, CRP requires model parameters that are detailed and difficult to obtain. In 1911, Frederick Taylor [Tay11] reported encountering resistance when trying to find out how long a manufacturing activity should take. Since then, the popularity of time and motion studies with shop floor personnel has not increased. The extraction of capacity parameters from historical records or engineering specifications is equally painstaking.

The amount of work required to run CRP must then be weighed against its usefulness. The CRP resource model is a large spreadsheet, and CRP as part of an MRP system competes with real spreadsheet programs running on personal computers. The most commonly used capacity planning tool in the United States today is probably not CRP but Lotus 1-2-3, with the CRP logic implemented on it by manufacturing and production control managers. Those managers have some knowledge of the capacities of a few critical resources in their departments.

When they are informed of a change in production volume or product mix, they use their own spreadsheets to find out whether their resources can withstand it. The CRP logic is adequate for such "quick and dirty" calculations, but, within the context of MRP, there is a mismatch between the roughness of the method and the level of detail it requires in input data.

16.2. THE INTERPRETATION OF STANDARD THROUGHPUT TIMES

Outside purchases have lead times. Upon receipt of an order, a supplier promises delivery for a certain time, by which he or she feels confident the goods will be ready. The standard throughput times of MRP can be viewed in the same light more easily than as estimates of the sum of queue times and processing times.The department responsible for an activity is treated like an outside supplier. The standard throughput time is a time budget within which the work is to be completed. How it is split between waiting for processing, processing, and waiting for shipment is up to the supervisor.

MRP's treatment of each department as a contractor requires standard throughput times sufficiently high to be sure that they can be met. An outside supplier would normally not promise a delivery date he or she is only 50% sure of meeting. The buyer, on the other side, plans to receive the goods when due and not before, letting the supplier pay the holding costs of goods finished early. If MRP plans are followed, with throughput times set so that they have a high probability of being met, then material will be often ready early after each activity, but it will not be withdrawn by the next activity until it is "due." The effect on actual throughput times and inventories is the same as that of maintaining a safety output buffer. This is discussed in the next paragraph.

16.3. INADEQUACY OF SINGLE PARAMETER MODELS

16.3.1. Throughput Time Modeling

The following example is a two-step fabrication process with a high variability in processing times. Orders for one unit arrive as a Poisson process at rate λ. By conservation of flow, λ is also the rate at which units leave all the operations and the finished goods buffer, as shown in Figure 16.1.

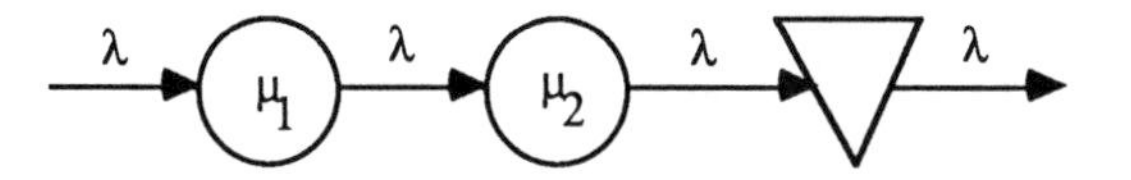

Figure 16.1. Two-Step Route with Exponential Processing Times

We assume operation i to have a processing time following an exponential distribution with parameter μ_i. The exponential distribution may be more random than manufacturing processing times, but it is used here because calculations with it are simple and it does show the effect of randomness on MRP performance.

If, when an order arrives, it is queued before the first operation and proceeds in FIFO sequence from then on without any control on releases to operation 2, the total time spent waiting and in process at each operation follows the exponential distribution with parameter $\mu_i-\lambda$, and the mean total time through the two operations is

$$W = \frac{1}{\mu_1 - \lambda} + \frac{1}{\mu_2 - \lambda}$$

With the hypotheses we made, we know from Kelly [Kel79] that the waiting times in the queues are independent. Therefore, the variance of the total time through the two operations is

$$\sigma^2 = \frac{1}{(\mu_1 - \lambda)^2} + \frac{1}{(\mu_2 - \lambda)^2}$$

We can be confident that the total time through the process will take no longer than $W+3\sigma$. If we take this value to be the standard lead time, by Little's law, the total inventory will be

$$L = \lambda \times (W + 3\sigma)$$

and the safety stock of finished goods will contain on the average

$$L_S = \lambda \times (W + 3\sigma) - \lambda \times W = 3\lambda\sigma$$

With $\lambda = 1$ and $\mu_1 = \mu_2 = 1.2$, $L = 31.2$ and $L_S = 21.2$ units.

How much can we improve on this result by using MRP to schedule through this process? Figure 16.2 is a modification of Figure 16.1 showing explicitly the inventory buffer between operations 1 and 2. We want to give both operations standard throughput times T_1 and T_2 that can reliably be met. To have the same degree of confidence as in the preceding, we set

$$T_i = W_i + 3\sigma_i$$

with

$$W_i = \frac{1}{\mu_i - \lambda} \text{ and } \sigma_i^2 = \frac{1}{(\mu_i - \lambda)^2}.$$

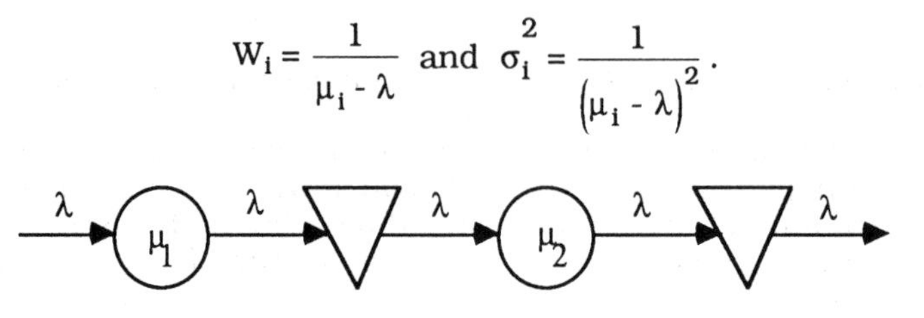

Figure 16.2. Two-Step Route with Exponential Processing Times and a WIP Buffer

With the same parameters as before, we get $T_1 = T_2 = 20$. The total throughput times and inventory are

$$W = T_1 + T_2 = 40$$

and

$$L = \lambda W = 40 \text{ units}$$

In this particular problem, the use of MRP logic worsens the throughput time and inventory performance by a good 30%, because it forces the planner to make worst case assumptions at every stage. When the dynamics of the series of simple queues is left to operate on its own, parts occasionally go faster than the average instead of slower.

16.3.2. Yield Modeling

Wight [Wig74] does not discuss yields. There are, however, many cases where part losses occur at such a rate that they cannot possibly be neglected in scheduling. The world leaders in manufacturing 256KDRAMs (Dynamic Random Access Memories) are in Japan, where the semiconductor industry yearbook for 1986 reported a yield of 54% [Han86]. In other words, nearly half of what the manufacturers set out to make is thrown away. This activity represents billions of dollars in sales, and cannot be considered pilot or experimental production.

If the standard yield of an operation in 95%, then MRP will plan as if 100 parts fed to this operation yielded exactly 95 good parts out. In reality, the number of good parts is a random variable with an expected value of 95. In many cases, we can model it with the binomial distribution — that is, assume that if N parts are processed, there will be n good parts with probability

$$P_n = \binom{N}{n} \times p^n \times (1-p)^{N-n}$$

The expected value of n is E(n) = Np, and the variance is

$$\sigma^2 = N \times p \times (1 - p).$$

With N=100 and p = 95%, n will be between 81 and 100 with a probability of 99.7%.

If the part is made on a regular basis, then a safety stock will be needed after the operation, which brings us back to the type of reorder point calculations MRP was supposed to avoid. If the part is rarely made, or is custom, then it is necessary to plan to build more than the use of the average yield would suggest, because using it would frequently lead to shipping less than promised.

16.4. THE TIME PERIOD DILEMMA

MRP views time as a sequence of periods, or buckets. If the elementary period is a month, then it is not possible to promise deliveries for any time but the end of a month. Shorter periods make for smaller, more frequent shipments, and more flexibility in responding to customers.

However, if the period is an hour, then a requirement that n units of A and m units of B be processed through resource X may be impossible to fulfill, because X takes 3 hours to process A's and has just started a load of 10×n units.

Thus the elementary time period must be short with respect to the lead times demanded by customers, and long with respect to the processing time of a load on any resource. Such a tradeoff is not always possible.

Let us examine in more detail the combined effect of modeling a manufacturing operation as a pure delay of length T, and of factory time as a sequence of periods of length p. Let N be the amount of inventory at the operation at the beginning of period i. Without making further assumptions on where that inventory is *within the operation*, we cannot say how many of these N units will have left it by the end of period i.

Figure 16.3 shows a model that is in use, in which the operation is modeled as a conveyor belt moving at constant speed and with a length proportional to its standard throughput time. At the beginning of the first period, the N units in inventory are assumed evenly spread over the conveyor, so that, by the end of the period, $\text{Min}\{N, N \times (p/T)\}$ of the original N units have left the operation.

In the absence of information as to how an operation is actually executed, this model seems no worse than any other. Figure 16.4 shows what happens when it is applied to a *sequence* $X_1, \ldots, X_i, \ldots$ of operations. Because this approach ignores capacity constraints, whatever peaks and valleys of inventory happen to be present at the beginning of period 1 are moved forward over time along the process without any damping.

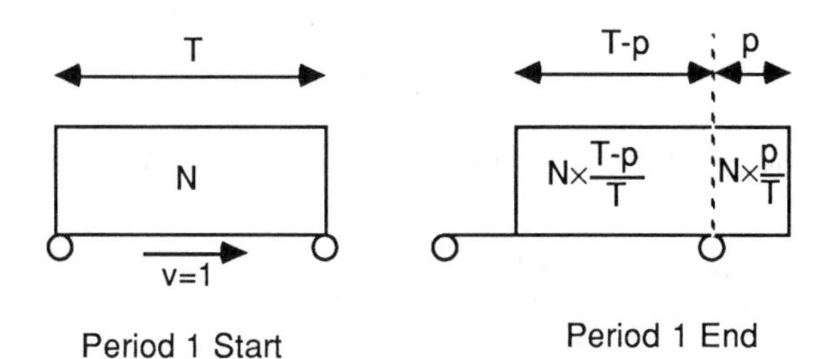

Figure 16.3. The Conveyor Model for an Operation

This logic excludes operations with a throughput time T=0, because spreading inventory over an interval of length zero involves a division by zero. From a shop floor perspective, an operation with T=0 cannot have any inventory because material leaves it as soon as it enters.

These considerations, however, do not stop users from defining operations to which they assign zero throughput times, for example, to model activities that are not individually visible because they are carried out by a subcontractor. The planning software does not always respond by an error message but sometimes by treating the inventory at the operation as a Dirac impulse of materials, which then moves forward as such through the whole process. In such a case, a combination of simplifying assumptions that are all individually reasonable can result in reports that are difficult to understand.

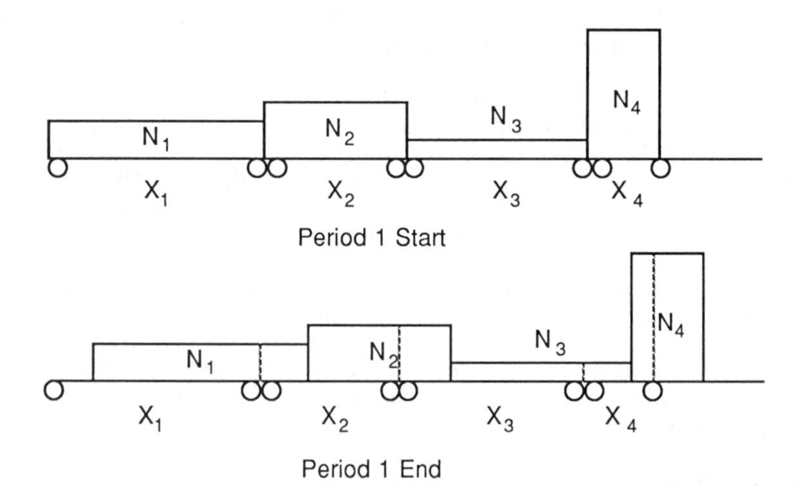

Figure 16.4. The Conveyor Model for a Route

16.5. THE 10,000-KNOB CONTROL PANEL

In spite of its simple logic, MRP is frequently perceived as requiring a huge implementation effort. Even if the routings and the bill of materials are set up for the CIM system as a whole, standard yields, throughput times, and processing times per part are needed for every activity. The number of those parameters easily runs in the thousands.

Various strategies may be used to reduce the number of decisions. It is possible, for example, to have users assign a throughput time and a yield to a whole route, and then to automatically assign to every operation on the route the arithmetic mean throughput time and the geometric mean yield: if a route with throughput time T and yield Y has n operations, each operation is assigned a throughput time of T/n and a yield of $\sqrt[n]{Y}$.

The number of parameters can also be reduced in another way if routes are constructed from standard operations and each operation is part of many routes. In this case, values for many thousands of route-operation combinations can be derived from a few hundred values set for standard operations.

Once the planner has somehow entered all the requisite parameters, the task remains of interpreting them in terms of the factory. What does it mean to change one of them before generating a pro forma schedule? If expedited throughput times are changed from 80% to 70% of standard, the resulting production plan will be different. But the planner does not have the power to make expedited material actually go that much faster.

If the dispatch system offers several algorithms, then that choice is a genuine control parameter. Likewise, management can and does frequently change the work calendar to adjust capacity. On the other hand, the planner has no direct control over throughput times. Those are, strictly speaking, a measure of scheduling performance, and not inputs to scheduling.

Finally, the sheer number of parameters makes driving a factory with MRP akin to using a control panel with 10,000 knobs. In the next chapter, we shall see how OPT goes about reducing that number to 14.

16.6. RESPONSE TO ACTIVITY LEVEL CHANGES

Let us assume that a factory that had hitherto been running to full capacity is reduced by a softening of its market to running at 50% of capacity. Let us also assume that it is a Class A MRP user. There is an MRP-II system in place and the actual manufacturing throughput times match the standards.

If the standard throughput times are unchanged and the output for every product is cut in half, then Little's law tells us that the work in process inventories will stabilize at $L = \lambda W$ where

- L is the work in process for a product at one operation.
- λ is the throughput of that operation for this product.

• W is the throughput time of this product through that operation.

If all the λ's are cut in half and all the W's are constant, then all the L's will also be cut in half. The planner, and the MRP system, will probably be credited for this spectacular reduction in inventory. The plant manager is however likely to apply pressure to reduce inventories further by cutting the standard throughput times for two reasons:

1. With its income from sales cut in half, the factory is short of cash and cannot afford to keep any more cash than necessary tied up in inventory.

2. Short throughput times make the factory more attractive to customers and thus can be part of a strategy to regain some of the lost sales volume.

The MRP system only contains one number to represent the capacity of each work center to process each product at each operation. The planner does not have a resource model such as the ones described in Chapter 13 to tell what level of inventory will sustain the required throughput and what throughput times will result from it.

The planner can play "what if" games with the standard throughput times, all 10,000 of them, but there isn't anything in the system to tell whether a particular waiting time is realistic. Meanwhile, if the production supervisors on the floor are measured on efficiency, they are unlikely to let their workers and machines be idle with inventory in front of them. The temptation to move it on even though the MRP system says it isn't necessary will be irresistible.

As material arrives in finished goods inventory ahead of schedule, the planner will have measured actual throughput times shorter than the standards and he or she can use those to update the standards. There may be more difficulties ahead in working off the bloated finished goods inventory, but the system should eventually stabilize at the new values.

If that is how the adjustments are made, the progress accomplished is really no thanks to the MRP system, but to the supervisors who disobeyed it. The results may have been achieved, but with the scheduling system following rather than leading.

16.7. WHEN DOES MRP WORK?

This chapter is critical of MRP as an approach to production scheduling, not because MRP is theoretically unattractive, but because it has so few satisfied users. After reviewing the method, we are now in a position to state conditions a factory must meet for it to work.

MRP is supposed to schedule production for a broad product mix without maintaining work in process inventory for every item in the catalog. However, its logic rests on the assumption that throughput times and resource capacities are independent of work load, which we have seen in Chapter 7 is never true. The flexibility of MRP is therefore limited to those product mix changes which do not affect resource work loads.

The combination of a sufficiently long planning period (for the flow model of capacity to be applicable) with the use of an available dispatching rule to schedule within each planning period must lead to meeting the schedule. We have no way to guarantee this.

The noble goal of "making whatever the customer wants" is at the root of many of MRP's problems. OPT, as we shall see in the next chapter, has a more elaborate resource model, but also uses standard yields and throughput times, and cannot avoid time-zero spikes.

Businesses whose main strength is manufacturing usually have the philosophy of selling what they make rather than making what they can sell. They recognize the need for flexibility but limit its scope. The highly successful Just-In-Time method is touted as a flexible manufacturing tool. Yet it requires maintenance of work in process for some parts made, at all stages of processing.

The question is not of flexibility versus rigidity, of the dedicated Model-T factory versus the job-shop that will take on anything, but of the balance between the two. MRP attempts to go too far in the direction of flexibility. It puts too much emphasis of what the customer wants, and does not embody a sufficient understanding of how factories work at the material flow level.

17

THE THEORY OF OPT

OPT is a system that schedules production off-line, like MRP, but takes into consideration utilizations and resource dependencies. Its theory has depth but is still evolving. Based on a factory analysis in terms of sales, inventory and operating expenses, OPT treats resources differently, depending on whether or not they have overcapacity.

OPT holds that dependent events and statistical fluctuations make it futile to seek balanced capacities. Unlike MRP, OPT distinguishes process batches, or parts processed between setups, from transfer batches, which move between resources. OPT claims to produce optimal schedules, but the optimization criteria and the bottleneck scheduling algorithm are secret.

17.1. OPT IN THE UNITED STATES

Touted by its supplier as revolutionary, more advanced than Just-In-Time, and capable of saving western industry from defeat at the hands of Japan, the OPT production scheduling system has been featured in *Fortune* and *Business Week*, and bought by such corporations as General Motors.

This achievement is remarkable for a system developed in the late 1970s by a physicist and that has yet to produce its first unqualified success story.

Besides the technical merits of OPT, other forces at work have been economic and political conditions, and creative marketing.

In a time when it is losing market after market to competition from Japan and newly industrialized countries of the Far East, U.S. manufacturing has responded mostly by layoffs and plant closings. The industrial northeastern part of the United States is dotted with ghost factories, while former manufacturing companies are turning into marketing organizations for products made elsewhere. Meanwhile, production management textbooks have come out with brave subtitles like "beating the Japanese at their own game" but with few ideas on how to go about it.

Goldratt, the inventor of OPT, preaches to manufacturing people whose jobs are in jeopardy. He declares himself on a crusade to save the western world from domination by Japan and his points are presented as revealed truths rather than proven.

The main OPT marketing tool is a novel entitled *The Goal*, and written by Goldratt and Jeff Cox. The hero, Al Rogo, is a midwestern plant manager threatened with closure within three months, who is saved by applying the ideas of a physicist turned production expert. In the end, Al Rogo saves the plant and is promoted to division manager. As of the spring of 1986, 25,000 copies of *The Goal* had been sold, and a sequel, called *The Race* was in preparation.

The Goal is based on the case of the Howmet lost-wax foundry of La Porte, Indiana. The plant manager, Danny LeFay, eventually became president of Creative Output, the company marketing OPT in the U.S. Unfortunately the OPT project was discontinued by his successor in the La Porte plant. This case illustrates how far OPT's record is behind that of Just-In-Time: the latter runs Toyota, the second largest automobile company in the world in sales, and, according to Ward's Auto World, the first in quality and labor productivity.

Unlike generic approaches like MRP and Just-In-Time, OPT is a theory bundled with a software product. As such, it is possible to evaluate it from the point of view of software engineering. Creative Output does not use structured analysis. The data flow and entity-relationship diagrams in this chapter reflect this author's understanding of the system and are not taken from any Creative Output publication. OPT's highly nonstandard terminology is not used here; it would both make the discussion unnecessarily obscure and impair comparison with other approaches.

MRP is based on a black box view of manufacturing. It doesn't want to know how machines within a work center are scheduled. By comparison, OPT has theoretical depth: it is an attempt to take into consideration relevant aspects of manufacturing ignored by MRP.

17.2. FROM BUSINESS ANALYSIS TO SCHEDULES

Goldratt insists that the only way to rate a manufacturing decision is by its impact on sales, inventory, and operating expenses. The soundness of this is not a revolutionary discovery but simply good financial analysis. Cost accounting is branded "number one enemy of productivity" because it does not support this approach. These questions were discussed at length in Chapter 12.

OPT aims to produce schedules that are (1) economic in the preceding sense, (2) realistic, in that they overload no resource, and (3) safeguarded against disruptions. The setting of a product mix to maximize profit while staying within the capacity of a factory is a constrained optimization problem not addressed by OPT. OPT starts from a list of orders, and strives to fulfil them.

An OPT schedule is safeguarded against disruptions in two ways:

1. By specifying slack time at non-critical resources to ensure timely delivery.
2. By maintaining an inventory buffer in front of critical resources to keep them busy.

Like MRP, and unlike Just-In-Time, OPT is an off-line system. As such, it cannot respond to disruptions *as they occur.* It has no choice but to make allowances for the fact that they do happen. Goldratt talks about "dependent events and statistical fluctuations" as an essential component of manufacturing, and uses dice games to get the point across to managers. Yet he and his organization see no value in the use of probability theory.

To model manufacturing disruptions, they invoke "Murphy's law" even in publications such as [Fox83] intended to be taken seriously. Manufacturing is disrupted by random events governed by the laws of probability, and those laws provide the only solid rules for estimating the impact of the disruptions and of corrective actions. An example is given in Appendix A. Randomness and probability inspire jokes and popular superstitions about "Lady Luck" or "Murphy" that have no place in engineering work.

The most concise formulation of the OPT philosophy, as taught to users, is in the form of nine "rules." Some are commandments: "Balance flow, not capacity!" Others are theorems: "An hour lost at a bottleneck is an hour lost for the entire system." The last one is a general principle providing little guidance: "The conjunction of local optima is not a global optimum." The following discussion is based on these rules.

17.3. THE OPT CONCEPT OF BOTTLENECK

In OPT, a bottleneck is a resource with a capacity that is less than or equal to market demand. This definition differs from more commonly used ones. To many MRP systems, a bottleneck is any resource loaded above a certain fraction of its capacity. In queueing network theory, the bottleneck of a network is the service center with the lowest capacity.

"An hour lost at a bottleneck is an hour lost for the entire system" says one of the "rules of OPT." No proof is given, and the example in Figure 17.1 seems difficult to reconcile with this assertion. We are considering a factory making a single product by a process involving two machines in succession and a market that would buy more than either of these machines can produce.

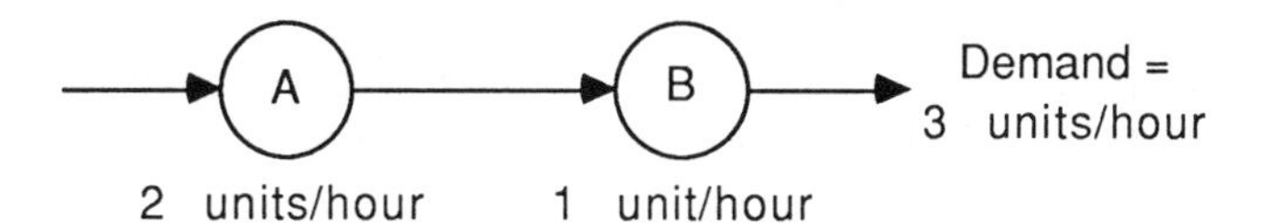

Figure 17.1. Bottlenecks with Unequal Capacities

A and B both meet the OPT criterion for bottlenecks, yet the rule applies only to B: if B stops for an hour, the factory misses an opportunity to sell one unit, but A has to be idle 50% of the time because it feeds B.

A resource which is not a bottleneck should be activated only as much as needed to keep the bottlenecks busy. Any activation above this level results in accumulation of inventory without generating any additional sales.

Theoretically, there is a set of bottlenecks for each product mix. In practice, there is little variation as to what resources are bottlenecks. The central idea of OPT scheduling is that different methods should be used to schedule bottlenecks and non-bottlenecks.

The Goal uses the analogy of a boy scout hike to show how bottlenecks affect production. The boys walk in a column and cannot pass each other. By walking, the column of boy scouts produces "walked trail" and has fulfilled the "order" when the last boy has reached the end of the hike. The length of trail between two successive boys is "work in process inventory."

There is one fat kid named Herbie in the line with an overloaded backpack who is too slow to complete the hike in the allotted time. His capacity to produce "walked trail" is less than the "market demand" and therefore he is a bottleneck. When the stronger boys in front of him walk faster than him, they create gaps — that is, accumulate inventory — but their efforts will not make the hike be completed any earlier.

The hike in *The Goal* is eventually completed on time: Herbie has been relieved of his excess luggage, which has been spread among his lighter companions. "Inventory" has been kept to a minimum by placing Herbie at the head of the line, so that nobody could walk faster.

The shop floor equivalent of these measures can be taken as soon as a resource is identified as a bottleneck:

- Rearranging shift changes and breaks to ensure that bottleneck processing is never interrupted. This may involve union negotiations (see Appendix A).

- Placing inspection stations in front of bottlenecks to keep them from processing already damaged parts.

- Activating whatever obsolete equipment may be available as a substitute resource to offload the bottlenecks, even if it is slower than current equipment.

In the medium term, bottleneck capacity can be increased by a setup time reduction program. Immediately, the process batch size should be set as large as possible on the bottlenecks, and only there. Schedules for bottlenecks should be developed first, and schedules for other resources derived from the bottleneck schedules.

17.4. FROM BOTTLENECKS TO CAPACITY CONSTRAINT RESOURCES

In OPT, there is no negative connotation to the word "bottleneck." Like the stars of a show, the bottlenecks are the resources on which success hinges. The role of the other resources is to serve the bottlenecks so that they deliver peak performance. OPT is in trouble in a factory that doesn't have any. What happens when every resource has a capacity that is superior to market demand?

One strategy, applicable for instance to a high volume factory whose market has suddenly gone soft, is to take the lowest capacity resource and use up all its idle time on setups. This cuts process batch sizes and lead times, and thus may attract quick turnaround business to the factory. This resource then becomes a bottleneck.

If bottlenecks cannot be created this way, then OPT falls back on "capacity constraint resources" — that is, resources that are overloaded at least half the time over the planning period.

A capacity constraint resource is not globally overloaded, as a bottleneck is, but its work load over time, as determined by classical capacity requirements planning, exceeds its capacity more than half the time. The concept of a capacity constraint resource is far from being as clear as that of a bottleneck. Why half the time and not 40% or 60%? There are no obvious interesting consequences following from this definition. Nonetheless, in the absence of bottlenecks, the capacity constraint resources thus defined are used to drive schedules.

If there are no bottlenecks and no capacity constraint resources, then Creative Output will admit that the factory is not a good fit for OPT. To OPT, either a resource is overloaded and it is a bottleneck, or it is a non-bottleneck. OPT makes no difference between resources loaded to 20% and 80% of capacity, although it is intuitively clear that there will be much more contention for the latter than the former.

17.5. THE UNBALANCED PLANT

17.5.1. Eli Goldratt's argument

The most common response to the discovery that a resource is a bottleneck is to want to increase its capacity to bring it "in balance" with the rest of the plant. It is a widely held notion that all processing resources should have the same capacity, and one that Goldratt disagrees with.

His argument is that the combination of dependent events and statistical fluctuations makes it undesirable to have capacity balance between the various resources. He supports it with the two-worker example shown in Figure 17.2.

From the point of view of the whole system, the slow speed of Worker 1 in the first hour is not made up by his fast work of the next hour, because Worker 2 runs out of parts to process. To make up for the lost time of Worker 1, Worker 2 would need a higher capacity.

According to Goldratt, the statistical fluctuations in worker capacities, combined with the dependence of Worker 2 on the output of Worker 1 show why a factory cannot and should not have a balanced capacity.

There is a strong case to be made that it is futile to try and balance capacities, but Goldratt does not make it. If there were an inventory buffer between the two workers, then Worker 2 would be protected against fluctuations in the speed of Worker 1. It is also often technically possible, if not economically justified, to eliminate the fluctuations in the speed of Worker 1 by automation. Goldratt's argument would lead one to believe that balanced capacity would be achievable and desirable in an automated factory, which is not the case.

17.5.2. External versus Internal Variability

Fluctuations in worker speed are an example of variability that is internal, in that it originates inside the factory. People and machines are factors of internal variability. Customers and raw materials, on the other hand, cause random changes in a factory *from the outside.* Internal variability can be reduced by management action; external variability is due to factors management has no control over.

The workload of any resource within a factory is a function of the current demand placed on it. A factory can only be said to have balanced capacity if all its resources have equal capacities to process any product mix. If it makes more than one product, then all products must have proportional requirements of every resource — that is, if product A requires 20% more of resource X than product B, then product A must also require 20% more of every other resource used for B. This condition is rarely met.

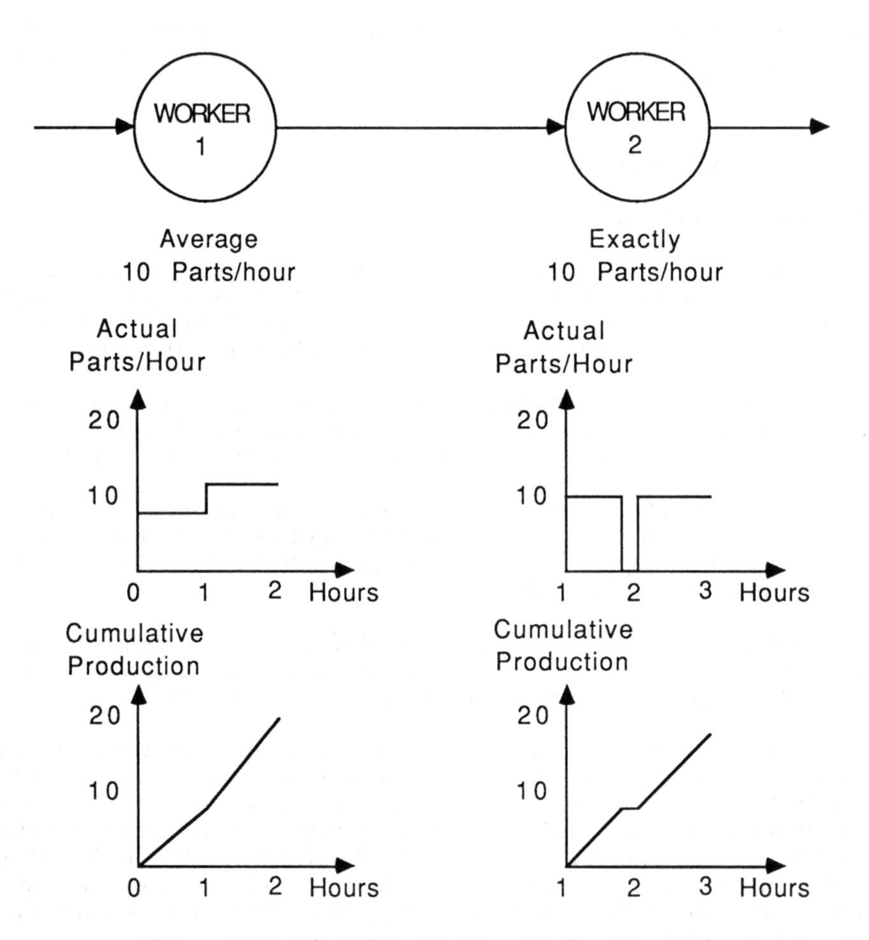

Figure 17.2. Eli Goldratt's Two-Worker Example

If A and B are two versions of the same car differing only in that A has a 2.0-liter engine and B a 3.0-liter engine. Then the assembly resource requirements for A and B are the same for everything except engines. Thus if the factory has a balanced capacity to produce A's, it is unbalanced for B's, and vice versa. In general, a factory with a balanced capacity for a certain product mix becomes unbalanced as soon as the mix changes.

17.5.3. The Cost of Capacity

A second argument *against* balanced capacities is the difference in acquisition costs between types of capacity. Every scheduling problem can be solved by "throwing hardware at it." With enough overcapacity, a resource will sustain its workload no matter how poorly scheduled it is. On the other hand, the same problem may be solved without any equipment purchase, by better scheduling of the already available capacity.

Within a factory, the costs of acquiring new capacity may vary by orders of magnitude between resources. In semiconductor wafer processing, $20,000 buys a cleaning sink with a capacity of 100 wafers/hour, while it takes $1,000,000 to get a lithography stepper that will do only 50 wafers/hour.

An apparent shortage of cleaning sinks may be solved by buying three more, but management will take a hard look at how available steppers are used before agreeing to buy a new one. If steppers become a bottleneck, it is not by accident but because they are expensive.

17.5.4. Flow Balance and Time

What does it mean to "balance flow" in a factory? On the one hand, some form of global flow balance is always satisfied, meaning that over a long period of time, flow is always balanced. Everything that goes *into* the factory eventually comes *out*, and there is an upper bound to how much can be inside. This can easily be expressed quantitatively.

Let

- $I(t)$ be the amount of inventory for a part at time t.
- $F_{in}(t)$ the rate at which units of this part flow into the factory.
- $F_{out}(t)$ the rate at which they leave in any form, including scrap.

Then

$$\frac{dI}{dt} = F_{in}(t) - F_{out}(t)$$

Let the length T of a time interval be long enough so that the volume

$$V_{out}(T) = \int_0^T F_{out}(t)dt$$

for this part out of the factory in the interval is large with respect to the maximum amount I_{max} of inventory that could possibly fit within the factory. If $V_{in}(T)$ is similarly defined, then

$$\frac{V_{out}(T) - V_{in}(T)}{T} = \frac{I(T) - I(0)}{T} \leq \frac{I_{max}}{T} \underset{T \to \infty}{\longrightarrow} 0$$

This is not what Goldratt has in mind when he says "balance flow" At the opposite extreme, is *instantaneous* flow balance, which could be expressed as

$$\forall t, \frac{dI}{dt} = 0$$

Arrivals and departures are synchronized so that the number of parts inside the factory never changes. Instantaneous flow balance is approximated more or less closely by assembly lines and the Kanban system, but that is not what OPT does either. Instantaneous flow balance cannot be maintained without knowing exactly when a part leaves and another one can be let in.

OPT schedules production from orders only. If, for a while, the requirements for a part drop to 0, the inventory may do so as well, which requires that dI/dt < 0 during the transition. Balancing flow, in the OPT context, means starting and processing no more than is required for the pending orders; it does not mean matching in- and outflow rates.

17.5.5. Process Batches and Transfer Batches

OPT makes use of these concepts, absent from MRP, to enable small quantities of parts to move through the floor while simultaneously minimizing the number of setups on bottlenecks. There may be logistical difficulties if an expensive transport system is involved, such as a fleet of forklifts. Other difficulties with small transfer batches arise if maintenance of lot integrity is an engineering requirement.

17.5.6. Considering All Constraints Simultaneously

The OPT literature on this subject is mostly negative. The conjunction of local optima is not a global optimum. Maximizing efficiencies or using "economic batch quantities" is pursuing local optima and does not necessarily contribute to the prosperity of the factory. But how do you get a global optimum? The OPT answer is: by using a secret formula, the algorithm for bottleneck scheduling that Goldratt refuses to publish.

17.6. THE SECRET ALGORITHM

The official line of Creative Output is that, in 1972, Goldratt discovered, but did not publish, the mathematical result on linear programming published in the Soviet Union by Khachian in 1979, and then refined it into a scheduling algorithm. He decided to forego the academic recognition he would have received from publication and build a business instead.

While broadcasting this unverifiable story, Creative Output also says that this algorithm is not that important after all and that what really matters is that users should understand the OPT philosophy.

Aside from generating what might be undeserved skepticism, the refusal to publish the algorithm has practical consequences. The economic function the algorithm maximizes is part of the secret; users don't really know in what sense the schedules are optimal and can't anticipate what they are going to be even in simple cases.

If the algorithm is everything that it is claimed to be, Creative Output stands to gain more from added credibility than it could lose from helping competitors.

Appendix A TIME LOST DUE TO BREAKS

A machine on the shop floor has just been identified as a bottleneck. Until now, management was not aware of any need to give this machine a special treatment; it hasn't been used to full capacity even though even its full capacity is below the market demand.

One of the causes of idle time on this machine is a clause is the union contract that guarantees operators a 30-minute lunch break and a 15-minute coffee break each shift. It also appears that some time is lost at each shift change.

By staggering breaks, providing relief operators and improving the tie-in between shifts, it is technically possible to eliminate this cause of time loss. However, it involves asking the union to agree to an exception in the contract, and therefore drawing on the short and precious capital of good will between management and the union. Is it worth it? To find out, it is worth spending two hours of a manager's time quantifying the problem.

The machine takes an average of 45 minutes to process a load. Of these 45 minutes, 21 are spent in activities requiring human attention: setting up and loading before processing, and unloading afterwards. The rest of the time, the machine can operate unattended.

Break start times strictly follow the union contract. If a break is scheduled at such a time that it would interrupt an unload-setup-load sequence, then that sequence is postponed until after the break. Variations in the processing time of a load make it impossible to predict the position of a break start time with respect to the processing cycle.

To formalize this is a general fashion, let

- 0 be a time at which unloading starts.
- M be the duration of the unload-setup-load sequence.
- D be the total length of the processing cycle.
- t be the scheduled start time of a break.
- B be the duration of the break.

The relation between the machine idle time caused by a break and the break start time is shown in Figure 17.3. There are three possible cases:

- If $t < M$, unloading is postponed until the end of the break, so that the machine time lost is $t+B$.

- If $M \leq t \leq D-B$, then the break is entirely contained within the period of unattended machine operation and no time is lost at all.

- If $D-B < t \leq D$, then the break starts within the period of unattended operation, but ends after unloading should have started. In this case, the time lost is $t+B-D$.

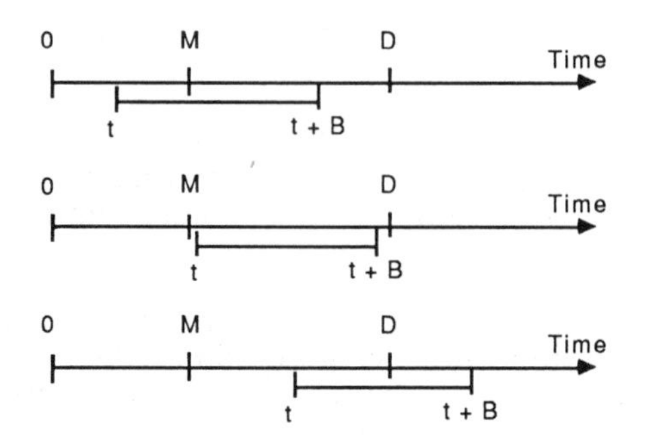

Figure 17.3. Time Loss Due to a Break

These results are summarized in the following table:

	Break Start Time	Machine Time Lost
Short Breaks	$0 < t < M$	$t + B$
$(B < D - M)$	$M \leq t \leq D - B$	0
	$D - B < t \leq D$	$t + B - D$
Long Breaks	$0 < t < M$	$t + B$
$(B \geq D - M)$	$M \leq t \leq D$	$t + B - D$

Let us assume that the small variations in processing time cause the scheduled break to be equally likely to start at any point between 0 and D. Then the expected machine time loss L for short breaks is given by

$$L = \frac{1}{D} \times \left[\int_0^M (t+B)dt + \int_{D-B}^D (t+B-D)dt \right] = \frac{M}{D} \times \left(B + \frac{M}{2} \right) + \frac{B^2}{2D}$$

and for long breaks by

$$L = \frac{1}{D} \times \left[\int_0^M (t+B)dt + \int_M^D (t+B-D)dt \right] = \frac{M}{D} \times \left(B + \frac{M}{2} \right) + \frac{D-M}{D} \times \left(B - \frac{D-M}{2} \right)$$

Applying these formulae with M = 21 mn and D = 45 mn, we estimate the time lost due to breaks every day:

	Break Length(mn)	Average Loss (mn)	Frequency (per Day)	Total Loss (mn/day)
Shift Change	0	9.8	3	29.4
Lunch	30	28.5	3	85.5
Coffee	15	14.4	3	43.2
TOTAL				158.1 = 2.6 hours

The capacity increase at stake in the coverage of breaks on this machine is 2.6/(24-2.6) = 12%. In this example, it can be obtained by taking shorter breaks more often and synchronizing these breaks with the period of automatic operation of the machine.

18

THE OPT PRODUCTION SCHEDULING SOFTWARE

OPT constructs a flow network from routings, bills of materials, orders, work in process, and resource descriptions. It then splits this network in two using the bottlenecks. A method similar to MRP then derives schedules for all other activities, using eight "scheduled delays" in place of the MRP standard throughput times.

Implementers of OPT are handicapped by its nonstandard terminology. Designed for portability, OPT has limited data management capabilities and is usable mostly with a preexisting database system. Its small track record and the inherent limitations of off-line systems do not support the inventor's claim that OPT is more advanced than Just-In-Time.

18.1. FUNCTIONS OF THE OPT SOFTWARE PRODUCT

18.1.1. External Context

Figure 18.1 is a logical external context diagram for OPT; it shows the final destination of the outputs and the origin of the inputs. How the OPT software actually interacts with its environment is another matter, discussed later in this chapter. Figure 18.1 shows Managerial-Parameters being decided by manufacturing or production control managers; this does not mean that those people enter them on an OPT input screen.

At this level, the differences beween OPT and MRP are not striking. Indeed, judging from their names, the outputs of OPT could be those of an MRP system. The differences lie in the content of the factory model and the algorithms used to produce the outputs. On the other hand, even at the level of Figure 18.1, there are significant differences between OPT and Just-In-Time in the way both systems communicate with the shop floor.

18.1.2. The OPT Factory Model

The relations between the factory model, orders, and inventory are shown in Figure 18.2. The term "work in process" is reserved for material that is part of the way through a routing. "Balance on Hand" is used to designate off-route material — that is, raw materials and components that have completed a routing.

To each entry in the bill of materials is associated an activity which can be either assembly or branching. A branching activity turns α units of X into β units of *one* of the $X_1,\ldots, X_n$ but not into β_1 units of $X_1,\ldots, \beta_n$ units of X_n simultaneously. Thus OPT does not model disassembly or binning.

Routings are sequences of activities involving neither convergence nor divergence. Material drawn from the balance on hand starts on a routing, is work in process at an activity along the routing until it completes it, when it is applied to an order if it is a finished good or otherwise returns to the balance on hand.

Activities do not exist independently of routings and bills of materials. The same activity cannot be part of several routings. The only commonality that can be expressed in the OPT data model is that of resources. Thus the OPT data model can represent two activities using the same machines, but not two routings containing the same activity. One activity definition can always be copied from one routing to another, but updates would need to be done explicitly on all copies.

Machine setup groups are used to model the general, worst case, in which the setup time depends both on the prior and the current activity. Setting up a machine is bringing it to a certain state, and, in general, this depends on the state it starts from.

Example: A furnace performs activity A at 800°C, B at 1000°C, and C at 1200°C. The setup time of the furnace includes a temperature ramp-up or cool-down and clearly depends on *both* the "from" and the "to" activities.

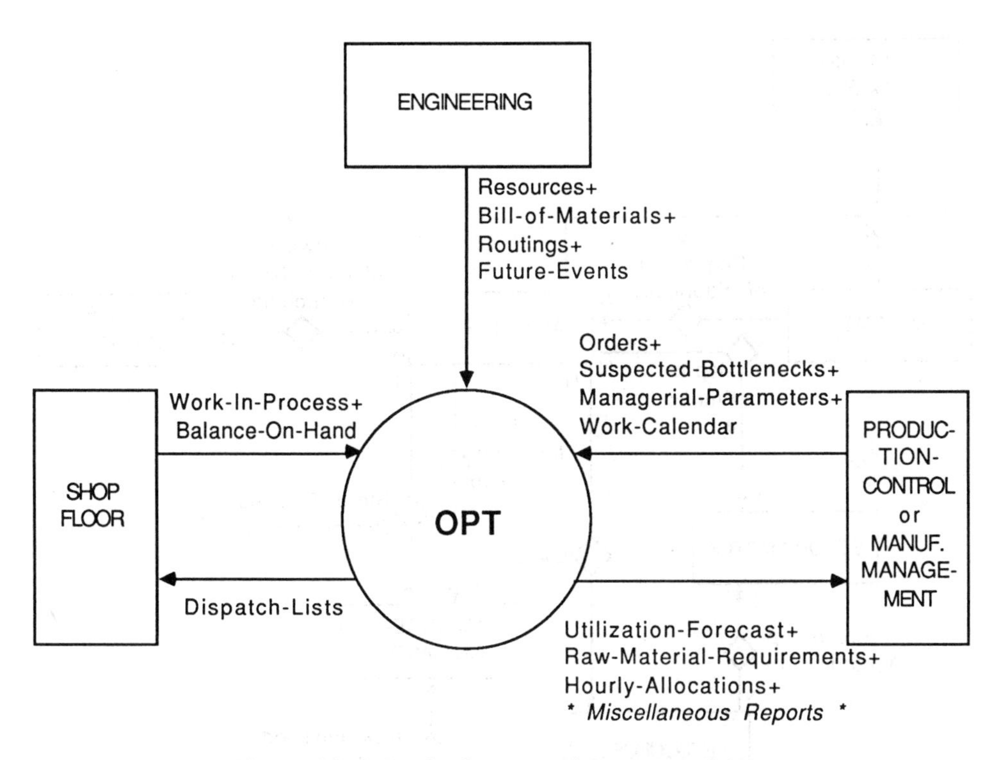

Figure 18.1. Logical External Context of OPT

The setup times of the furnace might be as in the following setup matrix:

To→		A	B	C	(Minutes)
From→	A		40	100	
	B	70		50	
	C	160	80		

Defining a separate setup matrix for every resource is usually both practically impossible and unnecessary. Resources are put in machine setup groups sharing the same matrix.

An interesting side of OPT is the wealth of its resource model, shown in Figure 18.3. Resources are groups of machines of the same make and model. To each activity is associated a main resource, with a time per part and a setup time, and, optionally, a substitute resource to be used when the main resource is unavailable. An efficiency ratio is specified to represent the differences in speed between individual machines in both resources.

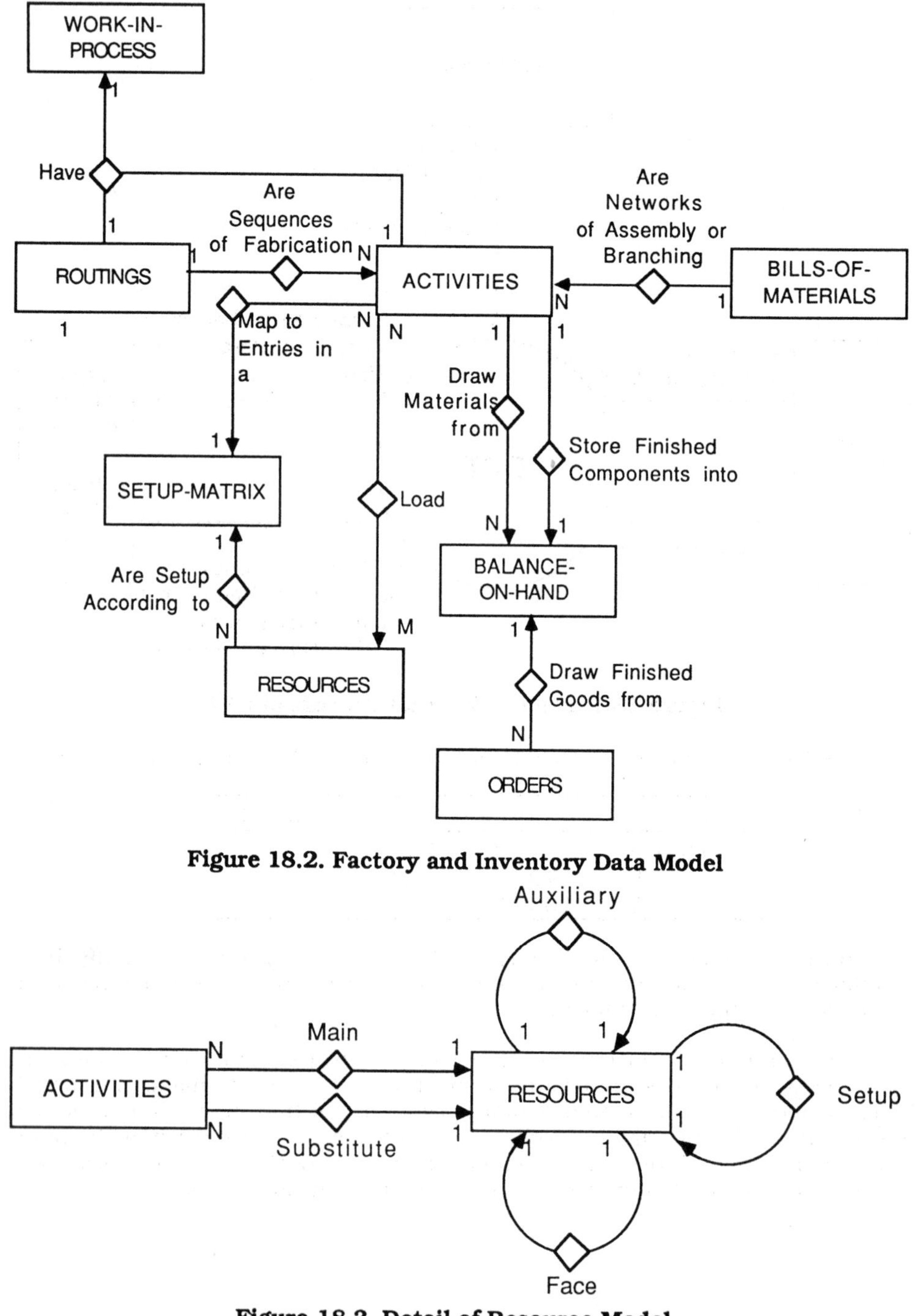

Figure 18.2. Factory and Inventory Data Model

Figure 18.3. Detail of Resource Model

If a machine requires an operator or a fixture, this is modeled by assigning it an auxiliary resource to be loaded at an appropriate level whenever the main resource is. If specialists are required to set up a machine, they are identified as setup resources. Setup resources are tied up for the duration of the setup time but not the processing time.

Finally, if an activity requires n units of a resource simultaneously, this set of n units is called a *face*, and defined by drawing n units from a resource called a *pool*.

18.1.3. The Essential Modules of OPT

The physical data flow structure of the OPT product is much more complicated than that shown in Figure 18.4. This is due in part to the way the software has been designed, but also to the omission from Figure 18.4 of modules called analyze, probe or refine. Those modules, though important, are not essential and would make Figure 18.4 difficult to understand.

1. The components of the factory model, the inventory status, and the orders are combined by BUILD NETWORK into a single Engineering-Network, such as illustrated in Figure 18.5. The Engineering-Network is a product network with resource assignments.

2. SPLIT first removes from the network those activities that have no load imposed on them by the Orders, and then uses the list of Suspected-Bottlenecks to turn the network into two networks, as shown in Figures 18.6 to 18.8.

   The OPT network contains the bottlenecks, the orders, and intermediate nodes called mules, hybrids, and raw material dummies representing blocks of nodes in the original network. The SERVE network contains all nodes, but the bottlenecks are marked as such.

3. "OPT" designates both the entire system and the module implementing Goldratt's secret algorithm. It is a finite forward scheduler, which means that it schedules tasks from the beginning of the planning period on the OPT network while taking into consideration the finite capacity of the resources. We will return to the issue of finite forward scheduling later. The output of the OPT module is a schedule for the bottlenecks and for order completion.

   There are six OPT *managerial parameters* setting the relation between transfer batch sizes at two successive stations, limits to the time a machine may be activated, and a "priority adjustment factor" affecting the secret priority function in some way not revealed to the user.

4. SERVE derives schedules for all the nodes of the SERVE network from the OPT schedule, using a backward scheduling similar to MRP. The SERVE instructions determine the planning horizon and eight scheduled delays, a concept discussed in detail in the section on SERVE.

5. REPORTS formats the schedules produced by OPT and SERVE into the outputs handed out to factory personnel.

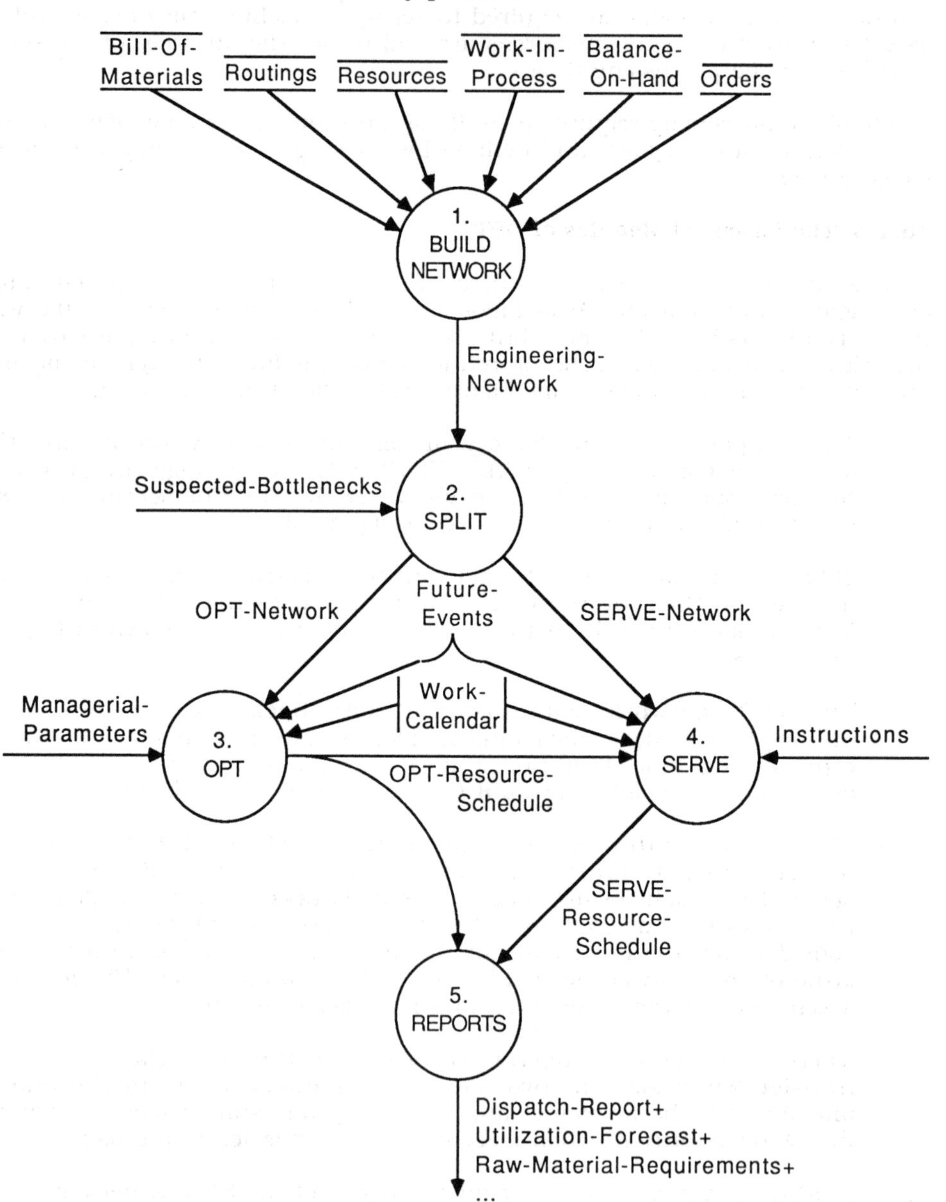

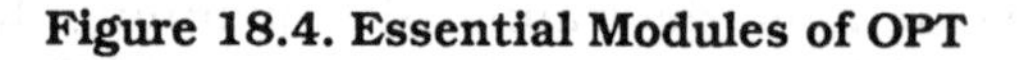

Figure 18.4. Essential Modules of OPT

Figure 18.5 shows the network graphic representation conventions of OPT, which are a variation on the ASME symbols. The material flows of OPT networks carry numbers called "arrow quantity" representing the average number of units of an input material flow needed to produce one unit of output. OPT models include an output buffer after each activity, which is omitted from the diagrams.

Figure 18.6 shows a sample output of BUILD NETWORK. A user-specified list of suspected bottlenecks is used by SPLIT to produce the networks shown in Figures 18.7 and 18.8. In Figure 18.7, the fabrication activities between raw material procurement stations and a bottleneck are lumped into *raw material dummies*, with parameters derived from those of the activities they summarize.

Chains of nonbottleneck fabrication operations between two bottlenecks are likewise aggregated into *mules*. Non-bottleneck assembly or branching activities remain as in the original network, and are called *hybrids*.

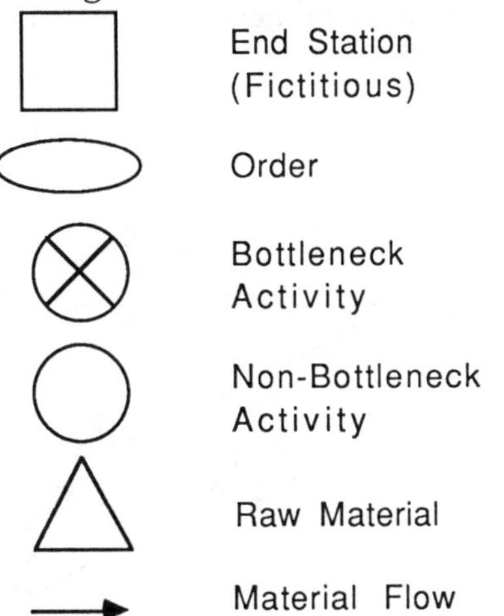

Figure 18.5. OPT Network Symbols

The SERVE network shown in Figure 18.8 contains all the nodes of the original network, but the bottlenecks and the orders are marked as stations with schedules produced by the OPT module, from which SERVE derives schedules for all other activities.

18.1.4. Backward Scheduling with SERVE

SERVE does backward scheduling in the sense that it deduces activity start times from desired completion times. The OPT module produces schedules for the blackened nodes of Figure 18.8. These schedules set desired completion times for the predecessor activities of the OPT nodes, and SERVE deduces from those times schedules for all other activities.

If an activity is to be completed at time t, SERVE schedules it to start at t-Δt, where $\Delta t = np + S$ and

- n is the transfer batch size.
- p is the processing time per part.
- S is a scheduled delay set in a manner discussed next.

This method makes SERVE differ from MRP in two respects:

1. Processing time and scheduled delay are separated rather than lumped into "standard throughput times."
2. There are only eight scheduled delays for a whole plant, as opposed to thousands of standard throughput times.

Which of the eight scheduled delays applies to a particular activity is deduced from the activity's position within the network, purely from topological considerations. The network is first divided into regions called A, B, and C, and the assignment of an activity to a region depends on its position with respect to the bottlenecks.

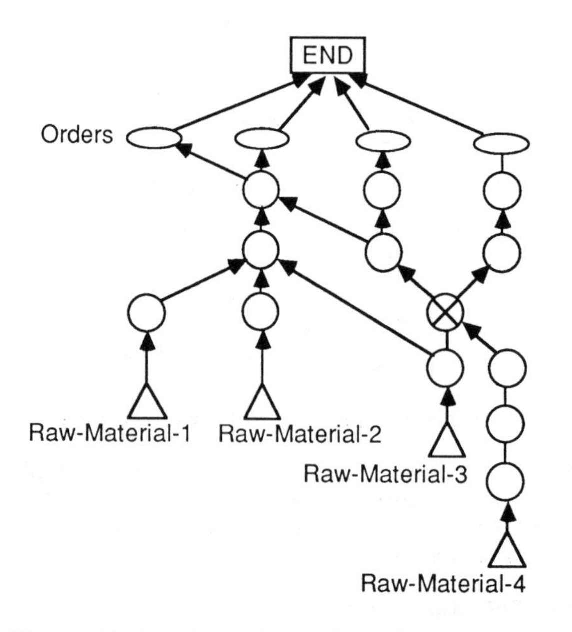

Figure 18.6. A Sample Engineering Network

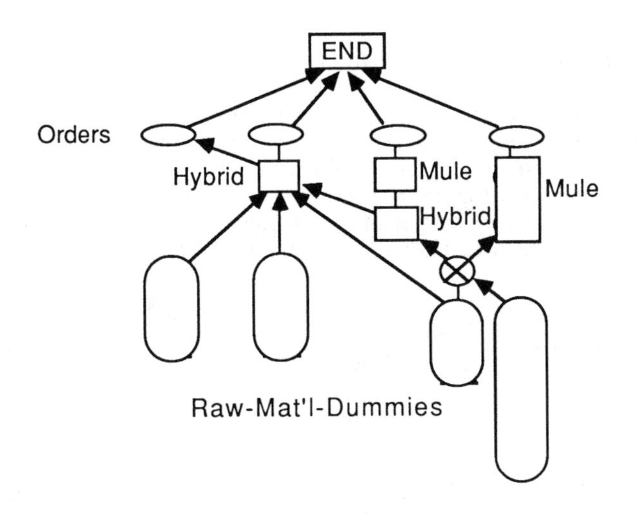

Figure 18.7. A Sample OPT Network

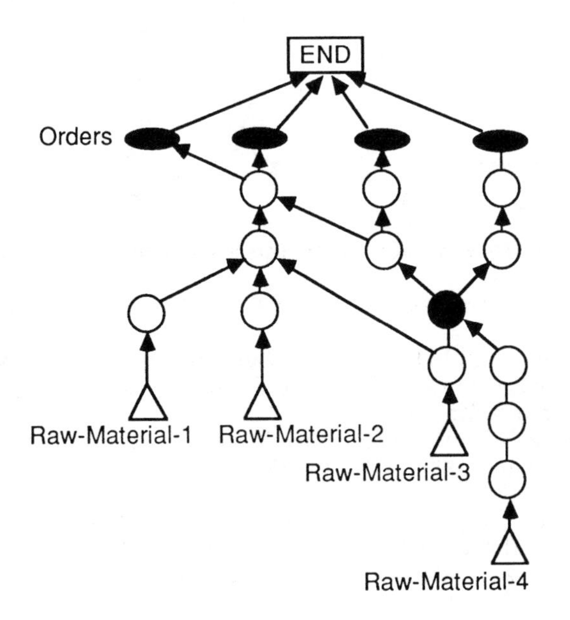

Figure 18.8. A Sample SERVE Network

The following discussion is not based on the OPT literature. We introduce the concepts of successor and predecessor subnetworks of a node. The successor subnetwork of a node contains its successors and their successors and so on up to the END node. The predecessor subnetwork likewise contains the predecessors and their predecessors, and so on all the way to raw material stations (see Figure 18.9).

The formal definitions take the following recursive form:

- The successor subnetwork of the END node is empty.
- The successor subnetwork of any other node contains both its successor nodes and all their successor subnetworks.
- The predecessor subnetwork of raw material stations is empty.
- The predecessor subnetwork of any other node contains both its predecessor nodes and their predecessor subnetworks.

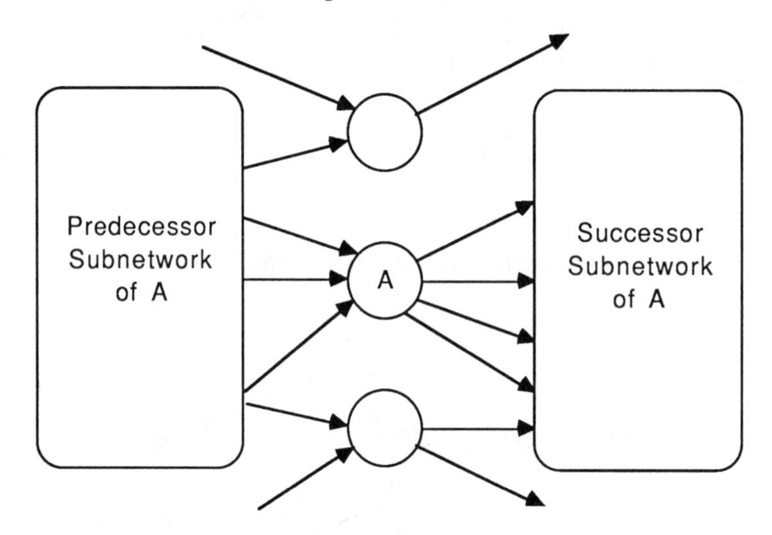

Figure 18.9. Predecessor and Successor Subnetworks

Using these two concepts, we divide the network into regions:

Region A: nodes with no bottleneck in their successor subnetwork.
Region B: nodes with bottlenecks both in their predecessor and their successor subnetworks.
Region C: nodes with bottlenecks in their successor subnetworks but none in their predecessor subnetwork.

Loosely speaking, region A contains the nodes between a last bottleneck and the orders, B the nodes between two bottlenecks, and C those between raw material stations and a first bottleneck. The region assignments can be expressed as a decision table:

Subnetwork	Contains Bottlenecks?			
Predecessor	Yes	Yes	No	No
Successor	Yes	No	Yes	No
Node Region	B	A	C	A

The position of each node within its region further determines its scheduled delay. An activity in region C, for example, may be a raw material station and be assigned the scheduled delay characteristic of outside purchases.

A region C node may feed a bottleneck or provide a region A or B node with parts to assemble with the output of a bottleneck. The scheduled delay it is assigned reflects the need to maintain a buffer in front of the bottleneck or the assembly station. A region C node may also feed only other region A or C nodes and have a scheduled delay reflecting the fact that it is not directly involved with bottlenecks.

After a similar analysis of region A and B nodes, the number of required scheduled delays reduces to eight. If the last four are omitted, then SERVE assigns them default values based on the first four.

18.2. GENERATING SCHEDULES WITH OPT

Figure 18.4 does not show the procedure to actually generate schedules. The initially available factory model is not a reliable source of bottleneck identification because of the frequent errors found in capacity data. As discussed in Chapter 7, depending on the type of factory, bottlenecks are found on the shop floor, by examining queue lengths, chronically short parts, or late orders.

18.2.1. Pure SERVE Run

The first step in using the OPT system is to run it with no bottlenecks defined. The OPT network is trivial and all the scheduling is done by SERVE. This "pure SERVE" run implements classical capacity requirements planning. Its Utilization Report should show as overloaded resources the bottlenecks discovered on the shop floor.

18.2.2. Focused Data Cleaning

Discrepancies between the lists of bottlenecks found by the software and on the shop floor must be resolved. The OPT philosophy is that it is vital to have accurate capacity data on bottlenecks but not on other resources. The factory's performance will be affected if a non-bottleneck is mistaken for a bottleneck or vice versa, but errors in the capacities of nonbottlenecks can be lived with.

18.2.3. Split-OPT-Serve-Reports

These modules are executed in sequence repeatedly until a satisfactory result is obtained. Unfeasible schedules result in "time zero" spikes in the SERVE schedules — that is, unrealistically heavy loads in the beginning of the planning period. From one run to another, the user modifies the OPT managerial parameters and SERVE instructions.

18.3. COMMENTS ON THE OPT PRODUCT DESIGN

18.3.1. The OPT Software — The Price of Portability

The designers of the OPT software have made portability a major goal. Figure 18.10 shows the physical external context of OPT — that is, where the software reads its inputs and produces its outputs. The factory model, inventory status, and orders are expected to come from another computer system. The outputs are printed reports to be handed out by the systems staff.

The end users of OPT interact with it not directly but through intermediaries. Systems engineers maintain the factory model, run the software, and distribute the reports. They are also frequently called upon to help in their interpretation.

The price of portability is the impossibility of making use of any feature of a given computer system which could not also be found on every other one the software might be run on. Sequential files pass this test; no database management system does. Therefore, the inputs to BUILD NETWORK are sequential files.

It is theoretically possible to write a set of BUILD NETWORK input files from scratch with a text editor, but the absence of any database management capability makes it practically impossible for any nontrivial example. On the other hand, any reasonable database system containing a factory model can be expected to provide these files as extracts.

BUILD NETWORK expects file names to follow certain conventions and each record within each file to contain certain fields, but it allows variations in formats. The column to field assignment structure of every file is communicated to BUILD NETWORK by means of a format file.

If the format file says that field X is between columns n_1 and n_2 of a file, then BUILD NETWORK will misunderstand any input that is off by one column. This type of brittleness is avoided in database systems by making the concept of a field independent from a physical location within a record. The absence of such facilities in OPT severely restricts the amount of editing that can be done on the inputs and makes it necessary to limit update access to these files to the trained systems staff. When there is no prior database system in place to extract the BUILD NETWORK inputs from, the most realistic approach may be to build one.

On the output side, every computer above a minimal size can drive a 132 column line printer. Therefore, the outputs of OPT are either printouts or printable sequential files.

All data transfers between modules are implemented by sequential files: each module writes output files which are read by others. These files must be explicitly named by users and are accessible to them. No control mechanism is provided to prevent a user from making changes in the engineering network with a text editor, only to see these changes reversed the next time BUILD NETWORK is run.

Although the use of primitive methods can be explained by the design goal of portability, the system does place requirements on users that have no obvious relation to its functions. For example, why is it necessary for the BUILD NETWORK input files to be sorted? Why must users run the OPT module on an empty network before they can do a pure SERVE run?

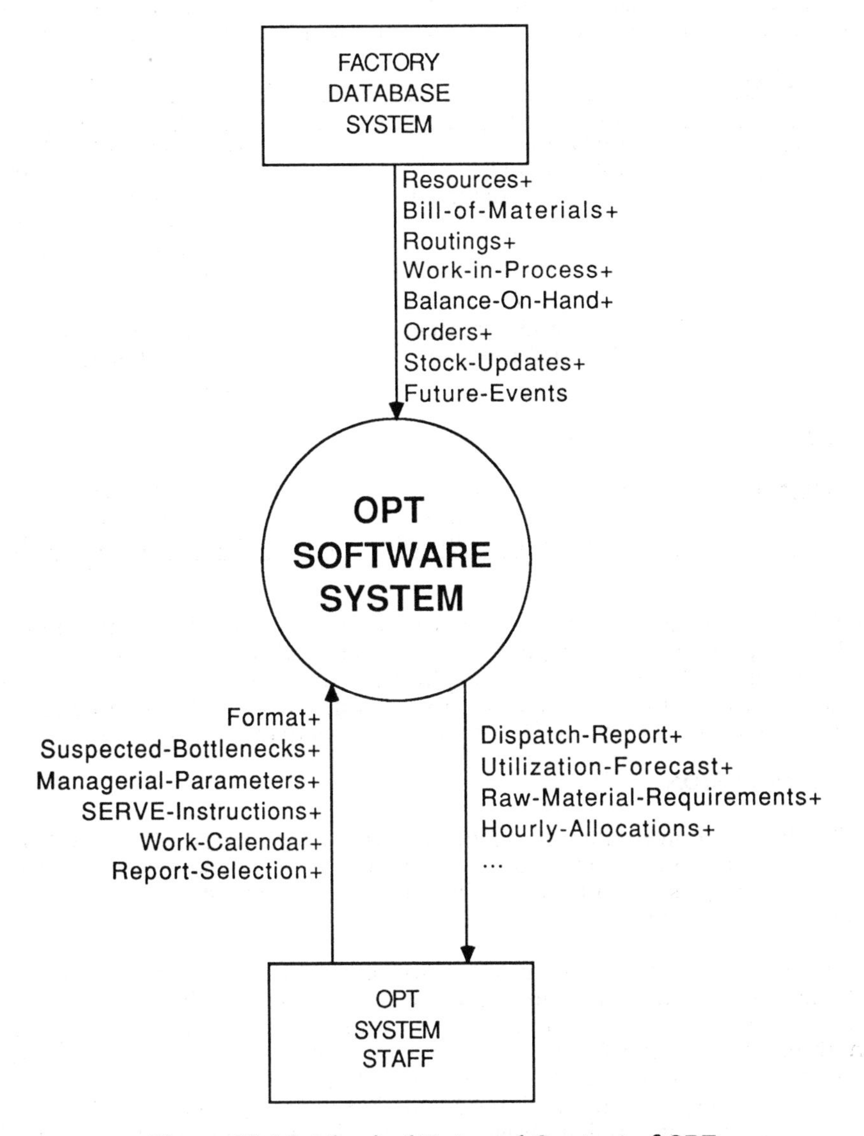

Figure 18.10. Physical External Context of OPT

18.3.2. OPT's Nonstandard Terminology

A reader interested in learning more about OPT from the Creative Output literature has to learn a unique terminology, which we have shunned in this chapter. It is not clear what is to be gained by forcing users to learn new words for old concepts or new meanings for old words. When addressing a manufacturing manager, the main effect of using familiar words in unusual ways is to be misunderstood. The following table is a short OPT glossary.

OPT Glossary

	OPT Meaning	Standard Meaning
Bottleneck	Resource with capacity ≤ market demand	1. Lowest capacity resource (Queueing Networks) 2. Resource exceeding a utilization threshold (MRP)
Master Schedule	Schedule produced by the OPT module	Finished Goods shipment schedule (APICS, MRP)
Thoughtware	Theory	(none)
Throughput	Rate at which a factory generates income through sales	Material flow rate out of a resource (Queueing Networks)

OPT's use of the word "bottleneck" causes a peculiar implementation problem: American workers perceive being labeled a bottleneck as a prelude to dismissal, and no amount of explaining can correct this impression. In practice, you have to use a more positive word. Everyone likes to be a "critical resource" as much as no one wants to be a bottleneck.

18.4. CONCLUSION

"The Japanese are killing a mosquito with a machine-gun," says Goldratt to describe setup reductions carried out on every resource regardless of utilization. He goes on to describe OPT as a more advanced approach because it focuses surgically on the bottlenecks.

This comment is based on two misconceptions. The main one is that implementers of Just-In-Time somehow would attack all problems at once without regard to urgency. Just-In-Time implementers known to this author do nothing of the kind [Abe86]: they carry out a business analysis of the factory along lines very similar to OPT, from which they decide whether the initial focus should be on machines (and which ones), materials, people, or the control system. The English literature on Just-In-Time describes completed implementations, not the implementation process.

The second misconception is that there is such a thing as *the* Japanese way to manufacture, when in fact there are many and Japanese plant managers are given conflicting advice by experts in total quality control, setup time reduction, or Just-In-Time.

As will become clear in the sequel, the domains of applicability of OPT and Just-In-Time overlap but are not identical. In particular, the Kanban system requires the maintenance of small stocks of work in process for every part at every stage. The wider the product mix, the more difficult it is to do. As the number of parts with no work in process maintained increases, the factory comes to resemble a pure job-shop, in that the scheduling precedence constraints are given by the routings and the bills of materials.

19

FINITE FORWARD SCHEDULING AND DISCRETE EVENT SIMULATION

"Finite Forward Scheduling" is the generation of schedules within given capacity constraints and forwards from a start time, as opposed to backwards from desired completion times. Production scheduling differs from general job-shop scheduling by the nature of factory resources, the opportunity to change tasks by adjusting batch sizes, and the presence of work in process.

In a "Nondelay Schedule," resources are allocated to tasks as soon as possible and never held open for forthcoming, urgent tasks. Timelines for resources and tasks are generated by maintaining a list of future events, the execution of which in turn causes other future events to be defined.

A finite forward scheduling engine can also be applied to discrete event simulation of factory activity, for example, to anticipate its average throughput time performance under given sequencing policies with random equipment failures.

19.1. FINITE FORWARD SCHEDULING

19.1.1. What is "Finite Forward Scheduling"?

The OPT module is described as a *finite forward scheduler*. *Finite* refers to its ability to respect given capacity constraints, and *forward* means that it deduces activity completion times from given start times and not the other way around.

The OPT module is secret. It is claimed to be an optimization algorithm, but the specific economic function it maximizes or minimizes is proprietary, and the users of OPT cannot precisely know what they are trying to accomplish when they run it.

Even though the suppliers of OPT are not helping us, we cannot ignore finite forward scheduling, because it is a fundamental approach and because a finite forward scheduling "engine" also is a discrete event simulation tool.

19.1.2. Job-Shop Scheduling versus Production Scheduling

Except in a few simple cases, it is not possible to find optimal schedules in one single forward pass in time. The contribution of each resource-to-task assignment decision to the performance of the schedule generally cannot be evaluated in isolation.

> Example: To schedule n tasks with due dates and precedence constraints on 1 machine so as to minimize the maximum lateness of any of the tasks, Lawler's algorithm starts out by finding out which task may be processed *last*, and thus proceeds backwards in time even though it is an optimization algorithm. A general optimization technique such as Branch and Bound makes tentative scheduling decisions which are later reversed if evidence is found that they could not possibly lead to an optimal schedule.

The general job-shop scheduling problem, as formulated in French [Fre82] is that of processing n jobs on m machines with technological constraints specifying the machine processing order for each job. In a factory, those constraints are given by the routings and the bill of materials, but scheduling production on a network such as that in Figure 18.6 differs from general job-shop scheduling in several respects:

- The resources are not necessarily single, indivisible machines, but work centers.
- The tasks to be performed can be defined in many ways by changing process and transfer batch sizes.
- The precedence constraints are relaxed by the presence of work in process inventory.

The scheduling precedence constraints are defined by the routings and the bill of materials only if all the raw materials are kitted and pegged to an order from the very start. Semiconductor wafer fabrication is an example of this, where silicon wafers are the only raw material, are lumped into lots, and are assigned lot due dates.

At the opposite extreme, if orders are satisfied by drawing from a finished goods inventory, then the task of replenishing the stock can start with the last operation as long as there is enough work in process inventory in front of it.

19.1.3. Nondelay Schedules

The method we are going to use produces nondelay schedules — that is, schedules which never leave idle a resource on which some task can be done. These schedules are not optimal: there are times when better overall performance is obtained by letting a specific resource wait a short time for an urgent job rather than do something else right away. The choice of nondelay schedules can be justified for resources that are bottlenecks in the OPT sense, because their time is so valuable that we want to spare them any idleness.

Our method will simulate the operation of the factory over the planning horizon, given a rule to choose what to load on a resource whenever there is an opportunity to do it. Some tasks, such as are modeled as mules, hybrids, and raw material dummies in OPT, are assumed to require no resource. They are modeled as simple delays with standard throughput times.

19.2. TASK AND RESOURCE MODEL

We shall consider two types of tasks:

- *Setups* — the number of which is an output of the schedule.
- *Processing a Transfer Batch*— a schedule is complete as soon as it includes all these tasks.

In this approach, the size of *process* batches is also an output, determined by the number of consecutive transfer batches processed on a resource between setups. Setup tasks require resources but consume no parts. Processing tasks on mules, hybrids, and raw material dummies consume parts but require no resources. Finally, processing tasks on bottlenecks require both parts and resources.

We want to build activity and resource timelines, as in Figure 19.1. With this in mind, we are calling "task" anything that eventually becomes a block in such a diagram, realizing that this is quite a heterogeneous concept, and that a data structure to model it has to accommodate variants (see Chapter 7, Appendix A).

The first step is to determine all the tasks to be performed in order to produce the finished goods on order and leave the work in process inventory at desired levels. We shall follow the OPT convention of considering the work in process at a node to be in its *output* buffer.

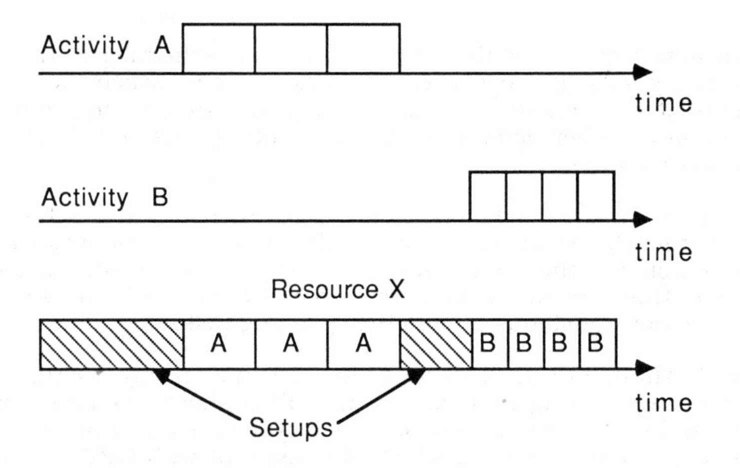

Figure 19.1. Types of Tasks

The net requirements at each activity are obtained by a backward pass through the network, starting with the END node and working towards the raw material stations. The gross outputs R_i required of a node are obtained by multiplying the net requirements out of its successors by the arrow quantities. The net requirements N_i for parts of type i out of a node are given by

$$N_i = R_i + W_i - C_i$$

where W_i and C_i are the desired and current amount of work in process for part i after the node.

We shall assume transfer batch sizes to be given and call them b_i. The number of transfer batches for part i needed out of the node is

$$n_i = \lfloor N_i/b_i \rfloor.$$

To each of these tasks, we can assign a processing time using the time per part of the node, but we would be hard pressed to give them a due date.

If the output part, for example, is a screw going into 20 different products, we have no way to decide which order it will eventually go into. We can not deduce a due date for the screws from the order due dates, because we would have to assume standard cycle times as in MRP. Our scheduling algorithm therefore cannot be based on due dates.

As we go forward in simulated time from the start time, all tasks will change state, from "pending" to "in-process" and finally "completed," as shown in Figure 19.2. However, to actually schedule them, we must break the "pending" state further down into

- "Waiting" for input parts to be available.
- "Queued" in front of resources once all the materials are available.
- "Ready" when all that is needed to get them started is the decision to allocate parts and resources.

We can now begin to see the outline of how the scheduler works. Whenever there are ready jobs, it must select one and start it, which means updating the inventory and resource states, carrying out the appropriate state transitions on all other pending tasks, and entering its completion time in a queue of future events.

When there are no ready tasks, the scheduler withdraws the next task completion from the event queue, moves the simulated time forward to match the task completion time, and ends the task. As in the previous case, this entails updating inventory and resource states, and carrying out the appropriate state transitions on all other pending tasks.

Before fleshing out this description, let us specify further the data structure we are calling a "task." Figure 19.3 shows its relationship with Resources and Part-Types, meant here as a type of work in process. One task produces only one type of part, which is consistent with OPT in not modeling disassembly. On the other hand, a task may require more than one input part and many resources.

All the resources needed perform a task are not treated the same way. We want to model all the resource relationships shown in Figure 19.3. Using De Marco's data dictionary conventions, we can refine the structure as follows:

Task = [Setup-Task
Non-Bottleneck-Production-Task
Bottleneck-Production-Task]

Setup-Task = Product + Operation + Duration + State + Resource-Configuration

Non-Bottleneck-Production-Task = Product + Operation + Duration + State + Output-Part + {Input-Part}

Bottleneck-Production-Task = Product + Operation + Duration + State + Output-Part + {Input-Part} + Main-Resource-Configuration + (Substitute-Resource-Configuration)

Input-Part = Product + Operation + Quantity

Output-Part = Product + Operation + Quantity

Main-Resource-Configuration = Resource-Configuration

Substitute-Resource-Configuration = Resource-Configuration

Resource-Configuration = {Machine + (Resource-Configuration)}

Machine = *Indivisible Resource*

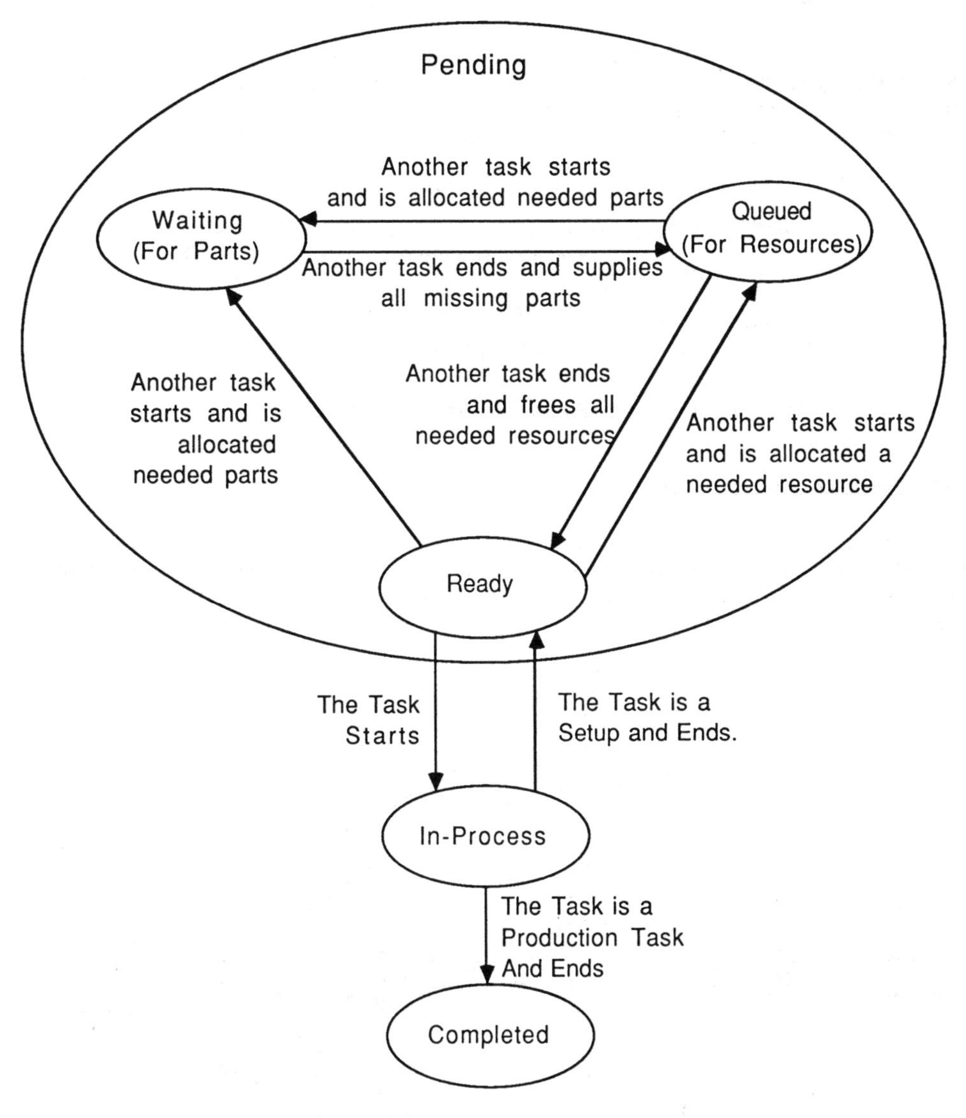

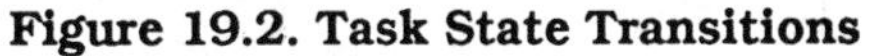

Figure 19.2. Task State Transitions

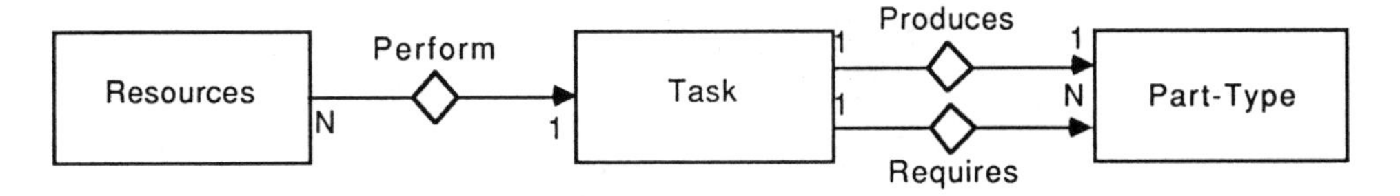

Figure 19.3. Relations between Resources, Tasks, and Parts

A Resource-Configuration is a tree of resources needed to perform a task. The "main machine," at the root of the tree, is the only one subject to setups. Only two resource states are considered, as shown in Figure 19.4. With the preceding task definition, there is no need for a separate "setup" state.

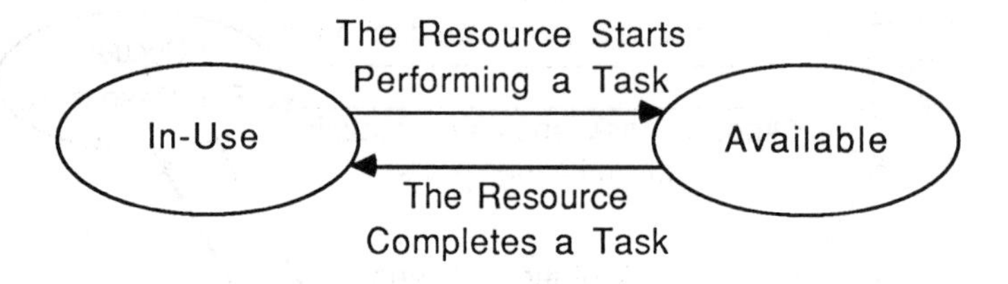

Figure 19.4. Resource State Transitions

19.3. Producing a Schedule

Figure 19.5 shows the three modules making up the actual scheduler, and their data flows. The output is the schedule data store, which is envisaged here as a journal of task and resource assignments. It contains the data needed to generate task and resource timelines.

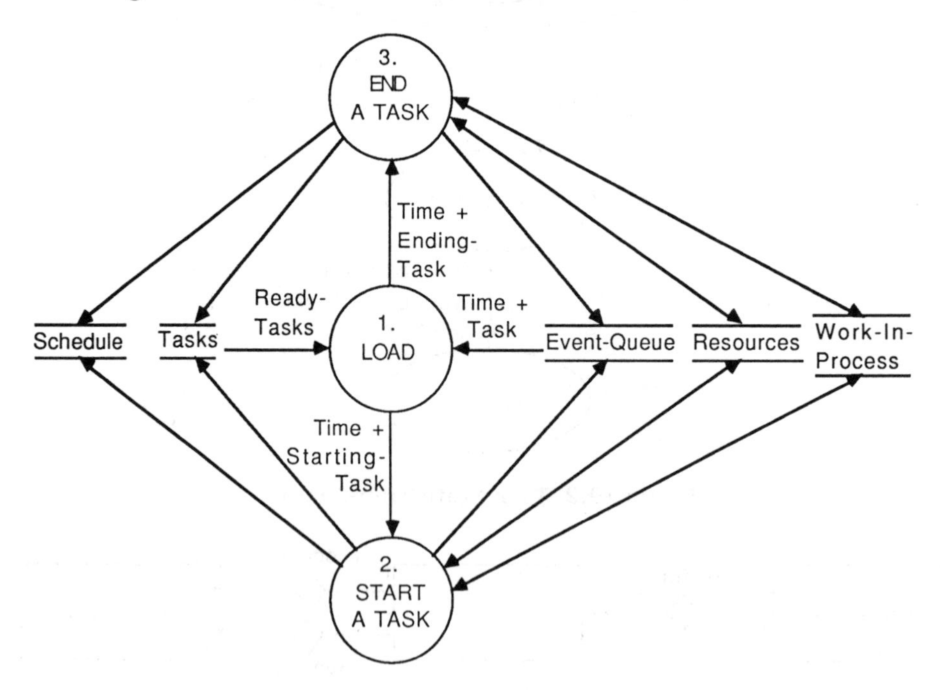

Figure 19.5. Finite Forward Scheduling Data Flows

We are assuming all resources to be free at the beginning of the planning period, and therefore all queued tasks are also ready. Thus the initialization work consists of setting all tasks in the waiting state and moving directly to the ready state those for which the work in process status indicates sufficient supplies of input parts. The operation of the bubbles of Figure 19.5 is described next.

1. LOAD
   Until all production tasks are completed, repeat the following:
   1.1. If there are Ready tasks,
        Select a Ready task (see discussion below),
        Start it.
   1.2. Otherwise repeat the following until there is a ready task
        or all production tasks are completed:
        Find the next task to complete in the event queue,
        Set the simulated time to the task completion time,
        End the task.

2. START A TASK
   2.1. Move the task from Ready to In-Process.
   2.2. Enter the task and completion time into the Event-Queue.
   2.3 Set all required resources to In-Use.
   2.4. Allocate Input-Parts to the task (i.e., decrement stocks).
   2.5. Set all Ready tasks that no longer have the requisite Input-Parts in sufficient quantities to Waiting.
   2.6. Move all Ready tasks requiring any of the resources just allocated to Queued.
   2.7. Append the Task Start to the Schedule.

3. END A TASK
   3.1. If it is a setup task, move it back to Ready,
   3.2. Otherwise move it to Completed.
   3.3. Delete the task completion from the Event-Queue.
   3.4. Free all the resources allocated to the Task.
   3.5. Increment the stock of Output-Parts.
   3.6. Move all Waiting Tasks that now have all their Input-Parts in sufficient quantities to Queued.
   3.7. Move all Queued Tasks with all required resources available to Ready.
   3.8. Append the Task completion to the Schedule.

When there is more than one ready task, how do we select one? The information we have about tasks in the data structures just described restricts the choice. The resources for which we have to make such choices are all in short supply, or they wouldn't be treated as bottlenecks. This is a justification for giving priority to tasks that don't require a setup to be performed over those that do.

When there is no choice but to perform a setup, we can choose the task with the shortest setup time, or the shortest duration or a combination of both. If there is a way to assign due dates, then the "earliest due date" criterion may also be used.

19.4. DISCRETE EVENT SIMULATION

There is no obvious economic function that is optimized by the scheduler we just described. The best that can be said for it is that it does not leave a bottleneck idle when it can find any way to keep it busy.

This algorithm works like a discrete event simulator, in that it goes forward in time and duplicates the effect of the events it schedules on a model of the factory. It differs from usual applications of discrete event simulation in that it assumes deterministic processing times and no equipment failures.

With the goal of producing activity and resource timelines for the bottlenecks of a factory over the next several weeks, there would be no sense in introducing random failures. A failure of machine X next Wednesday at 10:46 am produced by a simulation would have no reason to match reality.

The failure process of X is random, and, even if it is well modeled, the simulation matches reality on the average but not on an event-by-event basis: the real X may fail as often as the simulated one, but it will almost certainly not fail Wednesday at 10:46 a.m.

If a CIM system is in place, then its material tracking component knows the factory state at all times. In this case, the real-time application of the same task selection rules as the preceding algorithm results in the execution of the same schedule, but without anticipation of its final performance. This disadvantage is compensated for by the ability to respond to equipment failures.

20

MATERIAL FLOWS IN JUST-IN-TIME

The Just-In-Time system, developed at Toyota, is gradually spreading outside of Japan. Its implementation requires extensive industrial engineering work, starting with a search for the repetitive components of the manufacturing activity. The shop floor is reorganized into flow lines built from cells of different machines laid out in a loop, staffed by operators trained to run all of them, and with autonomous scheduling.

Setup times are reduced in several steps. First, equipment is dedicated by product families. Second, setup procedures are documented and standardized. Third, setup tasks are classified as "internal" if they require the equipment to be stopped and "external" if they can be done with the machine running. Finally, the equipment is modified to make internal tasks external.

Equipment availability is increased by training, by preventive maintenance and troubleshooting manuals, by involvement of production operators in periodic checks, and by small-group activities. The appearance of the shop floor is improved by identifying and discarding unneeded objects, assigning and labeling locations for tools and inventory, and enforcing stringent housekeeping standards.

In-line quality control is based on (1) integration of some form of 100% inspection into regular production work, (2) interruption of flow as soon as an anomaly is detected, and (3) fast problem reporting by such means as warning lights. Small changes called "Pokayoke" are made to the processes to make human errors difficult or impossible.

Statistical methods are confined to process capability studies. Human resources, as opposed to hardware, are viewed as the dominant factor of productivity and are managed accordingly, with workers being given stable employment, planned career paths, and intensive training. The production scheduling system rests on this foundation.

20.1. JUST-IN-TIME AS A HOLISTIC APPROACH

Just-In-Time (JIT) is a production system developed at Toyota in Japan, which has given a competitive edge to its practitioners in the automobile industry. Interest in this method outside of Japan has been rising in recent years, as evidenced by the publication of the books by Schonberger [Sch82] and Hall [Hal83], and the emergence of consulting companies offering Just-In-Time implementation help in Europe and in the U.S. The first international conference on Just-In-Time was held in April, 1986.

Although the goals of JIT can be expressed in the same terms as those of the systems described in the preceding chapters, the methods by which they are reached go far beyond resource allocation decisions. The purely production scheduling side of JIT cannot be understood without the industrial engineering foundations surveyed in this chapter.

20.1.1. Just-In-Time — What's in a Name?

What does "Just-In-Time" mean? The name has the intuitive appeal of work completed at the appointed time, no sooner and no later but it explains little. "Just-In-Time" is a goal, but its statement suggests no means to reach it. In fact, the common definition of Just-In-Time is almost identical to Wight's description of MRP as providing "the right parts, at the right time and in the right quantities." It can therefore not be expected to yield clues to the differences between the approaches.

In reality, as we will show soon, the techniques grouped under the label of Just-In-Time are consistent with the assembly line and mass production concepts developed in the American automobile industry in the early twentieth century. Just-In-Time can be viewed as the further development of a technology that has been neglected in the West for five decades.

"Just-In-Time" is a convenient label but it is losing meaning as its popularity rises. It is now being affixed for marketing reasons to products only remotely related to the ideas of those who coined the phrase. Like "zero defects," "just-in-time" is no more than a slogan. It could be interpreted to mean delivery on demand, or delivery on a promised date, no matter how remote. Toyota, the company that pioneered Just-In-Time, has a three-week lead time on customer orders placed with a dealer for a model that is not in stock. As a guaranteed performance, this lead time may be impressively short but it is not negligible.

"Zero Inventories" is another phrase, used by Hall to describe the same methods. It is also misleading if taken literally. In Little's "$L = \lambda W$," discussed in Chapter 14, the inventory L can only be zero with either the undesirable throughput $\lambda=0$ or the impossible throughput time $W=0$. What the "just-in-time," "zero inventory" approach truly aims for is to reduce L forever while maintaining λ. In Japan, it is also said of Toyota that "when they squeeze a dry rag, water comes out," which is not to be taken literally either.

The Kanban system is a means of controlling L and is the best known component of JIT. However, those who implement only the Kanban system soon find out that $L = \lambda W$ is maintained by reducing λ and not W. A backlog of orders promptly accumulates, leading to the conclusion that JIT does not work.

20.1.2. Seek Repetitiveness

Many believe that JIT is applicable only to repetitive, high volume production of a narrow product mix. This is a misreading of one of the basic tenets of the JIT approach, which is to seek out the repetitive aspects of a manufacturing activity. As the following example will show, JIT concepts can and have been adapted to such apparently nonrepetitive endeavors as shipbuilding or the construction of large newspaper printing presses [Wys86].

As K. Abe [Abe86] pointed out, ships have traditionally been built in horizontal layers from the keel up (see Figure 20.1). Technical reasons justified this method when ships were made of wood, and force of habit caused it to be carried over to metal construction, a technology which did not require it.

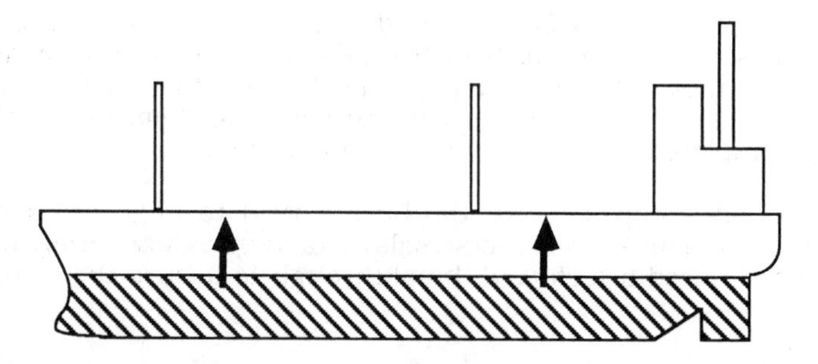

Figure 20.1. Bottom to Top Construction: The Traditional Method

Since few ships are built exactly alike, shipbuilding is not a repetitive process, but construction in horizontal layers compounded the problem because variability in the composition of adjacent layers caused an uneven flow of components to the shipyard.

On the other hand, two subsequent vertical slices of a ship are comprised of nearly identical quantities of metal sheets, pipes, wiring, and other parts. Therefore the flow of components to the shipyard can be smoothed by switching the sequence from a keel-to-bridge accumulation of layers to an assembly of slices as shown Figure 20.2.

We show construction proceeding from the center of the ship outwards because that is where the slices are most similar, but it could also bow-to-stern or stern-to-bow. Time is set aside to procure and assemble the more unique parts needed for the bow and the stern while assembly of the center is in progress. This is the key concept of MTJ's "Timetable Planning" system described in Chapter 21.

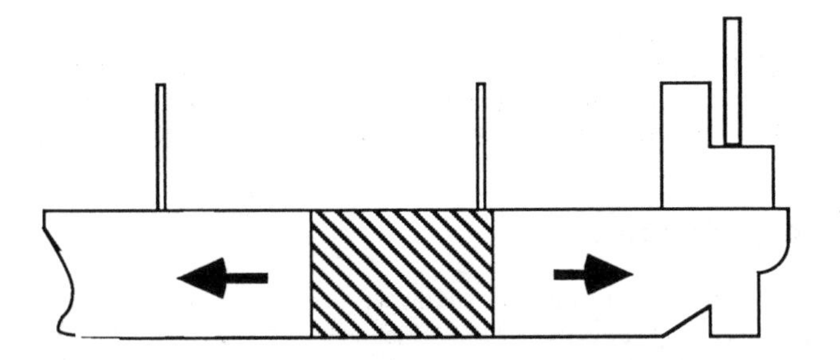

Figure 20.2. Construction in Vertical Slices: The Reverse Salami Tactic

This approach, now widespread in shipbuilding, makes it repetitive and susceptible to JIT. Wyss [Wys86] reports the application of this method to the not obviously similar case of newspaper printing presses at WIFAG in Switzerland. Like ships, these presses are large pieces of equipment made in small series and traditionally from the ground up, starting with a steel base, and adding the structure of rollers and feeders on top.

With Abe's help, WIFAG smoothed the flow of components to press assembly by switching to a front-to-back sequence, which Wyss called "reverse salami tactic" because it piles up the timelines of the various tasks like salami slices on a tray, as opposed to chaining them in such a way that one task is completed before the next one is started.

In the bottom-to-top method, the base served to align other components on top of it. Switching to the reverse salami tactic was not purely a scheduling decision, but required a *technical* breakthrough to assure the same alignment accuracy.

A better known method for enhancing repetitiveness is the type of schedule leveling, or order blending, described in the next chapter. A succession of small orders can be treated first-in-first-out without any transformation. However, if individual orders are large enough to tie up resources for weeks, then processing them FIFO would cause wide swings in the factory's activity and make it less repetitive.

The literature on Just-In-Time is focused on the applications to mixed model mechanical fabrication and assembly, for which the method was originally developed. This has two consequences. First, the emphasis on metal working and mixed-model assembly makes the discussions difficult to follow for readers involved in other types of manufacturing. Second, they may mislead the reader as to the breadth of applicability of JIT.

20.1.3. Manufacturing, Marketing, and JIT

Some economic activities are purely market-driven, in the sense that they consist of offering at every instant whatever goods or services a particular market wants. Other economic activities are production-driven, and consist of pushing available goods or services on markets that may or may not be eager to buy them.

MRP is an attempt to force factories to be market-driven when it is in fact against their nature, without supplying any technical means of changing that nature. JIT, on the other hand, is rooted in the production-driven reality of factories, and includes a set of tools to make them more responsive to markets.

Retail is market-driven. Retailers respond to every swing in consumer moods by stocking what moves. Their assets are primarily shelves on which to exhibit what they buy from producers, and there is no significant penalty to them in changing what they put on their shelves. Although manufacturers would wish the same to be true of factories, it clearly isn't. A factory represents millions of dollars invested in a particular set of products, which must be sold to sustain the factory's existence.

The stakes in flexibility are well illustrated by the rivalry between Ford and Chevrolet in the 1920s. In 1920, Ford's Model T accounted for 55% of the U.S. car market and was built in factories designed for it, with machines able to drill all the holes in its engine block in one pass, but unable to do anything else.

To compete with Ford, Alfred P. Sloan, head of then runner-up General Motors, chose to introduce yearly model changes, and thus "impose the laws of Paris dressmakers on the automobile industry." This could not be done by building factories on the Ford pattern.

The Chevrolet factories were designed by Ford alumnus William Knudsen to be changed over from one year's models to the next in three weeks. When obsolescence eventually forced Ford to switch from the Model T to the Model A in 1927, tooling changes shut down its factories for nine months and, by 1930, Ford's share of the car market had dropped to 30%.

This example does not show that it is necessarily a mistake to dedicate a factory to a single product. It is still being done today for some integrated circuits and it is the cheapest way to serve certain markets. What the example does show is that there are cases in which it is a dangerous decision. In these cases, flexibility must be engineered into the factory, but within limits, because it is acquired at a cost. Knudsen made Chevrolet factories support yearly model changes, but he did not make them able to accommodate thousands of options for every customer.

Many western reports about JIT in Japanese factories describe high productivity as being obtained at the expense of flexibility. At the same time, the *Japanese* literature on the subject describes JIT as a strategy to make factories flexible. In reality, the flexibility of a factory is a economic tradeoff between responding to the market and manufacturing cheaply. Nuances are easily lost in the search for catchy phrases.

Manufacturing organizations, on a day-to-day basis, are production-driven, in that they are always selling what their factories make. The sales organization is seeking buyers for the content of a catalog, to which changes are made slowly with respect to the manufacturing cycle. Increasing a factory's flexibility means reducing the time it needs to respond to changes in the market.

20.1.4. Job-Shops, MRP, and JIT

Pros and Cons of Job-Shop Layouts

The most flexible type of factory is the job-shop, with constraints neither on the set of resources needed for an activity nor on the sequence of activities. In other words, its workload may require its resources in any combination and any order. The job-shop layout described in Chapter 14 is, as its name indicates, best suited for job-shops. In flow lines, machines are laid out in processing sequence, which presupposes the existence of such a sequence.

The absence of coupling between processes and floor layout is not the only appeal of the job-shop structure. Some factories making only 30 varieties of roller bearings are laid out like job-shops in spite of their narrow product mix and fixed processing sequences.

Besides the ease of simply placing identical machines side by side, another perceived advantage is resilience in the face of equipment failures, as shown in Figure 20.3. Each machine in a flow line is a link in a chain and brings the whole line down when it fails (see Figure 20.4). With a job-shop layout, the workload of a failed machine can be shared among the other members of the same work center.

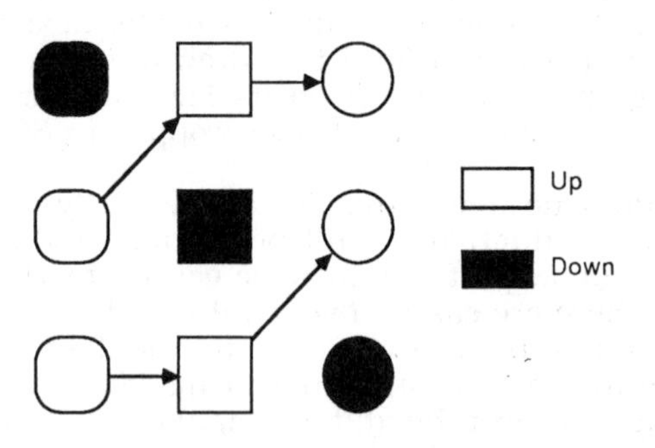

Figure 20.3. Coping with Equipment Failures in a Job-Shop

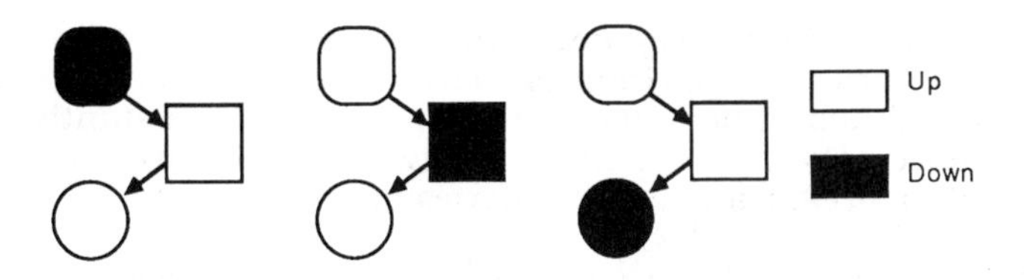

Figure 20.4. Equipment Failures Bringing Down Flow Lines

In actual operation, these "advantages" cause job-shop layouts to be associated with large inventories, excess capacity, loose maintenance practices, and poor quality, for the following reasons:

- Parts move around shop floors designed without regard to their flow patterns, and therefore in complicated paths with numerous buffers.
- Low equipment reliability is tolerable and therefore tolerated, while productivity expectations are lowered.
- Finally, since the successive operations comprising a process are performed in distant work centers, responsibility for the resulting quality is spread among too many groups.

The job-shop layout should be reserved for factories that are truly job-shops. Most factories are not, and an analysis of many of those that appear to be exposes them as flow shops at a deeper level, as was shown in the example of the Reverse Salami tactic.

All the scheduling systems discussed in the previous chapters take the factory as it is and tell it what to do. From the point of view of MRP, if most factories are laid out as job-shops, so be it. It is not the planner's concern. The basic construct of a job-shop is the work center, and it is modeled as such in MRP.

JIT and Factory Response to Change

JIT is a departure from this stance. The improvements at stake in purely scheduling decisions are too small a fraction of a factory's potential to be worth pursuing in isolation. The JIT philosophy is to solve the problems of job-shops by engineering factories into flow lines without jeopardizing their flexibility.

20.1.5. Holistic Manufacturing

What we have said so far can be summarized by saying that JIT is a holistic manufacturing technology, in that, unlike MRP or OPT, it affects every aspect of manufacturing. JIT is as much an industrial engineering (IE) and quality control (QC) system as it is a scheduling system. In attempting to treat scheduling as a separate activity, disjoint from IE and QC, the designers of MRP and OPT were simply following the organization structure of the factories to which they were trying to sell their systems.

In most American factories, production scheduling, IE, and QC are the purviews of separate departments with minimal communications. A production control department trying to implement JIT instantly runs into a territorial dispute. From the point of view of software marketing, only systems respecting organizational boundaries are salable.

Unfortunately, good software marketing does not necessarily make for good production scheduling. JIT is not a software product. It is a set of concepts affecting the life of many departments. The motivation to implement it is the superior performance of JIT manufacturers, and the decision to do so can only be made at a level in the hierarchy that is beyond territorial disputes between departments.

In factories with rigorous separation between departments, each one works to improve the factory under the assumption that none of the others can improve. Where coordinated actions are not encouraged, they are not undertaken, and the factory foregoes the benefits of synergy. In this environment, if a process change reduces the scrap rate of an operation, production control will notice it from production reports, and may update its planning parameters to reflect it. This can be done without a single meeting between the groups. On the other hand, a project to change the shop floor layout to simplify the planning model requires prior consultation.

20.2. FUNDAMENTALS OF FLOW LINES

It is tempting, but misleading, to say that "the first step" towards JIT is to simplify the scheduling problem by converting the factory to a flow line structure. This must take place, but it does not necessarily precede everything else. There may be a technically preferable order in which to implement the various components of JIT, but such considerations are overridden in practice by the politics of each factory. These activities are not steps each to be completed before the next one is started. They are instead neverending parallel efforts, comparable to the parts of various instruments in a symphony.

20.2.1. Demand Structure Analysis

To MRP or OPT, the work load of a factory *is* a Master Production Schedule — that is, a list of products, quantities, and due dates, all to be treated in exactly the same way. As could be expected of general-use software products, these systems make no assumption about the structure of the work load. This is tantamount to acting as if the next order was equally likely to be for any item in the catalog, in any quantity, with any due date.

To anyone with access to the demand records of a factory, this is a pessimistic assumption: most of the demand is accounted for by a small minority of items providing a stable core of production activity, supplemented by a large number of small orders for the others. Many, however, fail to draw the two following consequences from this structure:

- Different methods should be used to schedule the activities respectively needed to meet the core and the variable demands.

• The layout of the shop floor should reflect the demand structure.

The identification of the demand structure may not be trivial. In the example of the "reverse salami tactic," a stable demand for ship components only appeared after the ship assembly method was changed. In other cases, an apparent plethora of end items may be caused by variations on what is, from the point of view of manufacturing, a single product.

From the product made under a five-year government contract down to the spare part for a discontinued model that is ordered once a year, there is a continuum of levels of demand stability, from deterministic to random. However, to translate the results of the analysis to industrial engineering decisions, the products must be grouped into a small number of categories, albeit with some arbitrariness.

First, products with a stable, high demand deserve dedicated flow lines on the Model-T pattern — that is, each optimized to manufacture one of these products but inept at anything else. Setting up such lines will occupy the industrial engineers, but the result may eliminate the need for many planning decisions.

Second, there are products with a level of activity that is too low to justify dedicated lines but high enough to be kept flowing through the factory at all times. These products need common flow lines designed for flexibility rather than speed.

Finally, the numerous other products are each required so rarely that it would be pointless to retain any work in process inventory for them. Attempting to do so would fill up the available space with small quantities of many part types that would be difficult to track and subject to handling damage and obsolescence. Production of these items is started from scratch upon receipt of an order. They are the legitimate realm of the job-shop organization.

20.2.2. Flow Line Organization

As we have just seen, a large part of the shop floor needs to be organized into flow lines, a concept that has existed since the days of the Model T. In the West, authors such as Gallagher [Gal86], describe "Group Technology" as an analysis technique for the design of flow lines, even though, in practice, it is best known as a method for coding and classifying mechanical parts by process similarity. Furthermore, the JIT concept of a flow line is considerably richer than that in [Gal86].

A JIT shop floor is organized in heterogeneous groups of machines called *cells*, that are (1) laid out in series, (2) served by operators each qualified to run all of them, and (3) with autonomous planning and scheduling. One key concept of flow lines is to group machines in series rather than in parallel as in a job-shop. As just mentioned, "in series" only has a meaning in the context of a sequence of operations. A set of machines that is laid out in series for a process will have the output of each machine be the input of the next one *for this process* but not necessarily for another.

Comparison of JIT with Group Technology

Gallagher [Gal86] presents Group Technology as a general approach to flexible manufacturing, but he fails to address both the strategic issue of its range of applicability within a factory, and the tactical problems of laying out cells for flexibility and quality.

Through demand structure analysis, we are able not only to target products for flow line conversion, but also to identify those for which the job-shop structure is best suited. No such discussion can be found in [Gal86]. As described by him, Group Technology clusters parts by process similarity, in order to define machine sequences able to serve these parts. The analysis tools range from check sheets of machine requirements by process to statistical cluster analysis.

As an application example, Gallagher describes group technology in shipbuilding, but stops at classifying parts, based on whether they are made from plates or rolled shapes, with parallel or asymmetric sides, and internal or external to the ship. No such concept as the Reverse Salami Tactic is included.

Since it is usually not feasible to make a single chain of *all* the machines in a factory, group technology leads to a partition into several smaller chains, or "cells," but group technology stops at this. In JIT, cells are not only comprised of machines performing several functions, they are also laid out and managed in a definite fashion, rooted in the idea that human operators are the most critical resource.

JIT Cell Structure

JIT cells are laid out so that the entrance and the exit are contiguous. There are three reasons for this. First, quality problems occurring inside a cell are detected more effectively when arrival from upstream and shipment downstream are done at the same location. Second, the inflow and outflow of parts can be synchronized. Third, when the activity level of the cell is reduced, the operators must move between machines, and it becomes critical to minimize the time spent walking between the last and the first.

The flexibility of this structure is illustrated in Figure 20.5. A lone operator must rotate between all the machines. Two operators can either work the same way or split the machines among themselves. When and only when the production volume warrants it, will each machine be served by its own operator.

Operator awareness of the transformations occurring within a cell is heightened when incoming and outgoing parts are both visible. It is human nature to feel little concern for results one does not see. Experience shows that reorganization of a shop floor into JIT cells improves quality, even when part characteristics are "seen" only through measurement devices.

The cell structure is best suited to machines that process parts one at a time. Within a cell, the machines must be physically close, so that parts can easily be fed from one to the next through special-purpose fixtures with minimal buffers. The closeness minimizes materials handling and enables operators to move quickly between the machines when needed.

In many factories, operators *sit* on the *outside* of equipment modules. In a JIT cell, they *stand inside*, so that they can easily move around. The operators are mobile and each is expected to be proficient in all the operations of the cell. This structure is meant to support load-dependent staffing and gradual automation.

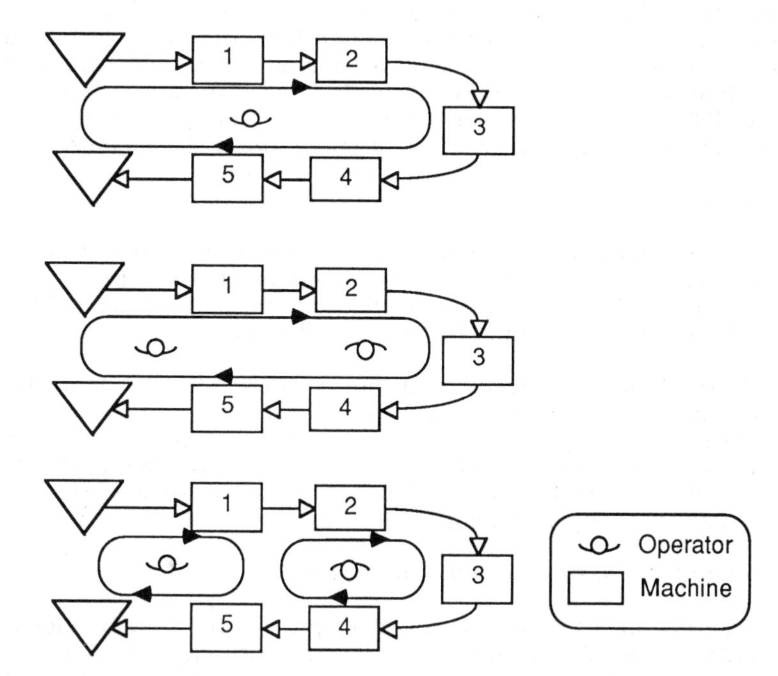

Figure 20.5. Three Operating Styles with One or Two Operators

Variations in the factory's work load will cause the level of activity in a cell to vary as well. A JIT cell requiring three operators at full load will only need two when loaded at two thirds of its capacity and one when loaded to one third of capacity. This is generally not the case in other types of cells. Operators would quickly be exhausted if they had to stand up, run 20 feet, and sit down again just to switch from one function to another.

The closeness of machines in a JIT cell makes it easier to develop cheap, automatic material transfer systems between them. This, combined with automation work on the machines themselves, makes it possible to gradually replace operators with robots.

From a scheduling standpoint, the key parameter of a cell is its *cycle time*, defined as the time between two successive part completions. If a part is needed on the average once every two minutes over a planning period, then the cell will be managed so as to come as close as possible to delivering one *exactly* every two minutes.

20.2.3. Setup Time Reduction

Next to the Kanban system, the technology of quick setups is the component of Just-In-Time that has received the most publicity. As we discussed in Chapter 14, frequent changeovers can make an indivisible resource behave as if it were infinitely divisible. Until setup time reduction is made a priority, changeover times can be 40 minutes for a integrated circuit tester or 40 hours for a diecast machine. The scheduling systems of the previous chapters will take these figures as inputs and work around them. The JIT approach is instead to engineer them down until they no longer interfere with the desired mode of operation of the factory.

Dedication by Families

Cell design results in equipment dedication by product family, one effect of which is the elimination of the most complicated setups. If a machine is used only on similar products, then the amount of work needed to change it over will be less than if it must be used on all products.

Carried to the extreme, this logic would lead to dedicate equipment to each product — an expensive and inflexible solution. The current practice, however, is the opposite extreme, where factories are organized as job-shops with no equipment dedication, even if they have few products and a stable demand. Finding the proper balance between these extremes is one of the purposes of group technology.

Standardization of Procedures and Training

The documentation of changeover procedures may be lacking, obsolete, or unintelligible. The traditionally low priority nature of this task makes it unlikely to have been pursued with vigor and with the type of tools described in the first half of this book.

Once the range of changeovers has been narrowed, the next step is to review the procedures for standardization and simplification. For setup time reduction, the main benefits are (1) the identification of unnecessary steps, (2) the elimination of hesitation, and (3) a broader diffusion of setup skills.

When a manufacturing process is debugged, it tends to be frozen in the first version that works. More often than not, this version contains steps specifying such things as a tighter temperature range than needed. While such steps are not harmful to parts, they are wasted effort. When the goal is to get a line up and running for the first time, chasing down these problems is not a good use of engineering time, but this research must be done later as part of process improvement.

The execution of undocumented procedures requires an effort of memory by operators, resulting in errors and in time lost wondering what to do next. Simple check sheets can eliminate this. By "simple," we mean that they must be lists of actions to be performed in sequence, with none skipped and no branching. None of the steps listed must require the operator to make a decision as to whether it should be executed, or else time is again lost and mistakes can be made.

Finally, the review of changeover procedures causes more employees to learn them and reduces the likelihood of delays due to the unavailability of specialized setup technicians.

Machine Modifications

The analysis of changeover procedures reveals that some activities — called "external setup" — can be performed while the machine is still running, while others — the "internal setup" — require it to be stopped. External setup activities may require time and a variety of resources, but they do not interrupt production the way the internal setup does.

The systematic conversion of setup activities from internal to external is the next field to explore. Shingo [Shin80] cites the example of die changing in diecasting. Heating a die by shooting hot metal into it is an internal step that is made external by preheating the die outside of the machine.

In every factory, one would expect to see the tools and fixtures needed for a changeover to be located near the machines within easy reach of the operators but, employees have no incentive to do it until setup time reduction is made a priority.

To speed up the individual activities, the tools should include jigs that are precisely positioned on the machine through guiding pins and gauge blocks to eliminate manual adjustments. All fixtures should be fastened by quick-release clamps, with handles rather than bolts, because the latter take several minutes to attach or detach.

20.3. AUTOMATION

Where MRP or OPT are neutral with respect to automation, JIT includes a strategy on how to do it. The first principle is that automation should not be viewed as a means of solving outstanding manufacturing problems. These should be solved first, and automation done on the basis of a smooth, high performance manual operation. There is no shortcut from a job-shop with poor quality and bloated inventories to a high quality automatic line. A sequence of incremental improvements must first make it susceptible to automation.

The view of automation as a problem solving panacea dates back to the development of the "American system of manufacture" for interchangeable parts in the 19th century as described in [Hou83]. It is now reinforced by the belief that feedback control can and should be relied upon to correct all deviations in manufacturing processes.

Feedback control is the appropriate tool to keep a plane on course when it is hit by gusts of winds, because no human action can keep the gusts from occurring. Manufacturing quality problems, on the other hand, are not acts of God but *consequences* of human actions. Other human actions therefore remove these causes, so as to make feedback control unnecessary and automation feasible with cheap, open-loop systems.

Gradual improvements on the shop floor, brought about by employees, raise their level of expertise about the factory beyond that of outsiders. It is therefore a mistake to rely on outside contractors for automation. The JIT philosophy is that the automated systems must be designed and implemented by the factory's work force.

20.4. INTEGRATION WITH EQUIPMENT MAINTENANCE

20.4.1. Motivation

More than a job-shop, a flow line is vulnerable to equipment failures. The machines are as links in a chain, and must all be up for the line to be up. Equipment reliability must therefore be assured at a higher level. Theoretically, a line of 10 machines, each up 95% of the time will be up only $.95^{10}$ = 60% of the time. The reliability needed to make the production system operate is only achieved by a determined effort.

At first sight, the distinction between production and maintenance appears as clear as that between driving a car and repairing it. A closer look, however, reveals gray areas. In a silicon wafer cleaning process, for example, sulfuric acid and aqua regia baths must be changed every four hours. This is a "production task," done by operators. In the same factory, aluminum targets used for sputtering metal onto wafers must be changed every 50 runs. That is "preventive maintenance" and performed by a specialized technician. One task may be more complicated than the other, but it is difficult to point to any feature that intrinsically classifies one as production and the other as maintenance.

Automation gradually reduces the amount of purely production work but increases the need for maintenance, by populating the shop floor with machines that are both more numerous and more sophisticated. This evolution causes productivity to depend less and less on the operators' diligence and dexterity, and more and more on their ability to keep the equipment up. The distinction between operations and maintenance is blurred to the point that NUMMI, the Toyota-GM joint venture, deliberately gave the responsibility for production in such areas as body assembly to people whose prior experience was in the maintenance of spot welding equipment.

Furthermore, the performance of all maintenance by a separate department runs into scheduling difficulties due to the mixture in its workload of prevention and repair tasks. The former can be planned in advance and can be done with a high technician utilization. The latter requires "firefighters" to be available on short notice, which means that their utilization must be low.

The temptation is rarely resisted to have both types of tasks performed by the same group, with repairs given priority over preventive maintenance. A vicious cycle ensues, as repairs on some machines delay the preventive maintenance of others, while these delays in turn cause further failures. This situation is lived with in job-shops with excess equipment, but is incompatible with JIT.

20.4.2 Training and Operation Standards

Aside from engineering better equipment reliability, the solution involves restricting the intervention of specialists by delegating small, periodic maintenance tasks to production operators. Given sufficient management support, the maintenance goals are reached by the aggressive application of IE to generate operations manuals and maintenance manuals, and train affected personnel in their use.

Operation standards cover equipment startup and shutdown, preparations before starting processing on a batch of parts, and activities to be carried out during processing. The maintenance manuals describe

- Periodic checks, to be made part of normal operations.
- Machine overhaul criteria.
- Failure analysis instructions.
- A list of spare parts.
- The engineering foundation of the process.

Scheduled training sessions disseminate the knowledge in the manuals in the work force, test their quality as communication tools, and convey the message that management intends for them to be used.

20.4.3 Housekeeping

Until this activity is initiated, the shop floor is likely to be cluttered with defective or obsolete inventory, unused machines, tools, or dies, and excess storage devices. The challenge is to get rid of them without accidentally discarding things that are truly needed.

One possible method is a "red tag campaign" along the following lines:

1. Plant management launches the campaign and forms a task force.
2. The task force then
   a. Decides the object types to target.
   b. Designs the red tags (that is, decides what data to record on them).
   c. Sets criteria for attaching the red tags.
3. Assigned groups of employees, *other* than shop floor operators, affix the red tags on objects meeting the criteria.
4. The red tag items are moved to an assigned area and disposed of.
   a. Dead inventory is thrown away.

b. Items that would be expensive to get rid of are allowed to remain in the red tag area until it is needed for another purpose.

5. Summary reports are collated to evaluate the results of the campaign.

At that point, everything that is on the shop floor is there for a known purpose. The next steps are (1) to assign locations to all objects and (2) to place enough labels on the shop floor to make these objects easy to find for everyone.

Location assignments should follow ergonomic principles, with frequently used tools and consumed parts placed within easy reach of operators. Tools used as a set should be arranged on a tray carrying the complete set, with cavities matching tool shapes to make absence obvious. At an operator's work bench, cords that pull power tools back into place when released simplify the task of replacing the tools after use.

Cleaning is the last dimension of the housekeeping effort, with standards that depend on the type of product being manufactured. The level of cleanliness needed for integrated circuits would be absurd in metal working. The direct effect of dirt on product quality is, however, not the only reason for wanting to maintain a clean environment. A grimy shop floor, littered with cigarette butts and styrofoam cups, is also damaging to employee morale. Since production gives many opportunities for liquids to spill and dust to fly, keeping the work place clean is a never-ending task and one which must be done largely by the operators themselves.

Once the preceding results have been attained, they must be maintained. Currently useful objects turn into clutter as they become obsolete or worn out. Under time pressure, people forget to return objects to their assigned locations or to keep the shop clean. A continuous effort of discipline is needed to maintain the shop floor in the desired state.

The "technology" of the actions described in this section does not exceed that possessed by anyone who looks after a family household. Yet it is clearly not universally applied in factories. Some American companies periodically subject their open office areas to "Mr. Clean" inspections in which a group of managers goes by every cubicle and chastises employees with cluttered desks. The most common employee reaction is resentment and the perception of these inspections as an invasion of privacy.

Labeling all objects and locations on the shop floor makes it easier to understand, but this transparency is not wanted by everyone. In companies that provide no job security, employees are prone to protect themselves by withholding information. In such environments, they are content with the shop floor appearing to management as an uncharted wilderness, in which only they know the way.

20.5. INTEGRATION WITH QUALITY CONTROL

20.5.1 Effects of Flow Lines on Quality

Although the subject of quality control deserves a thorough treatment elsewhere, it cannot be ignored in an overview of Just-In-Time, because improving quality is one of the key goals of the method. A JIT line is built for speedy detection of quality problems, and many of the actions we have already outlined serve this purpose.

The short throughput times associated with a high throughput, bare bones inventory environment translate to short problem detection times. Low inventories also means less handling and therefore less handling damage. Furthermore, a shop floor that is dense in process equipment and uncluttered with storage devices offers better visibility.

With operation sequences performed in one room instead of being split among several, managers and engineers can see the state of the process at a glance instead of examining multiple computer screens and walking over long distances simply to find out what the shop smells like or read the operators' body language. Simple alarm mechanisms compensate for the crudeness of the information they convey by their immediacy and low cost. Finally, the intense maintenance activity ensures that equipment is in a condition to produce good parts.

20.5.2 Just-In-Time and SQC

Further activities are carried out strictly for quality control. At cell gates, operators are expected to perform some form of inspection of every outgoing part, stop production when they observe any anomaly, and report the problem by means of warning lights called "andon." The role of statistical methods is limited to process capability studies. At each operation, inspections are performed either on all parts, on the first and the last within a batch sequence, or on none at all, but not on random samples. Control charts are not used as a daily process control tool.

JIT specialists criticize statistical quality control (SQC) for failing to catch a fraction of defective parts. For example, an $\overline{x}$-chart, which verifies the presence of sample averages within $\pm 3\sigma$ of a reference, will fail to detect some aberrant *individual* values. An SQC expert would counter that an $\overline{x}$-chart is simply a diagnostic tool. When a process is in "statistical control," the $\pm 3\sigma$ control limits of the $\overline{x}$-chart are tighter than the tolerances that must be met by good parts. In theory, control limit violations provide evidence of significant deviations *before* they have become severe enough to damage parts.

Unfortunately, the subtleties of SQC as described by Deming or Juran are not generally understood on the shop floor. JIT implies a rejection of SQC not as it is in theory, but as it is misapplied. SQC may solve the problems technically, but this is a moot point. If the work force cannot be trained to use SQC properly, then it is clearly not an appropriate technology.

20.5.3. Foolproofing the Process with Pokayoke

One last dimension of the industrial engineering work associated with JIT is a set of small changes to the manufacturing process intended to prevent mistakes and known as *pokayoke* (pronounced "pokah-yokay"). It is difficult to imagine a theory of pokayoke. It is however possible to provide inspiration by examples. This is what *Kojo Kanri* magazine did in July, 1986, when it published a special issue [KKA86] intended to be a "picture encyclopedia of fail-safe techniques."

Household electric coffee mills that are started by pressing down the lid offer a daily life example of pokayoke. This trick makes it impossible to have the lid come off while the mill is running. In consumer electronics goods, color coded cables and special-purpose jacks are also pokayokes in that they make it difficult or impossible to hook up components incorrectly. In the software world, strong typing of data structures can also be viewed as pokayoke, in that it helps programmers detect mistakes at compile time that would be difficult to diagnose in later stages.

Here, we can only give a idea of what pokayokes look like, through a few examples:

- A part that is nearly symmetrical has a high risk of being inserted into an assembly upside down. A pokayoke for this case is to redesign the part to be fully symmetrical, so that it doesn't matter in which direction it is inserted [KKA86].

- A limit switch is placed over a conveyor belt to detect parts that are too high because a machining operation has been skipped (see Figure 20.6, based on [KKA86]).

- At an assembly operation where operators must draw two types of small parts from boxes, a flip-lid box with two compartments prevents operators from drawing parts out of order, as in Figure 20.7.

20.6. PEOPLE AS THE MOST CRITICAL RESOURCE

Hardware changes on the shop floor can be shown off to visitors, but the results of changes in how people work are invisible at first. Equipment can be depreciated, so that its impact on accounting results is spread over time. On the other hand, training is an expense and must be written off in the period it is done. Machines stay in the plant, but the best people may be lured away by competitors. These and other similar justifications are used by managers who believe that the productivity of a factory is determined by its hardware and not by its people.

History shows how mistaken this point of view is. Germany's industrial eclipse after World War II lasted only as long as it took its skilled work force to rebuild factories, but Iran's massive equipment purchases of the 1960s and 70s failed to turn it into an industrial power. The need for grass roots acceptance of a productivity improvement program constrains the technology that can be used. Just as foreign aid agencies discovered that hand tools could help agriculture in parts of the world where tractors could not, so must those who want to improve factories realize that cheap, low technology may succeed where expensive, high technology fails.

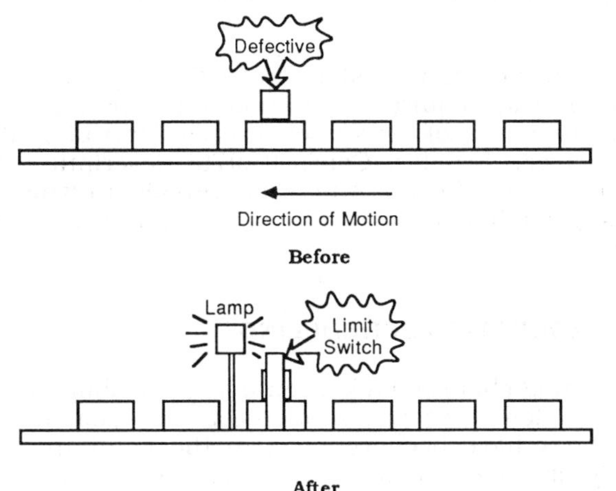

Figure 20.6. Limit Switch Pokayoke

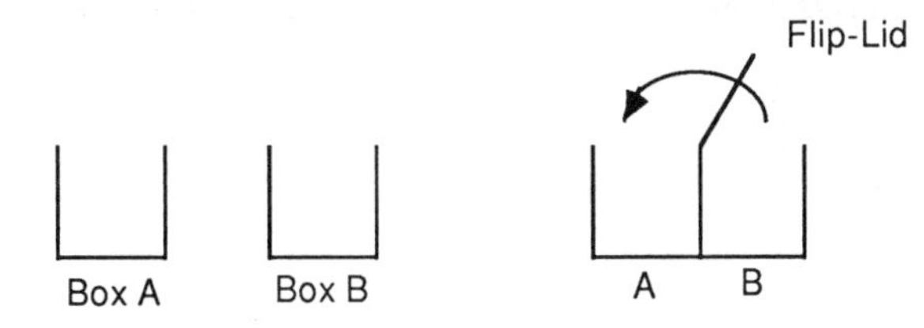

Figure 20.7. Flip-Lid Pokayoke

As just described, the changes made to a factory in the initial stages of Just-In-Time implementation require thinking, but no large outlays of funds. The improvement in factory performance brought about by these changes free up financial resources for further projects. These in turn gradually raise the level of technical sophistication of the work force and enhance its receptiveness to new methods.

In a high growth period, productivity improvements enable the factory to withstand a heavier workload without hiring, but in a low growth period, they make it technically possible to *shrink* the work force. Employees will refuse to cooperate if they feel that the whole effort will lead to layoffs, and that must therefore not be true. Whatever staff reductions occur once JIT implementation has started must be done by not replacing employees who retire or resign voluntarily.

The operators' ideas are needed to find ways of running each cell with fewer of them. When one operator is removed from a cell, it must be the best rated, and the transfer must be perceived as a promotion. Within this context, working oneself out of one's current job becomes the key to getting a better one. Automation will not eliminate people from factories but change their roles from operators to maintenance technicians and NC-machine programmers.

The acquisition of all the new skills needed by operators in a JIT context must be organized. This requires a systematic rotation between the different jobs, and the maintenance of personnel records describing all the tasks each operator has been certified for. Cole's [Col79] description of the personnel administration methods of Toyota Auto Body reveals a higher level of attention to career planning for line workers than most American firms give to their professional staff.

20.7. BACK TO PRODUCTION SCHEDULING

Once the preceding changes have been done, the shop floor is a network of partially dedicated flow lines based on the actual demand structure of the factory. Setup times have been reduced to the point that it is possible to emulate infinitely divisible resources by small batch processing on indivisible resources. Improvements in equipment availability have eliminated the need for large buffers of work in process, and the combination of short throughput times with pokayoke has reduced scrap and rework to negligible levels. Now is the time to discuss how to schedule this superb production tool.

21

JUST-IN-TIME PRODUCTION SCHEDULING

While internal causes of variability in a factory can be engineered away, external causes must be addressed by establishing stable, long term supplier relationships and leveling schedules to smooth material flows. Customers provide suppliers with demand forecasts that are gradually frozen as their term draws near, with the range of allowed adjustments becoming progressively narrower.

MRP's time-phased bill of materials explosion logic is used to predict part requirements, but with daily time buckets and without trying to prescribe actual materials movements. Shipments and movements between cells can be triggered by flows of cards as in the Kanban system, or specified in calendars as in the Timetable system.

21.1. DAMPING EXTERNAL VARIABILITY

As we discussed in earlier chapters, part of the variability in a factory's activity comes from its environment. While such engineering actions as setup time reduction enhance the factory's ability to cope with variability, scheduling can be used to mitigate its effect on the shop floor.

Possible actions include (1) developing stable relationships with suppliers of raw materials, (2) using MRP to anticipate requirements with increasing precision as the term draws near, and (3) turning large orders into interleaved, small ones to smooth requirements over time.

Actual part movements inside the shop floor are driven by other mechanisms, characterized by decentralization and simplicity. The best known is the *Kanban* system, in which the material flows are regulated by tokens detached from parts when they are withdrawn from buffers and giving permission to move or produce replenishments. The Kanban system is not the only option for Just-In-Time, and we will discuss Abe's *Timetable Planning* system as an alternative that is suitable in many cases.

Where some high level decisions are based on MRP logic and execution is controlled by Kanbans, discrepancies are bound to appear when actual requirements either exceed or fall short of expectations. Rules to resolve such conflicts are a necessary part of each supplier-customer relationship.

Consumers can be coaxed by advertisements and promotions, but the final decision on the volume and the timing of their orders remains theirs alone. Manufacturing products eventually sold to individual consumers have their collective whims as a factor of variability that can never be eliminated. Orders from consumers are outside the scope of control of a manufacturing organization.

Most factories, however, do not serve the general consumer market directly but are instead nodes in a network of factories where the final products of each are raw materials to some of the others. Factories that supply customers who are themselves factories are called OEMs, for *original equipment manufacturers*. Even those factories that are at the end of a chain often produce goods that end in the hands not of consumers but of businesses or government institutions with more stable buying patterns.

In this context, various steps can be taken to organize the flow of materials in the network so as to minimize the variability of the workload of each node. Before reviewing what those may be, let us examine, as an example of how not to do it, the supplier-customer relations that prevailed in the U.S. semiconductor industry prior to the recession of 1985, and that are partly to blame for it.

21.1.1. The Overbooking Model

To protect themselves against the unreliability of integrated circuit (IC) suppliers, printed circuit board (PCB) assemblers usually buy each type of IC from at least two sources. This practice has actually let semiconductor companies to license their technology to competitors for "second sourcing."

The PCB assemblers started placing orders for the same parts with different IC suppliers, and canceling all the outstanding orders as soon as one of the suppliers came through. Meanwhile, expecting cancellations, IC makers deliberately overbooked their capacity (see Figure 21.1).

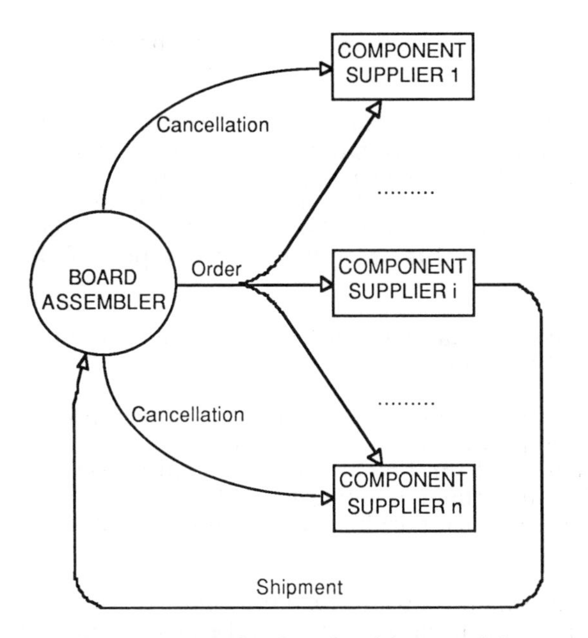

Figure 21.1. The Overbooking Model

The dangers of this mode of operation should be clear. In particular, future activity becomes hopelessly difficult to forecast. The word "order," in this context, loses the implication of commitment it has in normal business practice, as each one has to be weighed with a probability of coming through.

The root cause of overbooking is lack of trust. The supplier and the customer do not believe each other's word, and are both forced to justify the other party's worst assumptions. Because customers cancel their orders, suppliers have no alternative to overbooking their capacity. Conversely, suppliers are unreliable, in part because they overbook. Customers have no choice but to place redundant orders. Let us now examine the type of relationship that may be created to break this vicious cycle.

21.1.2. Demand and Production Forecasting

Figure 21.2 describes a possible alternative. Production Planning uses the same backward scheduling logic as MRP, but its outputs are used only to anticipate what will happen on the shop floor, *not* to prescribe it. A similar situation is that of a retailer, who bases stocking decisions on forecasts, but for whom actual sales are triggered by transactions that occur independently of the forecast.

Strictly speaking, the enclosed bottom part of Figure 21.2 is not part of the forecasting system, which simply relies on the availability of some form of shop floor control. In the specific pattern shown in Figure 21.2, all production activities are triggered and regulated by flows of cards called "Kanban" according to rules described later in this chapter. In the Japanese

IC industry, the forecasting system is used, but the shop floors are not run by Kanbans.

Changes to the Demand-Forecast are regulated by agreements with customers. Since they are more painful in the near than in the distant future, the agreement may limit allowed adjustments to ±5% of the volume for the first two weeks, to ±15% for the second month, and ±25% for the third month. The system works perfectly as long as the sequence of orders matches the forecast, but since this is unlikely, two types of deviations must be envisioned:

1. The customer's actual needs fall short of the forecast.
2. They exceed the forecast.

There may be as many ways to resolve these problems as there are supplier-customer relationships. One possibility in case of order shortfall is for the supplier to hold the finished goods for an agreed upon period of time, at the end of which the customer buys them regardless of need. In the opposite case of an order surplus, a resolution must be negotiated case by case.

In a JIT environment, inventory data come to production planning not from the shop floor but from marketing and sales, which owns all the finished goods. Since work in process inventory levels are determined by the Kanban system, production planning does not need to get them from the shop floor.

By comparing the Demand-Forecast with Finished-Goods-Inventories, Production Planning does the following:

1. It determines whether any production is necessary.
2. If it is, Production Planning does a CRP calculation to identify eventual capacity shortages. The capacities used take into consideration expansion or contraction plans over the horizon of the Demand-Forecast.
3. Through negotiations not shown in Figure 21.2, Production Planning attempts to resolve the capacity shortages by getting Production Management to plan for overtime or by borrowing capacity from outside contractors, refining the Production-Forecast in the process.

Under normal conditions, the Production-Forecast will require no action from production managers. Occasionally, it may require them to update the Kanban system by changing the numbers of kanbans in circulation or the numbers of parts a kanban is attached to, the latter being more difficult because of its impact on the physical transport system.

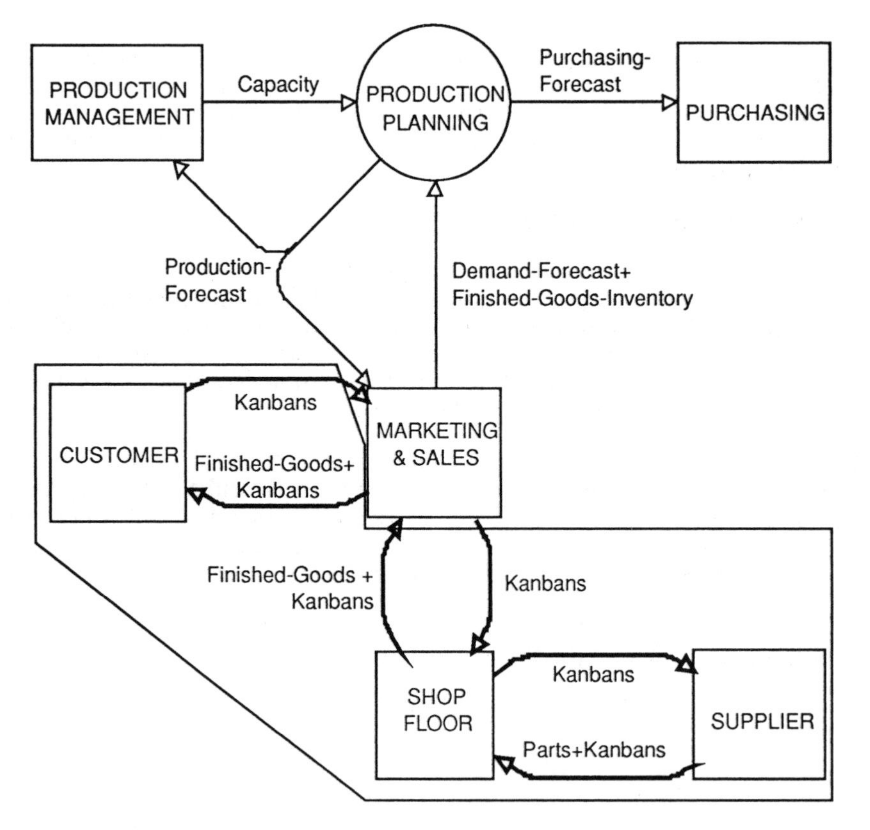

Figure 21.2. Demand and Production Forecasting

To give an accurate picture of the system described in Figure 21.2, we need to discuss *timing*. Following is a common but not universal pattern:

- Capacity planning is done every six months, resulting, from the point of view of Production Planning, in a schedule of updates of capacity values.
- Production Planning is done on the 20th of every month, giving daily forecasts for the following month and weekly forecasts further into the future.
- Kanban system maintenance is done when needed, which happens a few times a year.
- Normal kanban transactions occur many times a day.

21.2. GENERATION OF LEVEL SCHEDULES

In Chapter 20, we have seen, in the example of the reverse salami tactic, how an assembly process could be organized so as to smooth the demand for components. In this and in other cases, the component supplier does not need to take any additional action to level his work load and can process orders on a first-in-first-out basis.

On the other hand, an automobile assembly plant receives dealer orders for a variable mix of cars and needs to take special care to sequence their production so as to smooth the flow of components to assembly. This is also necessary for component suppliers who receive large orders from a small number of customers.

21.2.1. First Blender Algorithm

If the products were sequenced by drawing on their demand in a round robin fashion, there would be long runs of the products with the longest queues due to exhaustion of the shorter ones. This approach would clearly not work. Let us try to blend the orders in such a way that final assembly will appear to be dedicating a fixed fraction of its capacity to each product all the time — that is, will behave like a divisible resource.

Let us consider the work load at time t_0 and proceed to blend it without regard for the orders which might arrive between t_0 and the time t_1 by which all the orders in the queue at t will have been processed. At all times between t_0 and t_1 we want the fraction of the work load processed to be the same for all products. This is accomplished by the following method:

1. Let n_i be the quantity of product i on order at time t_0.
2. Build a sequence of the form

$$S = \left(\left[i,\ q_{mi}\right]\right),\quad m = 1 \text{ to } n_i \text{ and } i = 1 \text{ to } k$$

where $q_{mi} = \dfrac{m}{n_i + 1}$

3. Sort the sequence S by increasing q_{mi}.
4. The level schedule is then given by extracting the sequence of i's from the sorted S.

This approach is illustrated in Figure 21.3.

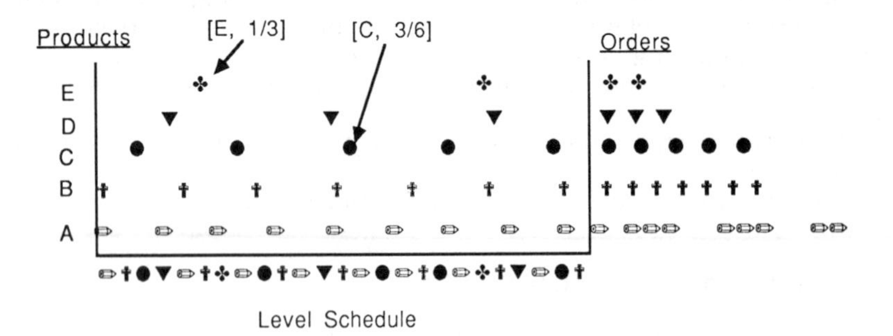

Figure 21.3. First Blending Algorithm

This simple algorithm has value as an illustration but relies on too many restrictive assumptions. It would be effective in fabrication, but, with non-trivial bills of materials, a schedule that is level at final assembly is not necessarily so for components. Monden's algorithm, described below, makes use of the bill of materials of each product.

The longest possible run length in a schedule processed by the First Blender is the ratio of the number of orders for the two highest volume products.

21.2.2. The Monden Algorithm

We now leave the realm of concept illustration for that of practical application. Yasuhiro Monden [Ono83] presents a blending algorithm he describes as a generalization of a heuristic developed for Toyota assembly lines by Koyu Terada. Like the First Blender, this method sequences a set of outstanding orders and must be run periodically.

Unlike the First Blender, it takes into account the bill of materials, which a significant part of the factory model. It is remarkable that Monden's method ignores not only due dates, but operation processing times as well. The focus on the bill of materials is the mark of assembly: when shipments from an assembly plant are delayed, it is more often for lack of components than for any other reason.

The Monden algorithm sequences orders so as to minimize the flow rate variation of *every* component. Following Monden's conventions, we introduce the following notations.

Let

- A_i, i = 1,..., N denote the products of the factory.
- Q_i, i = 1,..., N be the quantities on order for all products.
- $Q = \sum_{i=1}^{N} Q_i$
- a_j, j = 1,..., M designate the M components going into the N products.
- n_j, j = 1,..., M be the quantity of component j required to satisfy all the orders.

Then

$$m_j = \frac{n_j}{Q}$$

is the average quantity of component a_j needed per unit out over the order mix $(Q_1, \ldots, Q_N)$.

The vector

$$G_k = \left(k \times m_1, \ldots, k \times m_M\right)$$

gives what the cumulative consumption of each component for the first k units out would be if all units, regardless of product, consumed exactly the average quantity of each component. But, for each choice of k units to produce first, there is a vector

$$X_k = \left(x_{1k}, \ldots, x_{Mk}\right)$$

giving actual component requirements. One possible measure of the smoothness of component flow is the Euclidian distance D_k between G_k and X_k:

$$D_k = \left[\sum_{j=1}^{M} \left(k \times m_j - x_{jk}\right)^2\right]^{\frac{1}{2}}$$

Monden's algorithm is an approximate solution to the problem of minimizing

$$D = \sum_{\alpha=1}^{Q} D_\alpha$$

It works as follows:

1. Initialize
   1.1. b_{ij}= number of units of component a_j going into one unit of product A_i, for i= 1,..., N and j = 1,..., M.
   1.2. k = 1
   1.3. x_{j0} = 0, for j = 1,..., M
   1.4. S = {1,..., N} = set of all product indices with non-empty queues.
   1.5. L = () = list of already sequenced orders, initially empty.

2. Repeat the following until S is empty:
   2.1. $\forall i \in S$, set

$$D_{ki} = \left[\sum_{j=1}^{M} \left(k \times m_j - x_{j,k-1} - b_{ij}\right)^2\right]^{\frac{1}{2}}$$

   2.2. Select the product A_{i^*} such that

$$D_{ki^*} = \min_{i \in S} D_{ki}$$

   2.3. Append A_{i^*} to the list L of sequenced orders.

   2.4. Decrement the outstanding order count for product A_{i^*}:

$$Q_{i^*} = Q_{i^*} - 1$$

2.5. If $Q_{i}^{*} = 0$, set $S = S - \{i^*\}$

2.6. If S is non-empty, then do the following:
  2.6.1. $\forall$ j= 1,...,M, set $x_{jk} = x_{j,k-1} + b_{i^*j}$
  2.6.2. Set k = k + 1

Monden also gives several variants of this algorithm with higher performance in some circumstances, but not always applicable. Monden's algorithm reduces to the First Blender when all products have only one component.

21.3. FLOW CONTROL ON THE SHOP FLOOR

21.3.1 The Kanban system

Kanban Flows between Two Cells

The "level schedule" produced by the blender is not truly a schedule, in the sense that it does not specify when any order should be completed, but only the sequence in which they should be. In the kanban system, the responsibility for responding to the level schedule can be delegated to a system of cards and to rules governing their flows.

Before building a data flow model of the kanban system, it is necessary to view it as a material flow system in the sense of Chapter 10. The reason for this is that the most significant features of the kanban system are not visible in the data flow model.

The stores of cards used in the kanban system undergo insertions of arriving cards and withdrawals of departing cards rather than the read and write operations that lend themselves to data flow modeling. In a data flow model, insertions and withdrawals are both updates and therefore both represented by arrows pointing to a data store, which makes it impossible to represent a distinction that is fundamental to the kanban system.

For this reason, we start the analysis with a material flow model of the kanban system operating between two work cells L and M. The model presented in Figure 21.4 is derived from the description in Appendix A of [Sch82]. The differences with Schonberger's discussion are

- The use of an abstract terminology, the work cells being called L and M rather than "Drilling" and "Milling." The purpose is to avoid giving the impression that the technique is applicable only to metal working.
- The use of the standard ASME symbols for material flow models.
- The elimination of the distinction between the card collection box and the dispatch box.

The reason for the existence of two separate boxes is physical separation between the place at which the cards are collected from containers and that at which they are attached to other containers. This distinction is a constraint due to the use of physical cards; it is not essential to the logic of the kanban system. A mechanism automatically conveying the cards to their destination as soon as they are collected from containers would not change the logic by which the kanban system schedules.

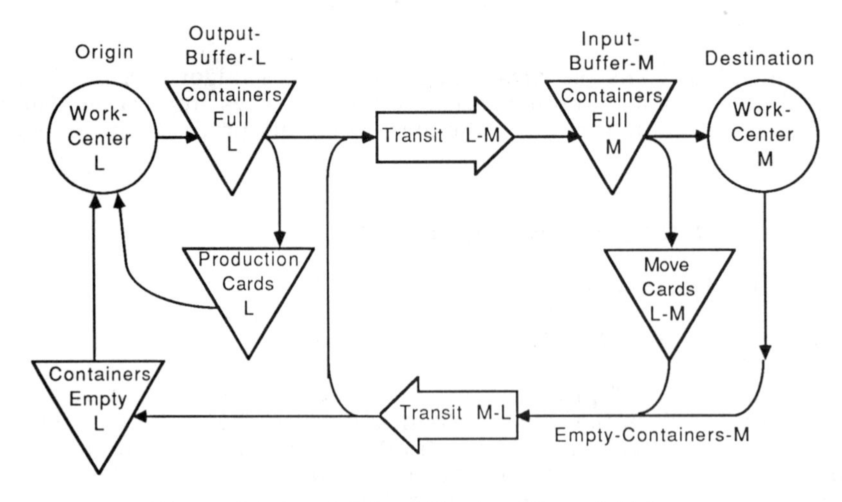

Figure 21.4. Dual-Card Kanban Material Flows

Each work cell has input and output inventory buffers. Move-Cards regulate the content of input buffers; Production-Cards regulate the content of output buffers. When a Full-Container of M parts is required at work cell M, it is withdrawn from the input buffer and its Move-Card goes into the Move-Card collection box.

As soon as work cell M produces an Empty-Container, a Move-Card is withdrawn from the collection box and attached to it. The Empty-Container with the Move-Card travels to work cell L, where the card is detached:

- The Empty-Container goes into the store of Empty-Containers ready to receive the output of work cell L.
- The Move-Card is attached to a Full-Container in the output buffer of L.
- The Production-Card of the Full-Container is detached and placed in work cell L's collection box.
- The Full-Container, with the Move-Card attached, is transferred over to work cell M.

Meanwhile, the Production-Cards at work cell L serve as a dispatch list. They are withdrawn from the box in first-in-first-out order. Based on a bill of materials for work cell L, the input Full-Containers required at L to produce one output Full-Container are withdrawn from the input buffers. This cycle is repeated upstream in the processing sequence all the way to raw material purchases.

An Empty-Container leaving work cell M always has a matching Move-Card in the collection box, because one was detached from it when it was moved into the work cell. An Empty-Container waits until the Transit system gets around to picking up the Move-Card from the collection box. Move-Cards may wait in the collection box for a matching empty container to leave the work cell.

On the side of work cell L, containers in the input Empty-Container store always have a matching Production-Card. They wait until the work cell picks it up. In the output buffer of work cell L, Full-Containers wait until an Empty-Container arrives back from work cell M with a Move-Card attached.

Exception Handling

The preceding discussion contains no provision for shortages:

- What does the kanban system do when there is no output Full-Container at L to attach to the Move-Card?
- What does it do when there is no input Full-Container at M to withdraw in response to a Production-Card?

Those situations are of particular interest because the philosophy of the kanban system is to uncover problems by reducing the numbers of cards in circulation until a shortage occurs. Then the management response is to stop the line and solve the problem. There is however a situation where all components are short even though there is no evidence of a problem: that is when a new factory is brought up for the first time directly on the kanban system. Figure 21.5 shows part of a possible response to these situations by identifying logical places to put the cards when shortages occur.

Kanban State Transitions

Figures 21.6 and 21.7 give another perspective on the operation of the kanban system. They are state transition diagrams for Move-Cards and Production-Cards, describing the states they can be in and the events causing changes. As with the preceding, additions for shortage handling are in bold.

Data Flow Model

The data flow model of Figure 21.8 shows two modules, respectively managing Production-Cards and Move-Cards. Unlike Figure 21.4 and 21.5, it represents not material flows on the floor but the functions of a hypothetical software system.

PRODUCE issues a Move-Card to MOVE when it receives an empty container in front of an upstream work cell L in the cell just used. Since the Move-Card was previously attached to an empty container, PRODUCE has to read the Empty-Containers file.

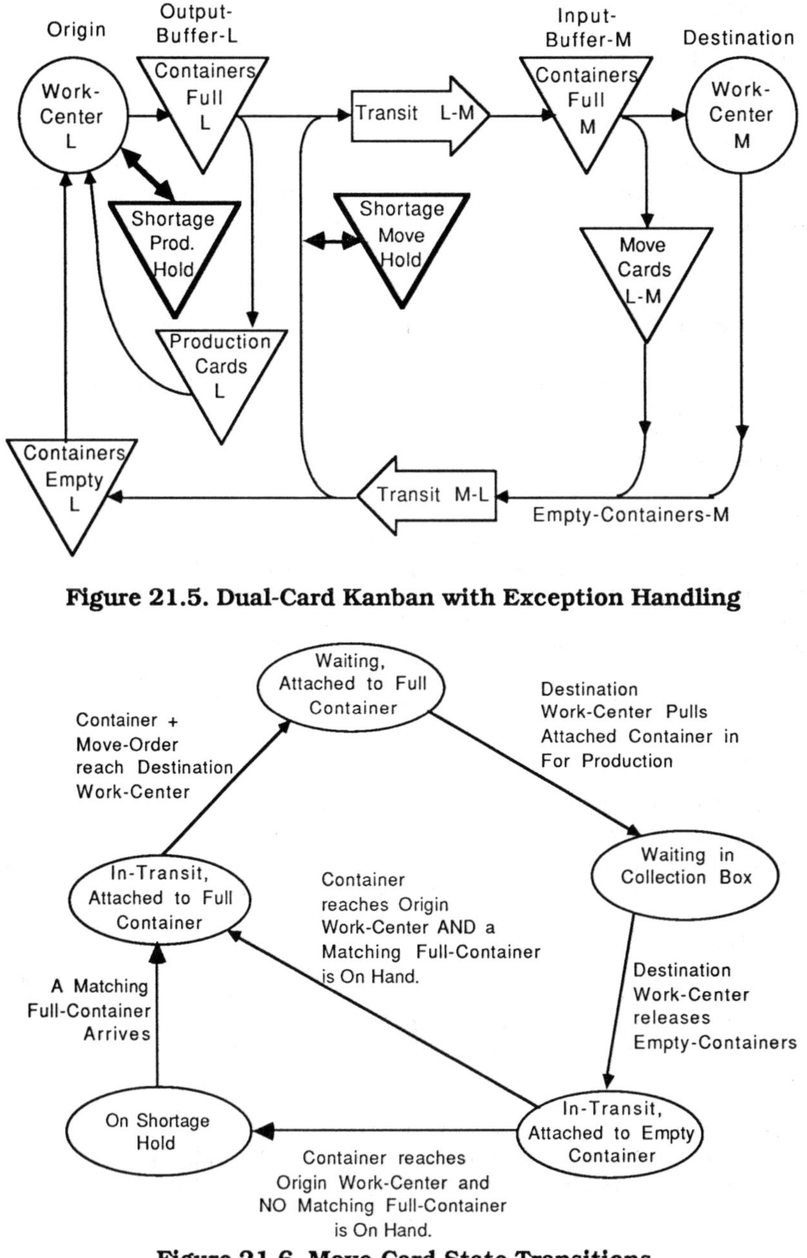

Figure 21.5. Dual-Card Kanban with Exception Handling

Figure 21.6. Move-Card State Transitions

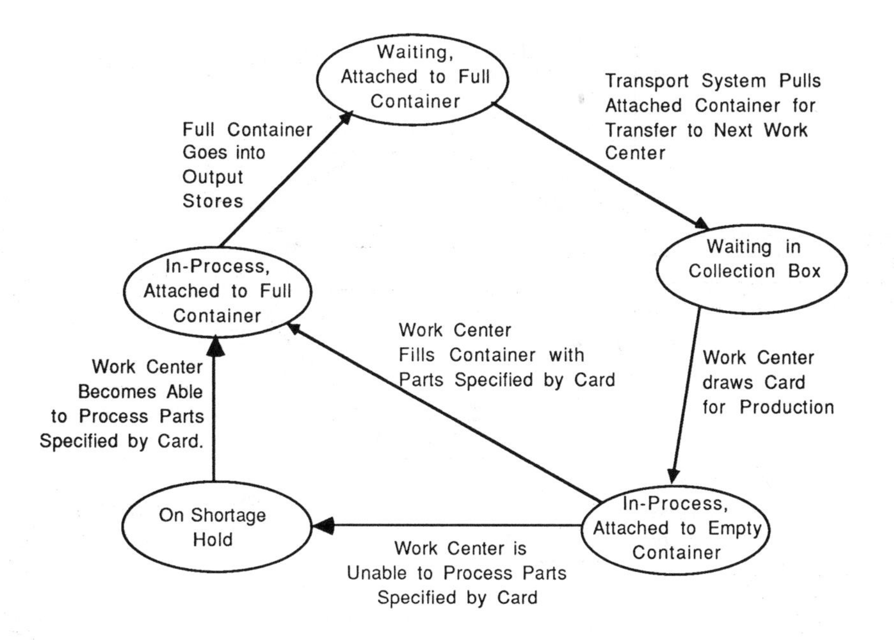

Figure 21.7. Production-Card State Transitions

MOVE likewise passes a Production-Card to PRODUCE when it withdraws a full container from the output buffer of the upstream work cell and, for this, must read the Full-Containers file.

The factory model is broken up into the components needed by PRODUCE and MOVE. To know what input buffers to draw from, PRODUCE needs the Bill-of-Materials. To know where to take full containers, MOVE needs the Routings.

Free Move- and Production-Cards are the means by which the kanban system tells the shop floor what to do, and this is why they are an output. A card is attached to material when it is being processed or transported, or when it should be left alone. The Attached-Card data flows are notifications by the shop floor to the kanban system that it has taken responsibility for material disposition.

Hall [Hal83] shows examples of cards bearing individual IDs, but no example of uses the information that a certain action on parts was authorized by a particular card. It does seem as if the cards are simply anonymous tokens, and that there is no need to differentiate two cards specifying the same action.

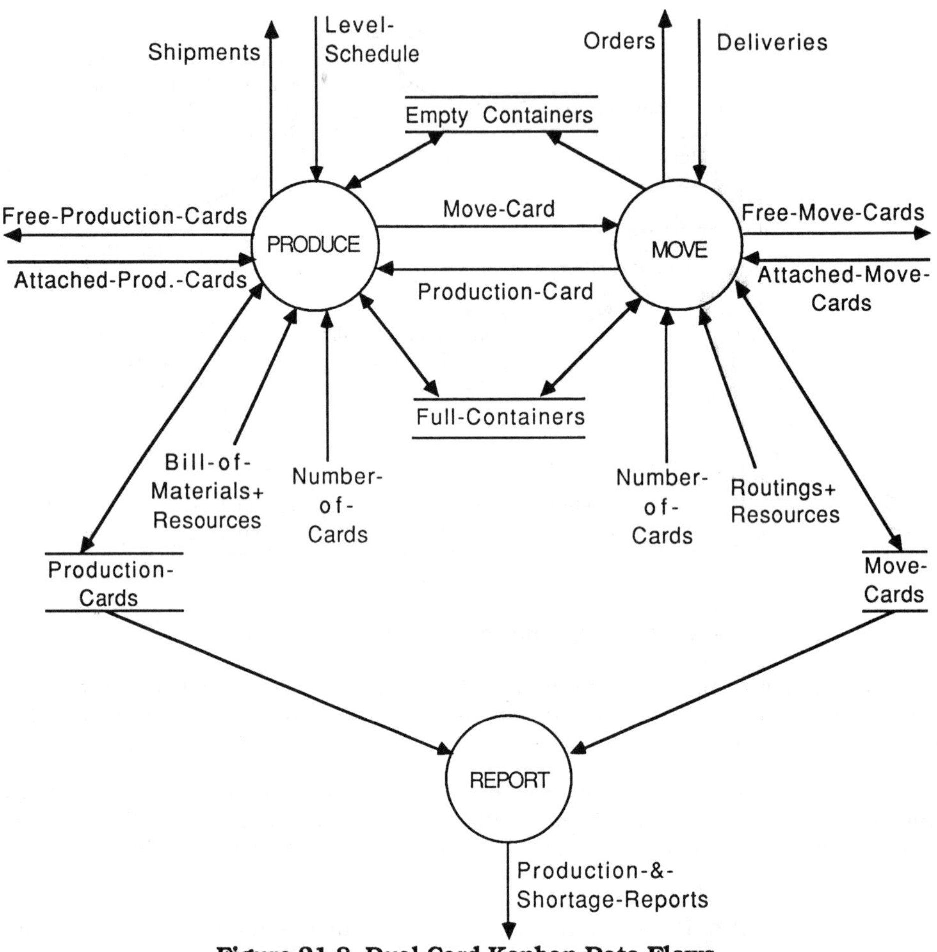

Figure 21.8. Dual-Card Kanban Data Flows

21.3.2. The Timetable Planning System

Forward Scheduling of Resources

Although both clever and simple, the kanban system can be overkill in factories such as are found among mechanical component manufacturers for the automobile industry and in factories with the "keystone" structure described in Chapter 10 for the example of silicon wafer slicing.

In this type of circumstance, MTJ's K. Abe recommends a form of finite forward scheduling which he calls "Timetable Planning." We have seen in Chapter 19 that, in general, finite forward scheduling was dauntingly complex. In some circumstances, however, it can be made simple enough for a cell leader to do *by hand* or with minimal computer support.

Unlike the kanban system, which coupled production actions with inventory movements as they occurred, Timetable Planning schedules resource allocation in advance. It works in a context where the industrial engineering actions of Chapter 20 and the use of small batches are sufficient to prevent the accumulation of inventory. Figure 21.9 shows a data flow model of the timetable planning system. The production planning function of Figure 21.2 is broken down by department.

By netting the Rolling-Forecast and Orders received from customers against Finished-Goods inventories, SALES produces a Master-Production-Plan from which MATERIALS PROCUREMENT deduces raw materials requirements by simple MRP logic as in the preceding, that are communicated to Purchasing as in Figure 21.2 and to Production Management in the form of a leveled Delivery-Schedule.

Monthly Production Calendars

SALES levels the Master-Production-Plan and, jointly with PRODUCTION MANAGEMENT, generates a Monthly-Production-Calendar for completions out of each of the flow lines on the shop floor, with the following structure:

Monthly-Production-Calendar = {Flow-Line + Date+Product-Detail+
Work-Load}
Product-Detail = {Product-Name +Order-Quantity+ Forecast-Quantity}
Work-Load = Number-of-Batches + Hours-Required

In Figure 21.9, the Monthly-Production-Calendar is shown as a data store rather than a data flow. The reason is that it is more than a message communicated once. It works like a bulletin board that remains permanently accessible to SALES, PRODUCTION, and PURCHASING.

The Monthly-Production-Calendar is formatted as shown in Figure 21.10. It is not generated once and for all but instead is dynamically updated as orders come in and override forecasts. This calendar can be viewed as a capacity reservation tool for SALES. The work load is estimated by classical CRP logic.

The schedules are formatted not as Gantt charts but instead like timetables of classes in a school. Anyone with a primary education anywhere in the world has spent at least six years in an institution scheduled this way and is therefore already familiar with it.

Weekly Production Calendars

PRODUCTION, or more precisely production supervisors, then generate Weekly-Production-Calendars for all the cells. The challenge is to build timetables that are not only consistent with the Monthly-Production-Calendar as it is originally written but can easily be made to reflect its updates. The data content of a Weekly-Production-Calendar is as follows:

Weekly-Production-Calendar = Flow-Line + Week-Name +
{Activity + Start-Time+End-Time}

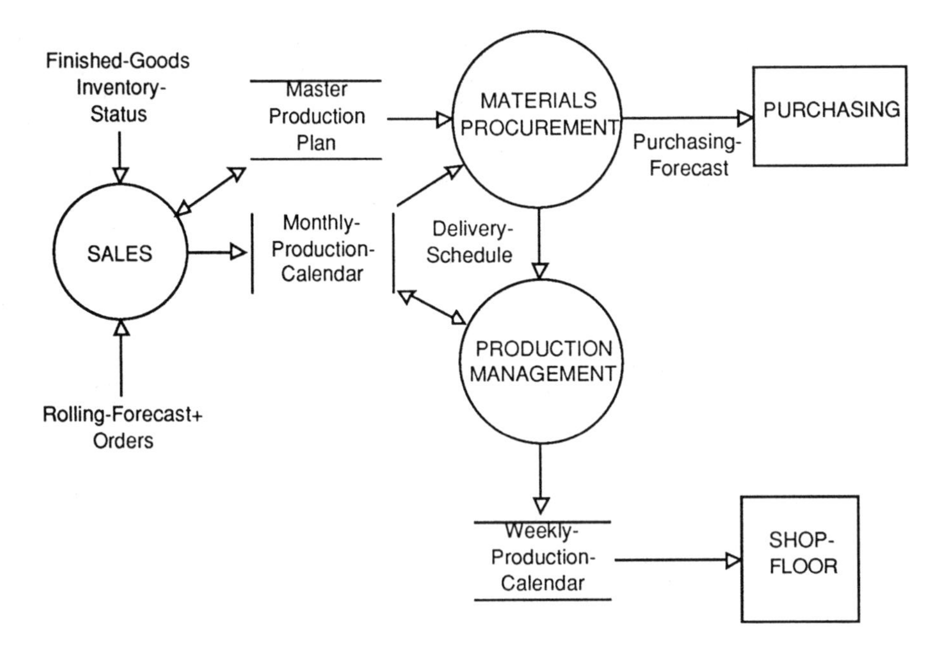

Figure 21.9. Timetable Planning Data Flows

Flow Line: Line-1

Product \ Date	5/23 (Mo)	5/24 (Tu)	5/25 (We)	5/26 (Th)	5/27 (Fr)	...
P1	1	2		1	1	...
P2	1		3	1	2	...
...	...	...	...	...	...	...
Pn		1	2	1		...
Total Batches	7	8	6	7	5	...
Work Load (Hours)	20	23	19	22	18	...

Booked Order (shaded): P1 5/23, P1 5/26, P2 5/25, Pn 5/26

Forecast (unshaded)

Figure 21.10. Monthly Production Calendar

but it is formatted as in Figure 21.11 or in a similar manner up to the user. The process of filling it out is explained next. Figure 21.11 is only an example. In some cases, Setups are not needed or are very brief.

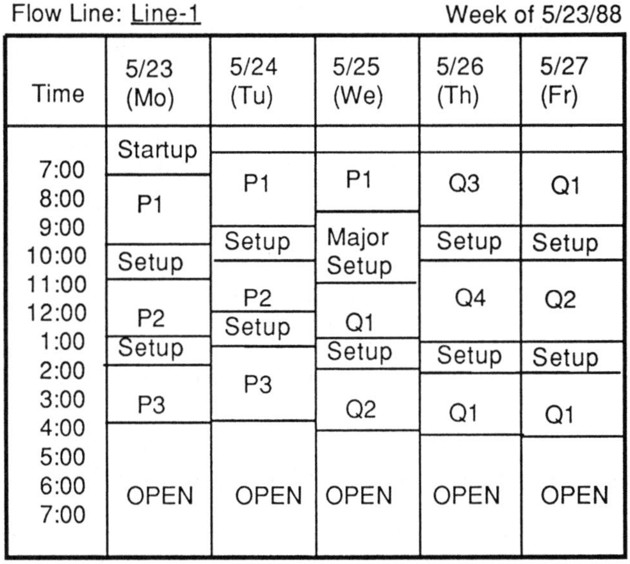

Flow Line: Line-1 Week of 5/23/88

Time	5/23 (Mo)	5/24 (Tu)	5/25 (We)	5/26 (Th)	5/27 (Fr)
7:00 8:00 9:00 10:00 11:00 12:00 1:00 2:00 3:00 4:00 5:00 6:00 7:00	Startup				
	P1	P1	P1	Q3	Q1
	Setup	Setup	Major Setup	Setup	Setup
	P2	P2	Q1	Q4	Q2
	Setup	Setup	Setup	Setup	Setup
	P3	P3	Q2	Q1	Q1
	OPEN	OPEN	OPEN	OPEN	OPEN

Figure 21.11. Weekly Production Calendar

The Weekly-Production-Calendar of a cell is filled out gradually, starting with what is known and predictable. For example, no matter what is produced, time must be allocated for equipment startup and shutdown. Then the setup structure of the cell, together with its forecast utilization, determines how many times it can be set up in the week and in what sequence, reserving the most time-consuming setups for the end.

As discussed in Chapter 20, in most factories, production activity has a stable core consisting of a small number of items for which work in process inventory is maintained. These can be allocated cell time right away, within the constraints of the setup sequence. Because there is work in process inventory this can be done independently for all the cells.

Slack time is set aside to accommodate changes in volume or unexpected orders for different products. In Figure 21.11, the slack time is concentrated in one *daily* lump at the end of the day. In another factory with other needs, it could be at other times, and in smaller or larger lumps. The key concept for Timetable Planning is that sufficient slack time be available, not *when* it is.

Scope of Applicability

The range of applicability of Timetable Planning is difficult to characterize in general terms. It works at WIFAG in the manufacture of large and complex printing presses, once the process has been made repetitive by the reverse salami tactic. On the other hand, it does not work in the fabrication of bolts, where high speed machines have long and variable setup times which throw off timetables.

The determining issue is planning depth — that is, the choice of what is to be controlled and how closely it should be controlled. If the critical processes done in a factory lend themselves to timetable scheduling, then these few timetables may be a sufficient synchronization tool for the whole factory.

The more stable and repetitive the demand is, the easier it is to produce the weekly calendars. If 10% or 20% of the demand is for custom parts on a job-shop basis, scheduling in the slack time will work. However, this mechanism breaks down if the job-shop component of the demand is too large.

Rather than dwelling at length on the circumstances in which Timetable Planning might not work, it is worth pondering the large number of cases in which it does. Custom integrated circuit fabrication and aircraft assembly may not be good fits, but the many factories making such products as roller bearings, spark plugs, faucets, or shock absorbers are.

21.4. TO WHAT EXTENT SHOULD JIT BE COMPUTERIZED?

The preceding discussion shows that the data manipulations associated with Just-In-Time are tightly coupled with the activity on the shop floor and are performed in real time for the kanban system, and on a frequent, periodic basis for materials requirements or timetable planning. The Just-In-Time system predates CIM, and the extent to which it should be computerized is a matter of debate.

The computer systems described in [Ono83] print physical cards which are attached to parts and read by optical character recognition (OCR). It is technically feasible to do away with the cards and replace them with equivalent software data structures.

There are several arguments against doing this. First, the physical cards provide a level of coupling between material flows and the scheduling system that could be lost when the cards are replaced with electronic images. Second, given the low cost of a system of physical cards, the justification of purchasing computer terminals and local area networks is far from obvious. The counter argument is that production scheduling is not the only function of a CIM system and that justification must take into account the other functions as well.

The following approaches can be considered:

1. No computerization — Level schedules and physical cards are generated and moved manually.
2. Computer-generated level schedules and physical cards — According to [Ono83], this is what Yamaha's PYMAC and the Kyoho system are doing.
3. Computer-generated level schedules and logical cards, moved by computer communication.

The two main drawbacks of 1 that 2 can remedy are the difficulties of communicating engineering changes to the shop floor and of keeping historical records of production. If the shop floor is to be emptied of people, some form of 3 will have to be developed.

3 is the only CIM option. It would provide shortage alarm notification to concerned managers who may be away from the shop floor, and record history automatically. The one disadvantage of 3 is that, since it would not provide physical tags to attach to material, there is always a danger that the connection could be lost.

22

PERFORMANCE MODELING AND JUST-IN-TIME

The throughput and throughput time characteristics of manufacturing systems are inferred through back-of-the-envelope formulas deduced from mathematical models, calculated using queueing network algorithms, or estimated from simulated histories produced by a discrete event simulator. The existing results are partial and limited to models based on questionable assumptions, but represent a start in a promising research direction.

When a demand coming from a large number of small, independent customers for a small number of products is leveled at increasingly longer intervals, the process batch size is asymptotically bounded by the ratio of the arrival rates of the two products with the highest volumes. On the other hand, when the leveling period shrinks to 0, level scheduling turns into a dispatch method in which the next order to work on is chosen FIFO from the longest product queue, and for which simulation runs do not show an even batch size distribution.

To the limited extent that a Kanban loop can be treated as Markovian, it can be modeled as a three-node cyclic queue and its behavior described by simple formulas. If the node capacities are balanced, the work in process in the loop is independent of throughput, and the throughput time drops as the throughput increases. If the nodes are not balanced, a type of cards only has a regulating effect on material flows if its downstream node is slower than its upstream node.

22.1. OBJECTIVES AND METHODS

22.1.1 Inferring Steady-State Throughput Parameters

The Just-In-Time approach described in Chapters 20 and 21 was developed not in operations research laboratories but on the shop floors of Toyota. Its effectiveness is known empirically, but there are nonetheless reasons to seek *theoretical* results to describe the performance of factories using it. From such results, we might gain insight into the reasons for the success of Just-In-Time and foresight about the future performance of a specific factory.

Given an order arrival pattern, we can infer bounds on the length of process batches in a leveled schedule. In a Kanban loop, we can infer stockout probabilities and average work in process inventories as a function of resource utilizations and numbers of cards in circulation. From these results, we gain some understanding of the conditions under which the full blown dual-card Kanban system can be simplified to a single-card system.

These results, however, are unsatisfactory for several reasons. First, their relevance is limited to a range of order arrival patterns that is small and not particularly realistic. Second, we are using a simplistic model in which resources have random, exponentially distributed processing times and no service interruptions. Finally, we model neither the synchronization of several resources nor assembly.

22.1.2 Three Approaches

Before explaining the specific results we have obtained, we will spend a few pages reviewing the general methodology of performance analysis. This will put into perspective forays made into this field in earlier chapters and the later sections of the present one.

There are three major types of methods, some of which we have already seen applied earlier in this book. They are (1) back-of-the envelope formulas, (2) queueing network performance estimation algorithms requiring a computer, and (3) discrete event simulation. The the properties of these approaches are summarized in Table 22.1.

Back-of-the-Envelope Formulas

By "back-of-the-envelope," we mean formulas that can be evaluated by hand or with a pocket calculator. We have seen back-of-the-envelope formulas earlier in this book: to anticipate the effect of reworks on throughput time in Chapter 13, or in Appendix B of Chapter 15. In these examples, the ease of evaluation was the result of applying advanced mathematics. Conversely, these mathematical tools require restrictive assumptions to hold, which narrow the formulas' range of applicability.

Table 22.1. Summary of Methodologies

	Back-of-the Envelope Formulas	Queueing Network Algorithms	Discrete Event Simul.
Mathematical Sophistication	High	High	Low
Modeling Flexibility	Low	Medium	High
Computer Requirements	None	Moderate	Large
Interpretation of Results	Easy	Easy	Difficult

Phipps' formula, for example, gives, as a function of processing time, the average queue time of a job in front of a single processor with the SPT discipline. It is no help in any other circumstance. Precious though they may be, such results are lucky finds, and we will see more later in this chapter. Other methods are needed for complex queue disciplines and for *networks* of processors.

Queueing Network Algorithms

The queueing network algorithms are philosophically related to the back-of-the envelope formulas, the main difference being that the estimation of performance parameters requires a computer and a program. CRP is a trivial but useful example of this approach. More sophisticated examples are found in Lazowska et al. [Laz84], not for factory problems but for the analysis of computer system performance.

Lazowska's methods work for the steady-state behavior of networks that are both separable and Markovian. To Lazowska, a "separable" network is one for which it is possible to carry out the analysis described in Figure 22.1 and in the attending data definitions. Although this structure is flexible, it is not able to represent parts waiting for other parts at assembly or multiple resources that are required simultaneously.

The Workload is divided into three classes as follows:

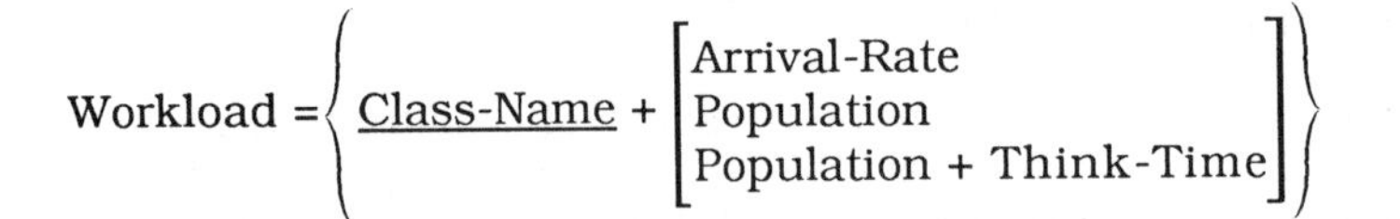

where, with Lazowska's concepts restated in manufacturing terms[1]:

- A demand coming from a large number of independent customers is represented by an Arrival-Rate.
- Customers with long-term contracts are represented by a constant Population.
- Regular but sporadic customers are represented by their Population and an average "Think-Time" between orders.

---

[1] In [Laz84], a computer's workload is comprised of *transactions* with an Arrival-Rate, a Population of concurrent *batch jobs*, and a Population of *terminals* with users who think over an output before requesting further processing.

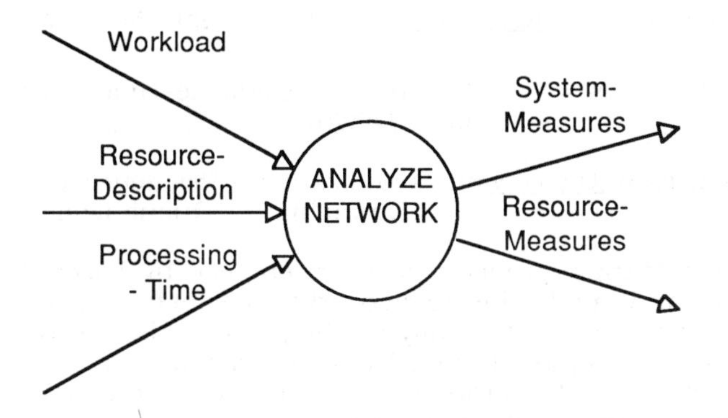

Figure 22.1. Separable Queueing Network Model

Resources are of two types. *Delay* resources are immediately available and have infinite capacity. *Queueing* resources have finite capacity and are indivisible. This results in the following definition:

$$\text{Resource-Description} = \left\{ \underline{\text{Resource-Name}} + \begin{bmatrix} \text{Queueing} \\ \text{Delay} \end{bmatrix} \right\}$$

The processing time for a particular class of work on a resource is

$$\text{Processing-Time} = \{ \underline{\text{Resource-Name}} + \underline{\text{Class-Name}} + \text{Visit-Count} + \text{Visit-Service-Time} \}$$

It is described by a number of visits and an average service time per visit. The throughput $\lambda_c(k)$ of resource k for class c becomes the product of the system throughput λ_c of class c by the visit count $V_c(k)$:

$$\lambda_c(k) = \lambda_c \times V_c(k)$$

The outputs are all expected values for the behavior of the system in equilibrium, as shown in the following definitions:

System-Measures = Aggregate-System-Measures + Class-System-Measures

The estimation of Aggregate-System-Measures is predicated on suitable units in which to add the throughputs and inventories of different classes — for example, turning "apples" and "oranges" into "tons of fruit."

Aggregate-System-Measures = Total-Throughput + Average-Throughput-Time + Average-Inventory
Class-System-Measures = {Class-Name + Class-Throughput + Class-Throughput-Time + Class-Inventory}

Resource-Measures = $\left\{\underline{\text{Resource-Name}} + \text{Aggregate-Measures} + \text{Class-Measures}\right\}$

Aggregate-Measures = Utilization + Residence-Time + Throughput + Queue-Length

Class-Measures = {Class-Name + Utilization + Residence-Time + Throughput + Queue-Length}

Although the Markov property is not referenced by Lazowska et al., their algorithms are only applicable to networks in which it holds. As stated in Chapter 5, it says that, conditionally on the present, the future of the network does not depend on its past. If, in a Markovian factory, we know that a part is in a machine at time t, the probability that it will leave it between t and t+dt is independent of how long it has been in it.

In manufacturing, this property is clearly not generally satisfied. However, the bulk of the literature on queueing is dedicated to Markovian systems and it is barely an exaggeration to say that if a system is non-Markovian, then it is mathematically intractable. In that case, we have nowhere to go but discrete event simulation. For this reason, we try to formulate models in such a way that they become Markovian, and use results based on the assumption of the Markov property even where we know it not to hold.

To understand the issues more clearly, let us examine Lazowska's use of the "arrival instant theorem" in a closed queueing network — that is, one with a workload made up of a fixed number of units (or tokens) circulating between the resources like the Production- and Move-Cards of Figure 21.4. A simple example is given in Figure 22.2, with a production line into which a new part is started when and only when one completes, so that completion is associated with the transfer of an authorization token back to the first operation.

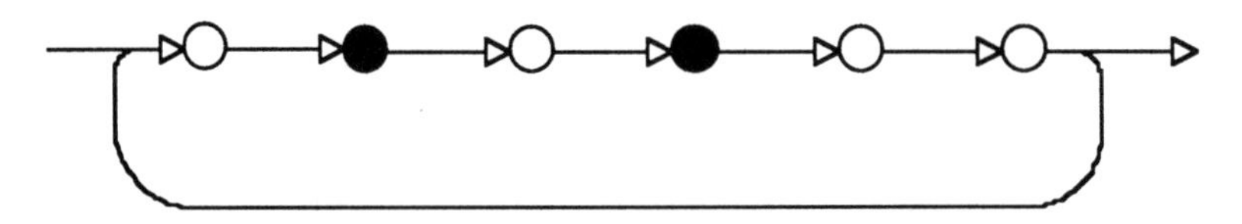

Figure 22.2. A Simple, Closed Queueing System

In a single class network with N units in circulation, the arrival instant theorem states that the expected length of the queue encountered by a unit when it arrives in front of resource k is the expected size $L_k(N-1)$ of that queue when there are N-1 units in the system. This theorem points to a method for obtaining the $L_k(n)$ by induction. Starting from an empty network, it consists of releasing units into it one at a time, waiting in between for equilibrium to be reached.

For N=1, the network characteristics are trivial. With Little's law and the arrival instant theorem, the network characteristics for any N are obtained by induction as follows:

1. Let
   - K be the number of resources.
   - D_k, k=1,...,K be the processing time through resource k.
   - $W_k(n)$, k=1,...,K, n=1,..., N be the throughput time of resource k when there are n units in circulation.
   - $\lambda(n)$, n=1,..., N be the throughput of the system when there are n units in circulation.

2. Initialize by setting

$$L_k(0) = 0, \ k=1,\ldots, K$$

3. For n=1, .., N do the following:

   3.1. For k= 1, ..., K, use the arrival instant theorem to get the throughput time of resource k:

$$W_k(n) = D_k \times [1 + L_k(n-1)]$$

   3.2 Use Little's law to get the system throughput

$$\lambda(n) = \frac{n}{\sum_{k=1}^{K} W_k(n)}$$

   3.3. For k = 1, ..., K, use Little's law again to get the average queue length of resource k:

$$L_k(n) = \lambda(n) \times W_k(n)$$

What Lazowska et al. do not say is that the arrival instant theorem holds only for Markovian networks. It is otherwise easy to build a counterexample. If the processors of Figure 22.2 have all the same deterministic processing time D, then the network is not Markovian. The completion time of a unit on one node is a function of its start time, and therefore the future depends not only on the present but on the past as well.

The releases of additional units into the network can be timed for movements to occur exactly at times t, t + D, t + 2D, and so on. Since the units move in lock step, as long as $n < K$, an additional unit can be released at the beginning of a cycle of length D when the first processor is free. Furthermore, in the absence of random variations, the pattern established by the releases endure forever. Instead of the values predicted using the arrival instant theorem, for $n \leq K$, Little's law gives us

$$\lambda(n) = \frac{n}{K \times D},$$

$$W_k(n) = D \text{ , and}$$

$$L_k(n) = \frac{n}{K}$$

While there are various ways to quantify randomness, intuitively, it appears that a real shop floor is "less" random than a Markovian network. On the other hand, it is also "more" random than the preceding deterministic model.

Discrete Event Simulation

We touched on the subject of discrete event simulation in Chapter 19, where we pointed out that the finite forward scheduling engine we described could be used as well for this purpose. Here, we discuss its application to performance analysis as a last resort, to be used only when answers can be found neither from back-of-the-envelope formulas nor from queueing network algorithms.

Discrete event simulation is free of the restrictions of applicability of the other methods. As long as the effect of scheduling decisions can be represented by events and state transitions as described in Chapter 5, this type of simulation is theoretically feasible, if not always practical.

Figure 22.3 shows how, unlike queueing network algorithms, performance analysis using a discrete event simulator breaks down into two parts. The first one is the simulator per se. It produces a *simulated* production history, which should have the same appearance as a real history gathered by shop floor transactions in a CIM system.

The second part is a module to infer performance measures from the simulated history. If the History-Analyzer can work both on simulated and observed histories, so much the better. This is, however, rarely possible.

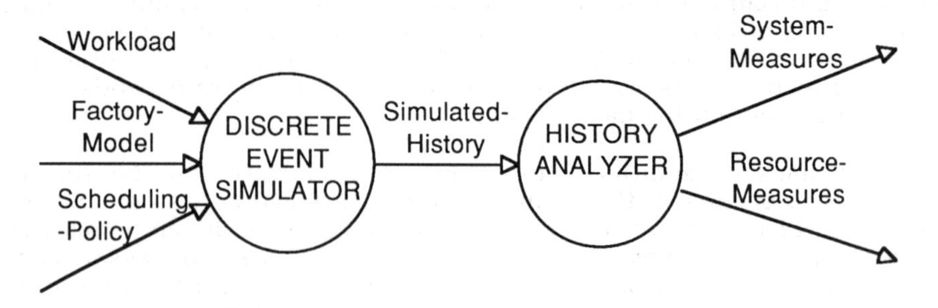

Figure 22.3. Discrete Event Simulation

The two stages both present major practical difficulties. The production of a simulated history requires enormous computer resources. One commercial package, for its demonstration case, needs a prohibitive 48 hours of a dedicated engineering work station to simulate one week of operations. In general, on a given computer system, discrete event simulation will take 100 to 1,000 times longer than a queueing network algorithm, *if* there exists one that can be applied to the problem.

The inference of performance characteristics from a factory history, real or simulated, is no trivial matter. Where the back-of-the-envelope formulas and the queueing network algorithms directly evaluate model parameters, the history analyzer must *estimate* them from "experimental" data. This method therefore has a dimension of statistical analysis that is not present in the other two. When simulations based on different Scheduling-Policies yield different results, the analyst must decide whether it is due to the policies or to the variability of the factory.

The statement in the summary table that discrete event simulation has low mathematical sophistication means only that it is a brute force method to produce simulated histories. The interpretation of this history, on the contrary, requires a rare level of statistical ability.

More often than not, this point is missed by the developers of discrete event simulation software, who make the desired outputs not performance measures but instead a cartoon showing parts moving on a shop floor on a graphic screen. Such cartoons have value as a design and debugging aid when generating the factory model, but producing them is not the essential activity of a simulator.

The preceding pages introduced a general methodology to analyze the performance of scheduling systems. We now explore what it can help us find out about to the specific subject of Just-In-Time.

22.2. EFFECT OF LEVELING ON PROCESS BATCHES

A leveled schedule is not needed in every application of Just-In-Time. If orders can be processed first-in-first-out without making waves in the material flows, no special treatment is necessary. To find out when that would be the case, let us examine a factory assembling a given set of products, receiving orders for single units from independent individual consumers, and pooling them into a single queue in order of arrival.

What will the process batch size distribution be at the final assembly work cell if no effort is made to level the demand? If this is an automobile assembly factory, long process batches of the same model will cause imbalances in the flow of components and subassemblies, eventually causing deliveries to be delayed by shortages.

22.2.1. Run Length in a Multiproduct FIFO Queue

Let us assume that there are k products, and that orders arrive as a Poisson process, for one unit at a time at rate λ_i per unit time for product i. The demands for all the different products are pooled First-In-First-Out into a single queue which has the Poisson arrival rate

$$\lambda = \lambda_1 + \ldots + \lambda_k$$

Any order within the queue is for product i with probability

$$a_i = \frac{\lambda_i}{\lambda}.$$

In this context, we have a process batch of n units if and only if n consecutive orders are received for the same product. The probability of having such a run of n orders for product i is

$$\begin{aligned} P_{in} &= \text{Prob(Exactly n-1 consecutive orders for i} \mid \text{Current order is for i)} \\ &= a_i^{n-1} \times (1 - a_i) \end{aligned}$$

A run of orders for product i will be at most of length n with probability

$$C_{in} = (1 - a_i) \times (1 + \ldots + a_i^{n-1}) = 1 - a_i^n$$

The first N such that $C_{iN} \geq 1 - \beta$ gives us a confidence interval for run lengths of product i.

It is the first integer N satisfying $a_i^N \leq \beta$, which, since $0 < a_i < 1$, is equivalent to

$$N \geq \frac{\ln \beta}{\ln a_i}$$

Example 1: There are five products, named A, B, C, D, and E with orders arriving respectively at rates 1, 0.5, 0.2, 0.2, and 0.2 per unit time, 99% of the time, the run lengths for all products will not exceed limits given by the following table.

Product	Arrival Rate	Average Run Length	Max. Run Length @ 99% Confidence
A	1.0	1.90	7
B	0.5	1.31	4
C, D, E	0.2	1.10	3

If such a pattern cannot cause imbalances in the flow of components and subassemblies within the plant, there is no need for blending.

The preceding discussion requires no assumption of any kind on how the factory responds to orders; it is entirely based on the order arrival process. We can say more about the system's behavior by making such assumptions, at the cost of a loss of generality.

It is generally not realistic to treat the final assembly resource of a factory as a single server with exponential service times, but the simplicity of the results obtained from this assumption make it worth making even if they are meaningful only as orders of magnitude. A back-of-the-envelope calculation does not have to be precise.

22.2.2. Effect of the Service Rate

A run of one product is ended as effectively when the queue goes empty as when the next order is for a different product, because the idleness of the final assembly work center also gives suppliers of components and subassemblies an opportunity to replenish stocks.

If, in the absence of any component shortage, the final assembly time follows the same exponential distribution of mean μ^{-1} for all products, its total utilization is

$$\rho = \frac{\lambda}{\mu}.$$

The discussion in Appendix A shows that the run length distribution is as it is here, but with

$$a_i = \frac{\rho\lambda_i}{\lambda} = \frac{\lambda_i}{\mu}$$

The average lead time

$$W = \frac{1}{\mu-\lambda}$$

is the same for all products, and thus the factory does not behave like an infinitely divisible resource (see Chapter 13). On the other hand, the average queue length of product i is

$$L_i = \frac{\lambda_i}{\mu-\lambda}$$

which is product-dependent. Leveling reverses this situation, causing product-dependent lead times.

<u>Example 2</u>: The product mix is as in Example 1, but μ = 2.625, so that ρ = 0.8. The common average lead time is W = 1.90. The product-dependent parameters are shown in Table 22.2. The maximum run lengths are shorter than in Example 1 because a run is considered over if the queue goes empty.

Table 22.2. Product-Dependent Parameters

Product	Arrival Rate	Average Queue Length	Average Run Length	Max.Run Length @ 99% Confidence
A	1	1.90	1.61	5
B	0.5	0.95	1.23	3
C, D, E	0.2	0.38	1.08	2

22.2.3. Maximum Run Length in a Level Schedule

The longest possible run length in a schedule processed by the First Blender is the ratio of the number of orders for the two highest volume products. If the process of order arrivals is, as just shown, a Poisson process with intensity λ_i for product i and the blender is run periodically with period T, then the number $N_i(T)$ of orders for product i arriving in an interval of length T satisfies

$$E[N_i(T)] = \text{Var}[N_i(T)] = \lambda_i T$$

For paramater values above 30, the Poisson distribution can be approximated by a Gaussian. If A and B are the two highest volume products as in Example 1 and the schedule is leveled every 100 units of time, we have $\lambda_A T = 100$ and $\lambda_B T = 100$, and we can use for $N_A(T)$ and $N_B(T)$, the approximation of two independent Gaussian variables, as shown in. Figure 22.4. Their ratio bounds the run length distribution in the leveled schedule, and the shaded area in Figure 22.4 shows where that ratio exceeds c.

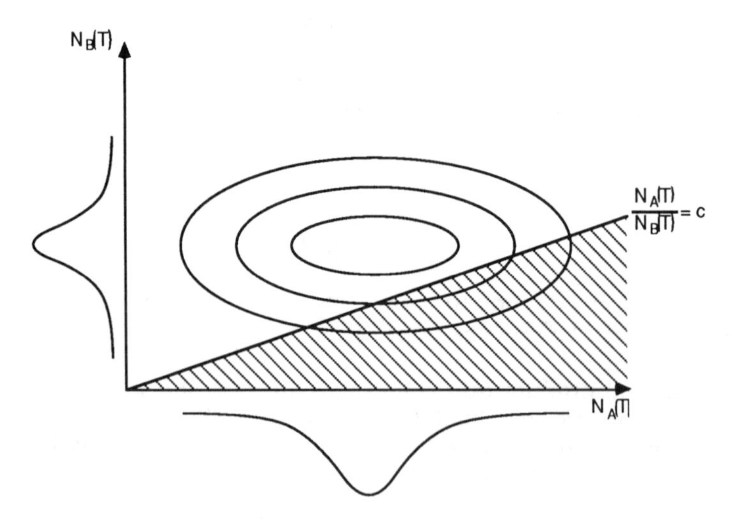

Figure 22.4. Ratio of Bookings for Two Highest Volume Products

$N_A(T) - c \times N_B(T)$ approximately follows the Gaussian distribution with mean

$$\mu = \lambda_A T - c \times \lambda_B T$$

and variance

$$\sigma^2 = \lambda_A T - c \times \lambda_B T$$

Therefore, if F is the cumulative distribution function of a standard Gaussian variable, and we want to find a c value that is going to be exceeded by the ratio less than 1% of the time, all we need to do is solve the following for c:

$$\text{Prob}\left[N_A(T) - c \times N_B(T) \le 0\right] = F\left[\frac{0 - \left(\lambda_A T - c \times \lambda_B T\right)}{\sqrt{\lambda_A T + c^2 \times \lambda_B T}}\right] = 99\%$$

From a table of the normal distribution, we find that this is equivalent to

$$\frac{\lambda_A T - c \times \lambda_B T}{\sqrt{\lambda_A T + c^2 \times \lambda_B T}} = -2.33$$

or, for $\beta = 2.33$,

$$c = \frac{\lambda_A \lambda_B T + \beta \times \sqrt{\lambda_A \lambda_B \left[T\left(\lambda_A + \lambda_B\right) - \beta^2\right]}}{\lambda_B \times \left(\lambda_B T - \beta^2\right)}$$

For

$$\lambda_A T = 100$$

and

$$\lambda_B T = 50$$

we get $c = 3.13$, so that, at the same level of confidence as in the pure first-in-first-out case, the maximum run length has been cut down to 3. Calculating c as a function of T, we get Table 22.3.

Table 22.3. Maximum Run Lengths

Leveling Period	Run Length Upper Bound
50	3.95
100	3.13
200	2.71
300	2.55
500	2.41
1000	2.28

22.2.4. The Zealot's Paradox

In the preceding section, we examined the effect of generating level schedules at increasingly *lower* frequencies. Here, we go to the opposite extreme. Instead of leveling the schedule periodically waiting a time T, it is tempting to revise the level schedule *every time* a new order arrives.

A simple variant of this idea is to level the schedule every time an order *completes* processing, whether or not new orders have joined the queue in the meantime. This is in fact a simplification, because it amounts to always drawing from the longest queue, and there is no need to actually generate an *explicit* level schedule.

Staying with the same example, we have the structure shown in Figure 22.5. Every time the processor is freed by completing work on a unit, the next order to work on is chosen first-in-first-out in the queue with the largest number of outstanding orders.

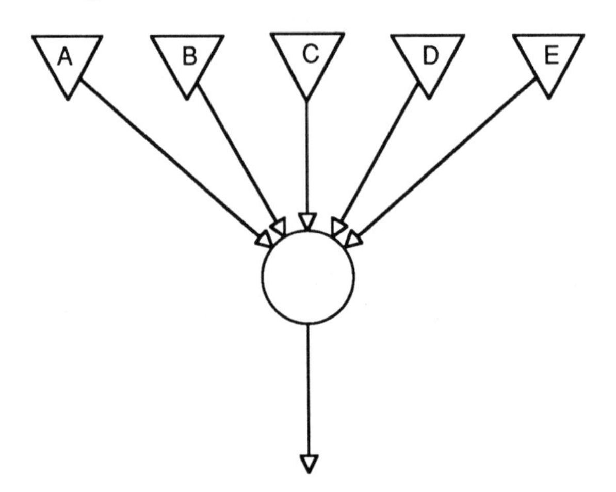

Figure 22.5. Five Products and One Processor

The resulting run length distribution of this method is not easily obtained, but some of its characteristics can be obtained by discrete event simulation. Table22.4 shows such simulation results. Ties on queue length are broken by using the product index — for example, if A and B have queues of equal lengths and longer than those of any other product, A takes priority over B. The simulation covered 100 units of time.

Table 22.4. Simulation Results

Product	Ave. Que. Length	Ave. Lead Time	Product. Volume	Number of Runs	Average Run Length	Max. Run Length
A	1.06	0.89	119	58	2.05	8
B	0.76	1.52	49	34	1.44	3
C	0.63	2.51	25	22	1.36	3
D	0.51	3.20	16	14	1.14	2
E	1.07	5.23	20	20	1.0	1
Overall	4.04	1.74	229			

Performance in terms of run lengths is not particularly good. In contrast with the plain first-in-first-out method, there are no large queue lengths differences between products with different order arrival rates. On the other hand, the lead times are now product dependent. Let us compare these lead time figures with those derived from the divisible resource model.

If the whole final assembly resource were dedicated to product i, a unit of i would be processed on the average in a time W_i satisfying

$$W_i = \frac{1}{\mu - \lambda_i}$$

units of time. The fraction of the resource going to product i is λ_i/λ, so that the lead time of product i in the divisible resource model is

$$W_i' = \frac{\lambda}{\lambda_i \times (\mu - \lambda_i)}$$

Simulation results and the divisible resource model are compared in Table 22.5.

Table 22.5. Lead Time Comparison

	Average Lead Times	
Product	Simulation	Div. Res. Model
A	0.89	1.29
B	1.52	1.97
C	2.51	4.33
D	3.20	4.33
E	5.23	4.33

22.3. PREFORMANCE ANALYSIS OF THE KANBAN SYSTEM

We focus here on the dynamics of the elementary dual-card Kanban cell shown in Figures 21.5 and 21.6, with the goal of finding answers to such questions as

- How often will shortages occur?
- Given an average work load, what is the average level of work in process inventory actually maintained by the Kanban system?

The literature describes the choice of numbers of production and move cards as "a key management decision." In practice, it does not seem to be much of a problem. The number of cards in each work center is initially set high enough not to exert a controlling influence on work in process. It is then gradually reduced.

22.3.1. What the Literature Says

Hall [Hal83] gives a formula for selecting numbers of Move-Cards and Production-Cards which, after some modification of terminology, reduces to

$$N_M = \lambda \times T_M \times (1 + A_M)$$

$$N_P = \lambda \times T_P \times (1 + A_P)$$

where:

- N_M and N_P are the numbers of Move-Cards and Production-Cards.
- λ is the production rate out of work cell M.
- T_M and T_P are the throughput times of Move-Cards and Production-Cards;
- A_M and A_P are factors meant to account for disruptions.

Similar formulas are also found in other references, but they all are in fact variants of Little's law. The most disturbing aspect of these formulas is that they are based on standard throughput times! Where are T_M and T_P supposed to come from? The modeling of variability by fudge factors is also disturbing.

22.3.2. Towards a Queueing Network Model

We propose here to infer the steady-state behavior of the elementary dual card Kanban cell of Figure 21.6 from a queueing network model incorporating the variability induced by orders. This will require mathematical work, explicit hypotheses about the material flow system, and the development of a probabilistic model of the container flow. Difficult to establish though they may be, the resulting formulas are both easy to evaluate and enlightening about the dynamics of the Kanban system.

Simplifying Assumptions

The containers are divided into three populations L, P, and M , as shown in Figure 22.6:

- Population L has the empty containers before work cell L and the one eventually being filled at work cell L;
- Population P has the full containers waiting for or in transit from work cell L to work cell M;
- Population M has the containers waiting full in front of work cell M or being emptied at work cell M.

The state of the system is the triplet (n_L, n_P, n_M) where n_i is the size of population i.

We make the following additional assumptions:

1. The services times of containers at Work-Cells L and M are random variables following exponential distributions with parameters μ_L and μ_M.
2. The transport time from L to M follows an exponential distribution with parameter μ_P.

3. The arrival of orders for parts out of M form a Poisson process with intensity λ such that

$$\rho_L = \frac{\lambda}{\mu_L} < 1$$

$$\rho_P = \frac{\lambda}{\mu_P} < 1$$

$$\rho_M = \frac{\lambda}{\mu_M} < 1.$$

1 characterizes the process of container departures from populations L and M. If the work cells are single, indivisible machines taking a fixed amount of time to process each part, 1 will be a pessimistic assumption even though departure times from the work cell are randomized by contention for the machine with other products.

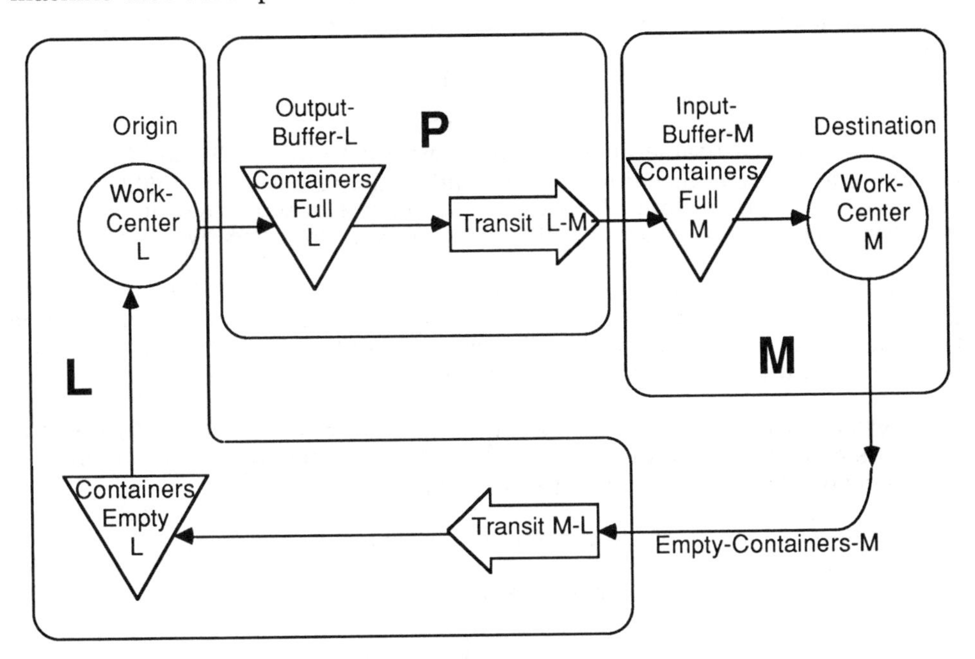

Figure 22.6. Reduction to Three Queues

3 expresses the requirement that the utilization of both work cells and the transport system be less than 100%. We will come back later to the issue of determining the μ_i from resource models.

The Cyclic Queue Model

Since we are looking here at movements of containers and not cards, the events to take into consideration are

M-L: An Empty-Container arrives at the buffer in front of work cell L.
$(n_L, n_P, n_M) \rightarrow (n_L + 1, n_P, n_M - 1)$

L-P: A Full-Container is released by work cell L.
$(n_L, n_P, n_M) \rightarrow (n_L - 1, n_P + 1, n_M)$

P-M: A Full-Container arrives at the buffer in front of work cell M.
$(n_L, n_P, n_M) \rightarrow (n_L, n_P - 1, n_M + 1)$

As long as there are no part shortages in front of M, the system behaves as if it were a single queue at M responding at rate μ_M to orders arriving at rate λ, and the lead time of the orders given from outside the system to work cell M is

$$T = \frac{1}{\mu_M - \lambda}$$

For this to be a usable approximation, we must choose N_M and N_P so that stockouts at M are sufficiently rare in steady-state to have a negligible impact. The second reason to aim for this is so that those stockouts that do occur can be interpreted as statistically significant departures from steady-state.

Whenever $n_L > 0$ and $n_P < N_P$ there is either a pending Production-Card for L or a Move-Card on shortage hold. In either case, a container at L is due for processing as soon as work cell L gets around to it. Therefore the mean rate at which L-P events occur when $n_L > 0$ and $n_P < N_P$ is μ_L.

Likewise, whenever $n_P > 0$ and $n_M < N_M$, there are pending Move-Cards for transfer of Full-Containers to M. Therefore the mean rate at which P-M events occur when $n_P > 0$ and $n_M < N_M$ is μ_P.

The allowable states of the system are characterized by

$$n_L + n_P + n_M = N_P + N_M$$
$$n_L \geq 0$$
$$0 \leq n_P \leq N_P$$
$$0 \leq n_M \leq N_M$$

With the above hypotheses, the following results are proven in Appendix B.

If we set

$$\alpha_{LM} = \frac{\rho_L}{\rho_M}$$

and

$$\alpha_{LP} = \frac{\rho_L}{\rho_P}$$

the stockout probabilities are

$$P(n_M = 0) = \begin{cases} \dfrac{1-\alpha_{LM}}{1-\alpha_{LM}^{N_M+1}} \times \alpha_{LM}^{N_M} & \text{if } \alpha_{LM} \neq 1 \\[2ex] \dfrac{1}{N_M+1} & \text{if } \alpha_{LM} = 1 \end{cases}$$

$$P(n_P = 0) = \begin{cases} \dfrac{1-\alpha_{LP}}{1-\alpha_{LP}^{N_P+1}} \times \alpha_{LP}^{N_P} & \text{if } \alpha_{LP} \neq 1 \\[2ex] \dfrac{1}{N_P+1} & \text{if } \alpha_{LP} = 1 \end{cases}$$

and

$$P(n_L = 0) = \begin{cases} \dfrac{1-\alpha_{LM}}{1-\alpha_{LM}^{N_M+1}} \times \dfrac{1-\alpha_{LP}}{1-\alpha_{LP}^{N_P+1}} & \text{if } \alpha_{LM} \neq 1 \text{ and if } \alpha_{LP} \neq 1 \\[2ex] \dfrac{1}{N_M+1} \times \dfrac{1-\alpha_{LP}}{1-\alpha_{LP}^{N_P+1}} & \text{if } \alpha_{LM} = 1 \text{ and if } \alpha_{LP} \neq 1 \\[2ex] \dfrac{1-\alpha_{LM}}{1-\alpha_{LM}^{N_M+1}} \times \dfrac{1}{N_P+1} & \text{if } \alpha_{LM} \neq 1 \text{ and if } \alpha_{LP} = 1 \\[2ex] \dfrac{1}{N_M+1} \times \dfrac{1}{N_P+1} & \text{if } \alpha_{LM} = 1 \text{ and if } \alpha_{LP} = 1 \end{cases}$$

and the expected values of the numbers of containers at M, P, and L are

$$En_M = \begin{cases} \dfrac{N_M + 1}{1 - \alpha_{LM}^{N_M + 1}} - \dfrac{1}{1 - \alpha_{LM}} & \text{if } \alpha_{LM} \neq 1 \\ \dfrac{N_M}{2} & \text{if } \alpha_{LM} = 1 \end{cases}$$

$$En_P = \begin{cases} \dfrac{N_P + 1}{1 - \alpha_{LP}^{N_P + 1}} - \dfrac{1}{1 - \alpha_{LP}} & \text{if } \alpha_{LP} \neq 1 \\ \dfrac{N_P}{2} & \text{if } \alpha_{LP} = 1 \end{cases}$$

and

$$En_L = N_P + N_M - En_M - En_P$$

Interpretation

The L, P, and M populations do not play symmetrical roles, because the containers queued at L are empty, whereas the containers queued at M and P are full. Expressed in number of containers, the work in process in the loop is therefore $n_P + n_M$.

If the service rates at L, P, and M are equal, we have $\rho_L = \rho_P = \rho_M$, and we can say that the capacity is balanced. In this case, since

$$En_M = \frac{N_M}{2},$$

$$En_P = \frac{N_P}{2}, \text{ and}$$

$$En_L = \frac{N_M + N_P}{2}$$

we find that the work in process L in *independent* of the utilizations ρ_i. From Little's law, $L = \lambda W$, and the throughput time W actually drops as the throughput λ increases!

When the utilizations are not equal, the formulas are functions of the ratios of the utilizations of two nodes. If L is much slower than M, then $\rho_L >> \rho_M$, and

$$En_M \approx \frac{1}{\alpha_{LM} - 1} = \frac{\frac{\rho_M}{\rho_L}}{1 - \frac{\rho_M}{\rho_L}}$$

N_M vanishes from this approximation. This means that, if the downstream cell is much faster than the upstream cell, the move cards are not necessary to regulate the transfers of parts. Likewise, N_P vanishes from En_P when $\rho_L >> \rho_P$.

Furthermore, since

$$P(n_M = 0) \approx 1 - \frac{\rho_M}{\rho_L}$$

the buffer in front of M will frequently be empty.

If, on the other hand, L is much faster than M, then $\rho_L << \rho_M$ and

$$En_M \approx N_M + 1 - \frac{1}{1 - \alpha_{LM}} \approx N_M - \frac{\rho_L}{\rho_M}$$

so that, the move cards do play a major role when a fast cell feeds parts to a slow one.

Appendix A RUN LENGTH WITH MULTICLASS SINGLE SERVER

We want to prove that if the service time follows the same exponential distribution of mean μ^{-1} for all products, the run length distribution is

$$P_{in} = a_i^{n-1} \times (1 - a_i)$$

where

$$a_i = \frac{\rho \lambda_i}{\lambda} = \frac{\lambda_i}{\mu}$$

and

$$\rho = \frac{\lambda}{\mu}$$

Proof: The single server processing the pooled output of k Poisson queues with arrival rates $\lambda_1,\ldots,\lambda_k$ at rate μ lends itself to modeling as a two-state Markov process as shown in Figure 22.7.

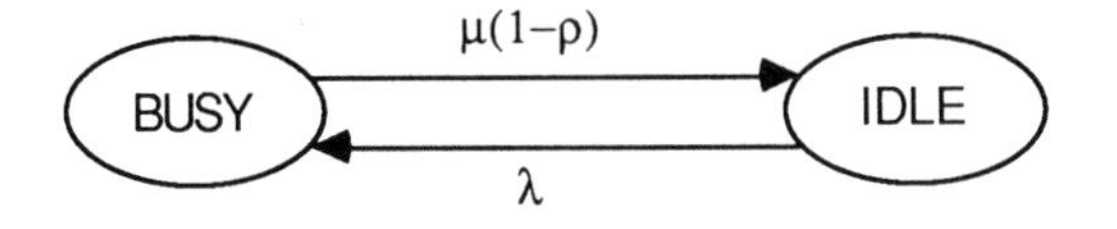

Figure 22.7.

Transitions from IDLE to BUSY occur by arrival of an order for any of the products, and thus at rate

$$\lambda=\lambda_1+\ldots+\lambda_k.$$

The equilibrium probability that the server is BUSY is

$$\Pi(\text{BUSY}) = \rho = \frac{\lambda}{\mu}$$

and therefore the equilibrium probability that the server is idle is

$$\Pi(\text{IDLE}) = 1-\rho$$

The transition rate q(BUSY, IDLE) from BUSY to IDLE is deduced from the balance equation

$$\lambda \times \Pi(\text{IDLE}) = \Pi(\text{BUSY}) \times q(\text{BUSY, IDLE})$$

which yields

$$q(\text{BUSY, IDLE}) = \mu(1-\rho)$$

When the utilization ρ is low, the work center goes from busy to idle almost at rate μ — that is, almost every time an order is completed — because there are rarely any other orders waiting. On the other hand, as ρ gets close to 1, transitions to the IDLE state become increasingly rare.

Figure 22.8 shows a refinement of this model where the BUSY state is broken down by product being processed. Thus when the server is in state A_i it means that it is working on an order for one unit of product A_i. The transition rate from any of the A_i states to IDLE is

$$q(A_i, \text{IDLE}) = q(\text{BUSY, IDLE}) = \mu(1-\rho)$$

because those transitions are not affected by what product is currently being worked on.

A transition from state A_i to A_k occurs when the following conditions are met:

1. An order completion occurs without leaving the work center IDLE, which happens at rate
$$\mu - \mu(1-\rho) = \mu\rho$$
2. The next order in the queue is for product A_k, which happens withprobability λ_k/λ.

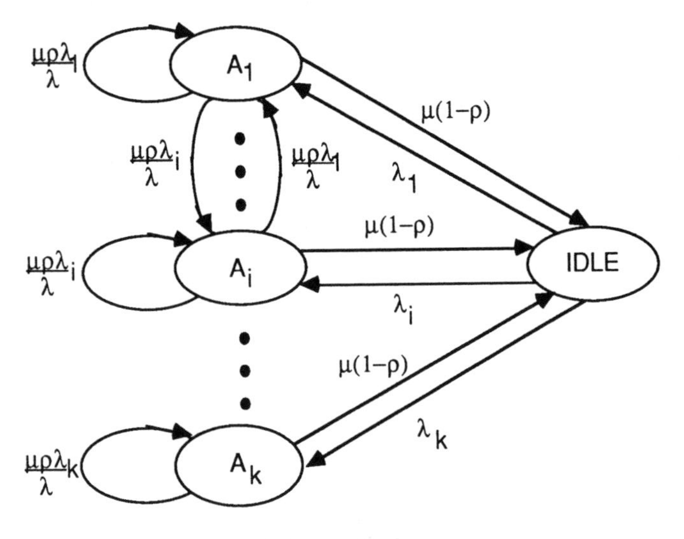

Figure 22.8.

Therefore

$$q(A_i, A_k) = \frac{\mu\rho\lambda_k}{\lambda}$$

It can easily been seen that the total transition rate out of state A_i is $q(A_i) = \mu$. The run length distribution is a function of what Kelly [Kel79] calls the "jump chain" of this process — that is, the sequence of states the process goes through without regard to how long it stays in each one. The transition probabilities of the jump chain we are looking at are

$$\forall i,j \in \{1,\ldots, k\},\ p(A_i, A_j) = \frac{q(A_i, A_j)}{q(A_i)} = \frac{\rho\lambda_j}{\lambda}$$

$$\forall i \in \{1,\ldots, k\},\ p(A_i, \text{IDLE}) = \frac{q(A_i, \text{IDLE})}{q(A_i)} = 1 - \rho$$

$$\forall i \in \{1,\ldots, k\},\ p(\text{IDLE}, A_i) = \frac{q(\text{IDLE}, A_i)}{q(\text{IDLE})} = \frac{\lambda_i}{\lambda}$$

The explicit form of the equilibrium distribution Π^J of the jump chain is not required to obtain the run length distribution, but its existence is. The global balance equations for the jump chain are

$$\Pi^J(\text{IDLE}) = \left[\sum_{i=1}^{k} \Pi^J(A_i)\right] \times (1 - \rho)$$

and $\forall\, i = 1,\ldots, k$,

$$\Pi^J(A_i) = \left[\sum_{j=1}^{k} \Pi^J(A_j)\right] \times \frac{\rho\lambda_i}{\lambda} + \Pi^J(\text{IDLE}) \times \frac{\lambda_i}{\lambda}$$

the solutions of which are

$$\Pi^J(\text{IDLE}) = \frac{1 - \rho}{2 - \rho}$$

and

$$\forall i \in \{1, \ldots, k\},\ \Pi^J(A_i) = \frac{\lambda_i}{\lambda \times (2 - \rho)}$$

When the utilization ρ is near 0,

$$\Pi^J(\text{IDLE}) \approx \frac{1}{2}$$

which means that nearly one state in two in the system's succession of states is IDLE. When ρ is near 1,

$$\Pi^J(\text{IDLE}) \approx 0.$$

The probability P_{in} of a run of n orders for product A_i is, conditionally on being in state A_i to begin with, the probability of observing n -1 transitions each leading back to A_i, followed by a transition to another state. Therefore

$$P_{in} = \frac{\Pi^J(A_i) \times p(A_i, A_i)^{n-1} \times \left[1 - p(A_i, A_i)\right]}{\Pi^J(A_i)}$$

$$= \left(\frac{\rho\lambda_i}{\lambda}\right)^{n-1} \times \left(1 - \frac{\rho\lambda_i}{\lambda}\right)$$

Q.E.D.

Appendix B. EQUILIBRIUM OF A TRUNCATED CYCLIC QUEUE

The containers circulate between populations L, P, and M, as shown in Figure 22.9, at the throughput rate λ of orders for the product under consideration.

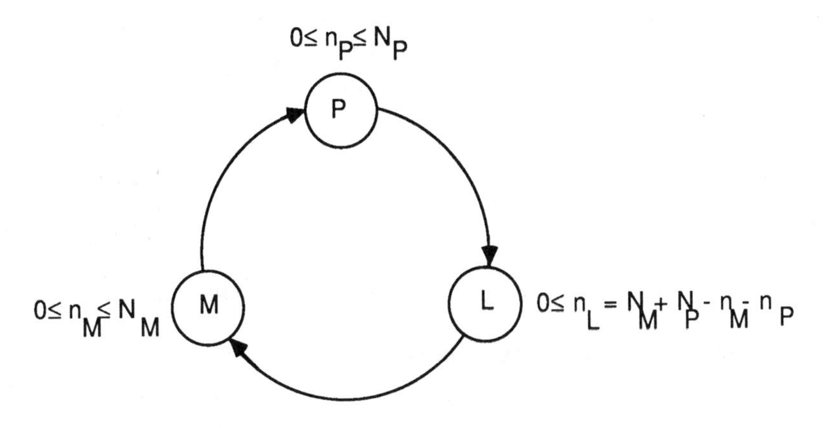

Figure 22.9. Truncated Cyclic Queue

If the processing times at L and M and the transportation time at P respectively follow exponential distributions of parameters μ_L, μ_P and μ_M, we have what Kelly [Kel79] calls a cyclic queue . In it were a general cyclic queue, the number of containers would be N_M+ N_P but there would be no restriction as to where these containers can be within the cycle.

By imposing $n_M \leq N_M$ and $n_P \leq N_P$, the Kanban system truncates the set of allowed states. Kelly shows that this does not alter the form of the equilibrium distribution of the system except for a normalizing constant. The utilizations of L, M and P are respectively

$$\rho_L = \frac{\lambda}{\mu_L}$$

$$\rho_M = \frac{\lambda}{\mu_M}$$

$$\rho_P = \frac{\lambda}{\mu_P}$$

in terms of which the equilibrium distribution takes the form

$$\Pi(n_L, n_P, n_M) = B \times \rho_L^{n_L} \times \rho_P^{n_P} \times \rho_M^{n_M}$$

where the normalizing constant is given by

$$B^{-1} = \sum_{n_M=0}^{N_M} \sum_{n_P=0}^{N_P} \rho_P^{n_P} \times \rho_M^{n_M} \times \rho_L^{(N_P+N_M-n_P-n_M)}$$

which simplifies if we introduce the function $f(x,y,N)$ for $0 \le x, y \le 1$ and $N>0$ defined by

$$f(x,y,N) = \begin{cases} \dfrac{x-y}{x^{N+1} - y^{N+1}} & \text{if } x \neq y \\ \dfrac{1}{(N+1)x^N} & \text{if } x=y \end{cases}$$

Then

$$B = f(\rho_L, \rho_P, N_P) \times f(\rho_L, \rho_M, N_M)$$

Thus the stockout probability at M is

$$P(n_M = 0) = \sum_{n_P=0}^{N_P} \Pi(N_M + N_P - n_P, n_P, 0) = B \times \rho_L^{N_M} \times \frac{1}{f(\rho_L, \rho_P, N_P)} = f(\rho_L, \rho_M, N_M) \times \rho_L^{N_M}$$

or, setting $\alpha_{LM} = \dfrac{\rho_L}{\rho_M}$

$$P(n_M = 0) = \begin{cases} \dfrac{1-\alpha_{LM}}{1-\alpha_{LM}^{N_M+1}} \times \alpha_{LM}^{N_M} & \text{if } \alpha_{LM} \neq 1 \\ \dfrac{1}{N_M + 1} & \text{if } \alpha_{LM} = 1 \end{cases}$$

Likewise, if $\alpha_{LP} = \frac{\rho_L}{\rho_P}$

$$P(n_P = 0) = \begin{cases} \frac{1 - \alpha_{LP}}{1 - \alpha_{LP}^{N_P+1}} \times \alpha_{LP}^{N_P} & \text{if } \alpha_{LP} \neq 1 \\ \frac{1}{N_P + 1} & \text{if } \alpha_{LP} = 1 \end{cases}$$

and

$$P(n_L = 0) = P(n_M = N_M \;\&\; n_P = N_P) = \Pi(0, N_P, N_M) = B \times \rho_P^{N_P} \times \rho_M^{N_M}$$

or

$$P(n_L = 0) = f\left(1, \frac{\rho_L}{\rho_P}, N_P\right) \times f\left(1, \frac{\rho_L}{\rho_M}, N_M\right)$$

which reduces to

$$P(n_L = 0) = \begin{cases} \frac{1 - \alpha_{LM}}{1 - \alpha_{LM}^{N_M+1}} \times \frac{1 - \alpha_{LP}}{1 - \alpha_{LP}^{N_P+1}} & \text{if } \alpha_{LM} \neq 1 \text{ and if } \alpha_{LP} \neq 1 \\ \frac{1}{N_M + 1} \times \frac{1 - \alpha_{LP}}{1 - \alpha_{LP}^{N_P+1}} & \text{if } \alpha_{LM} = 1 \text{ and if } \alpha_{LP} \neq 1 \\ \frac{1 - \alpha_{LM}}{1 - \alpha_{LM}^{N_M+1}} \times \frac{1}{N_P + 1} & \text{if } \alpha_{LM} \neq 1 \text{ and if } \alpha_{LP} = 1 \\ \frac{1}{N_M + 1} \times \frac{1}{N_P + 1} & \text{if } \alpha_{LM} = 1 \text{ and if } \alpha_{LP} = 1 \end{cases}$$

More generally, the marginal equilibrium distribution of n_M is $\Pi(k) = P(n_M = k)$, which satisfies

$$\Pi(k) = \sum_{n_P=0}^{N_P} \Pi(N_M + N_P - k - n_P,\, n_P,\, k)$$

$$= B \times \rho_M^{k} \times \rho_L^{N_M - k} \times \sum_{n_P=0}^{N_P} \rho_L^{N_P - n_P} \rho_P^{n_P} = f(\rho_L, \rho_M, N_M) \times \rho_M^{k} \times \rho_L^{N_M - k}$$

The expected value of n_M is

$$E(n_M) = \sum_{k=0}^{N_M} k\Pi(k) = f(\rho_L, \rho_M, N_M) \times \sum_{k=0}^{N_M} k \times \rho_M^k \times \rho_L^{N_M - k}$$

From this point on, the calculations are cumbersome but straightforward. If $\rho_M = \rho_L$,

$$E(n_M) = \frac{1}{(N_M + 1) \times \rho_M^{N_M}} \times \frac{N_M \times (N_M + 1)}{2} \times \rho_M^{N_M} = \frac{N_M}{2}$$

Otherwise, we have to use the following formula, which holds when $x \neq y$:

$$\sum_{k=0}^{N} k x^k y^{N-k} = \left[\frac{1}{f(x, y, M)} - (N + 1) \times x^N\right] \times \frac{x}{y-x}$$

It yields

$$En_M = \left[1 - (N_M + 1)\, \rho_M^{N_M} f(\rho_L, \rho_M, N_M)\right] \times \frac{\rho_M}{\rho_L - \rho_M} \quad \text{if } \rho_L \neq \rho_M$$

and the final formula is

$$En_M = \begin{cases} \dfrac{N_M + 1}{1 - \alpha_{LM}^{N_M + 1}} - \dfrac{1}{1 - \alpha_{LM}} & \text{if } \alpha_{LM} \neq 1 \\[2ex] \dfrac{N_M}{2} & \text{if } \alpha_{LM} = 1 \end{cases}$$

The expected value of the number of containers at P is obtained in a similar fashion:

$$En_P = \begin{cases} \dfrac{N_P + 1}{1 - \alpha_{LP}^{N_P + 1}} - \dfrac{1}{1 - \alpha_{LP}} & \text{if } \alpha_{LP} \neq 1 \\[2ex] \dfrac{N_P}{2} & \text{if } \alpha_{LP} = 1 \end{cases}$$

And finally, the result of the number of containers at L is deduced from the preceding two:

$$En_L = N_P + N_M - En_M - En_P$$

REFERENCES

[Abe86] Abe, Kei, *Personal Communication*, 1986.

[Bau85-1] Baudin, Michel, "Experience Curve Theory: A Technique for Quantifying CIM Benefits," CIM Review, Volume 1, Number 4, Summer 1985, Auerbach Publishers, 1985.

[Bau85-2] Baudin, Michel, "Manufacturing Yields, Reworks, and Cycle Times," 39th Annual Quality Congress, Baltimore, 1985.

[Bee83] Beeby, W.D., "The Heart of Integration: A Sound Data Base," IEEE Spectrum, May 1983, pp. 44-48

[Cod72] Codd, E. F., "Relational Completeness of Data Base Sublanguages," *Courant Computer Science Symposia Subseries*. Vol. 6, Englewood Cliffs, NJ: Prentice-Hall, 1972.

[Col79] Cole, R. E., *Work, Mobility and Participation*. Berkeley, CA: University of California Press, 1979.

[Cox74] Cox, D. R,. and Hinckley, D. W., *Theoretical Statistics*. London, UK: Chapman and Hall, 1974.

[Dat81] Date, C.J., *An Introduction to Database Systems*. Reading, Mass.: Addison-Wesley, 1981.

[Dat83] Date, C.J., *An Introduction to Database Systems*. Volume II, Reading, Mass.: Addison-Wesley, 1983.

[DeM79] De Marco,T., *Structured Analysis and System Specification*. Englewood Cliffs, NJ: Prentice-Hall, 1979.

[DeM84] De Marco,T., *Controlling Software Projects*. New York, NY: Yourdon Press, 1984.

[Doy85] Doyle, L. E., et al., *Manufacturing Processes and Materials for Engineers*. Englewood Cliffs, NJ: Prentice-Hall, 1985.

[Ebb83] Ebbinghaus, Flum and Thomas, *Mathematical Logic*. New York, NY: Springer, 1983.

[Fox83] Fox, R., "OPT vs. MRP: Thoughtware vs. Software," *26th Annual International Conference Proceedings*. APICS, 1983.

[Fre82] French, Simon, *Sequencing and Scheduling*. Chichester, UK: Ellis-Horwood, 1982.

[Gal86] Gallagher, C.C. and Knight, W.A., *Group Technology Production Methods in Manufacture*. Chichester, UK: Ellis-Horwood, 1986.

[Gol81] Goldratt, Eli, *The Unbalanced Plant*. Preprint, Creative Output Inc., Milford, CT, 1981.

[Gol84] Goldratt, Eli, and Cox, Jeff, *The Goal*. New York, NY: North-River Press, 1984.

[Hal83] Hall, Robert W., *Zero Inventories*. Homewood, IL: Dow Jones Irwin, 1983.

[Har73] Harrington, Joseph Jr, *Computer-Integrated Manufacturing*. Robert E. Krieger Publishing Company, 1973.

[Har85] Harrison, J. Michael, *Personal Communication*, 1985.

[Hel77] Helfert, E. A., *Techniques of Financial Analysis*. Homewood, IL: Dow Jones Irwin, 1977.

[Hen72] Henderson, B. and the B.C.G. Staff, *Perspectives on Experience*. Ann Arbor, MI: University Microfilms International.

[Hil80] Hillier, F. S., and Lieberman, G. J., *Introduction to Operations Research*. San Francisco, CA: Holden Day, 1980.

[Hir87] Hirano, Y., "JIT Seisan no 90 no Q&A," *Kojo Kanri Magazine*. Tokyo, Japan: Nikkankogyo, 1987.

[Hou84] Hounshell, D., *From the American System to Mass Production 1800-1932*. Baltimore, MD: Johns Hopkins University Press, 1984.

[Ishi85] Ishikawa, K., *What is Total Quality Control?*, Englewood Cliffs, NJ: Prentice-Hall, 1985.

[Jon88] Jones, V.C., *MAP/TOP Networking*. New York, NY: McGraw-Hill, 1988.

[Kap85] Kaplan, R. S., *Financial Justification for the Factory of the Future*. Harvard Business School, working paper 9-785-078, 1985.

[Kel79] Kelly, F.P., *Reversibility and Stochastic Networks*. Chichester, UK: John Wiley and Sons, 1979.

[KKA86] *Pokayoke Daizukan (The Big Picture Book of Failsafe Techniques)*. Special issue of Kojo Kanri monthly, July 1986, Tokyo, Japan: Nikkan Kogyo Shimbunsha, 1986.

[Laz84] Lazowska, E.D., et al., *Quantitative System Performance*. Englewood Cliffs, NJ: Prentice-Hall, 1984.

[McM84] McMenamin, S., and Palmer, J., *Essential Systems Analysis*. New York, NY: Yourdon Press, 1984.

[Moo77] Moore, F.G., and Hendrick, T.E., *Production/Operations Management*. Homewood,. IL: Richard D. Irwin 1977., pp. 303-305

[Mye79] Myers, Glenford J., *The Art of Software Testing*. New York, NY: Wiley Interscience, 1979.

[Ono78] Ono, Taiichi, *Toyota no Seisan Hoshiki*. Tokyo, Japan: Diamond Sha, 1978.

[Ono83] Ono, Taiichi, and Monden, Yasuhiro, *Toyota Seisan Hoshiki no Shin Tenkai*. Tokyo, Japan: JMA 1983.

[Pet83] Peters and Waterman, *In Search of Excellence*. 1983.

[Rei85] Reisig, W., *Petri Nets — An Introduction*. Berlin, Germany: Springer Verlag, 1985.

[Riv75] Riveline, C., *Evaluation des Coûts*. Paris, France: ENSMP Course Notes, 1975.

[Sch82] Schonberger, Richard J., *Japanese Manufacturing Techniques*. Free Press, 1982.

[Shi80] Shingo, Shigeo, *Toyota Seisan Hoshiki no IE teki Kosatsu*. Tokyo, Japan: Nikkan Kogyo 1980.

[Shi81] Shingo, Shigeo, *A Study of Toyota Production System from Industrial Engineering Point of View*. Tokyo, Japan: JMA, 1981.

[Slo64] Sloan, A. P., *My Years With General Motors*. New York, NY: MacFadden Books, 1965.

[Tan81] Tannenbaum, Andrew S., *Computer Networks*. Englewood Cliffs, NJ: Prentice-Hall, 1981.

[Tay11] Taylor, F. W., "Shop Management," (1911) in *Scientific Management*. Westport, CT: Greenwood Press Publishers, 1947.

[Tow64] Townsend, Peter, *Up the Organization*. 1964.

[VLS82] *VLSI Process Data Handbook*. Tokyo, Japan: Science Forum, 1982.

[Vol84] Vollman, T. E., et al., *Manufacturing Planning and Control Systems.* Homewood, IL: Dow Jones Irwin, 1984.

[War85] Ward, Paul T., and Mellor, Stephen J., *Structured Development for Real Time Systems.* New York, NY: Yourdon Press 1985.

[Wig74] Wight, Oliver, *Production and Inventory Management in the Computer Age.* Cahners Publishing, 1974.

[Wilde] Wildemann, H., *Flexible Werkstattsteuerung durch Integration von Kanban Prinzipien.* Germany: CW-Edition

[Wir76] Wirth, N., *Algorithms + Data Structures = Programs.* Englewood Cliffs, NJ: Prentice-Hall, 1976.

[Wys86] Wyss, A., *JUST-IN-TIME in der Einzelfertigung — JIT Anwendung beim Großmaschinenbau.* Bern, Switzerland: WIFAG Maschinenfabrik internal report, 1986.

[Yan79] Yanai, Hikiyoshi and Nagata, Minoru, *Shusseki Kairo Kogaku (Integrated Electronics).* Tokyo, Japan: Corona, 1979.

[Your77] Yourdon, E., and Constantine, L., *Structured Design.* Englewood Cliffs, NJ: Prentice-Hall, 1979.

INDEX

A

ADA 47
ASME 121
ASME Symbol 129
  Inspection 122
  Material Flow 123
  Operation 122
  Queue 122
  Transport 123
  Waiting 123
Acceptance Specification 18
Activity
  Custodial 113
  Fundamental 113
Administration 46, 114
Algorithm 106
Aloha 149
Analysis Process 23
Analyst 18
Application Programmer 47
Aristotle 88
Artificial Intelligence 27
Assembly 124
Attribute 112
Automation 295
  Maintenance 296

B

Back-of-the-envelope Formulas 323
Batch
  Process 182, 276
  Transfer 182, 276
Bill of Materials 26, 93, 102, 275
  in OPT 260
Binning 125
Boilerplates 36

Bottleneck 250, 272
Break
  Effect on Capacity 256
Buffer Analysis
  Age Based 200
  Schedule Based 200
Bulk Supplies 138

C

CIM 2
COBOL 47
CRP 210, 221, 239
Capacity Constraint Resource 252
Capacity Requirements
    Planning 210, 221, 239
Cell 291
Cells
  Group Technology 292
  JIT 292
Computer Integrated
    Manufacturing 2
Concurrent Processes 100
Consultant 19
Consumer Market 304
Contractor 240
Control Flow 47, 100
Control charts 299
Corporate Culture 22
Cost
  Decisions 158
  Events 158
  Labor 170
  Materials 170
  Overhead 170
Cost of Goods 157
Crossed Arrows 50
Customer Order 135
Cycle Time 180
Cyclic Queue 338

D

DBMS
  Hierarchical 88
  Network 88
  Relational 88, 102
Data
  Sink 140
  Source 140
Data Communications 7, 144
Data Dictionaries 36
Data Dictionary 78, 84
  De Marco's Conventions 84
  Definition 88
Data Flow 42, 47
  Diagram 44
  Modeling 41
Data Flow Architecture 144
Data Flow Diagrams 36
Data Flows 134
Data Object 112
Data Reconciliation 5
Data Stores 136
Database
  Administration 81
  Atoms 72
  Schema 81
  Size 73
Database Design 69, 70
De Marco 42
Decision Tables 36
Decision Tree 103
Decision Trees 36
Dedicated Hardware 112
Definition 88
  Atom 88
  Conceptual vs. Operational 90
  of a Function of Other Terms 88
  of a Primitive 88
Demand 290
  Backward Explosion 217
  Dependent 212
  Independent 212
Diagnostics 45
Diagram
  Data Flow 69, 100
  Entity-Relationship 78, 79, 87
    Set Inclusion 80
Diagram Aesthetics 50
Disassembly 125
Discounted Cash Flows 173
  Discount Rate 173
  Present Value Index 174
  Yield 174
Discrete Event

Simulation 274, 323, 328
  Cartoon 329
Dispatch 211
  Off-Line 226
  On-Line 226
  Real-Time 226
Dispatch List 312
Dispatching 225
  Critical Ratio 230
  EDD 228, 232
  Earliest Due Date 228
  FIFO 227
  First-In-First-Out 227
  SJN 228
  Shortest Job Next 228
  SPT 228
  Shortest Processing Time 228
  Slack 229
Distributed Computation 144
Distributed Databases 16, 144
Documentation of Military Parts 135
Due Date 179
Due Time 179

E

EDD 228, 232
Electronic Spreadsheet 75
End Signal 48
Engineering Analysis 32
Engineering Changes 65
Engineering Parameter Value 135
Entity 55
Entity Relationship Diagrams 36
Equipment 128
Essential Memory Model 69
Essential Model 108
Essential Systems Analysis 108
  Custodial Activity 111
  Event Partitioning 109
  Object Partitioning 110
Established Usage 90
Ethernet 148
Event 58
  Event Partitioning 109
  External 58
  Induced by History 59
  Internal 59
  Temporal 59
Experience Curve 161
Expert System 92
Exponential Distribution 241
External Context Diagram 33
External Entities 140
External Variability 303

F

FORTRAN 47
Facilities Monitoring 32
Factory Model 11, 18
Fields 78
File 42
Finished Goods 117, 135
Finite Forward Scheduling 275
First-In-First-Out 312
Fixed Costs 171
Flag 48
Floor Scheduling 32
Flow Conservation
  Data 135
  Materials 135
Flow Line 189
Foolproofing 300
Funds Flow 150

G

Group Technology 292

H

Henderson, B. 161
Hierarchical Explosion 100
Hold Time 180
Holistic Manufacturing 176, 289

I

IE 289
ISO 143
ISO-OSI Model
  Application Layer 144
  Link Layer 147
  Network Layer 146
  Physical Layer 147
  Presentation Layer 146
  Session Layer 146
  Transport Layer 146
Icon 129
Industrial Engineering 27
Insertion 136
Inspection 121, 122, 127
Interface Ring 116
Inventories 136
Inventory
  Buffer 121
  Finished Goods 140, 276
  Level Crossing 140
  Stockout 140
Inventory Buffers 129
Islands
  Automation 7
  of Information 5
Iteration 101

J

JIS 122
JIT 284
Job-Shop 189, 273, 275, 288
Juran, J.M. 151
Just-In-Time 7, 225, 249, 284
  as opposed to OPT 260

K

Kanban 212, 285, 323
  Dual-Card vs. Single-Card 323
Kanban System 273, 304, 311
  Move Card 312
  Philosophy 313
  Production Card 312
  Shortage 313
Kit 182
Kleinrock's Conservation Law 237

L

LISP 47, 97
Labor Productivity 161
Lawler's Algorithm 275
Lead Time 178
Learning curve 161
Level Schedule 286
Level Scheduling 308, 329
Little's law 192, 236, 327
Load 182
Logical Core 116
Lot 181

M

MAP 144
MIS 5
MODULA-2 47
MOS 118
MRP 209, 284, 289, 303
  as opposed to OPT 260
  Capacity Requirements Planning 210
  Class A user 210
  Closed-Loop 210
  CRP 210
  Manufacturing Resource Planning 210
  Materials Requirements Planning 210
  MRP-II 210
MRP-II 239, 305
  Dispatching 225
Management Commitment 10
Management Information Systems 5
Manufacturing Engineering 2
Manufacturing Organization 21

Manufacturing Specification 129
Mapping 71, 74, 76
  As Lookup Table 76
  As Rule 76
Market-Driven Activity 287
Markov Property 326
Master Production Schedule 290
Master Schedule 272
Material Flow 123
  ASME Symbols 121
  Hierarchical Modeling 132
  Model 117
Material Flows 117, 134
Materials Handling 144
Materials Management 2, 138
Memory Model 86
  Emphasis on Retrieval 70
  Equipment 69
  Product 69
  Route 69
  WIP 69
  Work in Process 69
Memory Modle
  Operation 69
Menu 49
Monden 309
Move Schedule 212
Multiple Sourcing 125

N

Name
  Deliberate Obscurity 91
  Encoded Information in 91
  Euphonic 91
  Unique Identification 91
  User-Definable 91
Name Field(s) 85
Need To Know 11
Non-Delay Schedule 276

O

OEM 304
OPT 248, 289
  Balance on Hand 260
  Build Network 263
  Functions 259
  Hybrid 265
  Machine
    Setup Group 260
    Setup Matrix 261
  Mule 265
  Resource Model 261
  Routing 260
  SERVE 263
    Scheduled Delay 266
OSI 143
Object 55
  Events Affecting 58
  State
    Description 55
    Space 56
Object Modeling 54
Object-Oriented Programming 112
Operation 122
Operations Research 323
Optimization 275
Order
  Arrival Pattern 323
Overbooking 304

P

PASCAL 47, 95
PCB 304
PDIP 137
PL/1 47
PROLOG 47, 97
Payback Period 172
Pegging 221
Perfect Technology Assumption 112
Performance Analysis 323
Performance Measures 6, 8
Petri Nets 38
Phipps' Formula 237, 324
Physical Flow Analysis 130
Planar Technology
  Diffusion 121
  Photolithography 121
  Thin Films 121
Poisson Process 235
Pokayoke 300
  Coffee Mill 300

Pragmatics 25
Precedence Constraints 275
Present Value Payback 174
Private Component File 110
Procedure 100
Process 42
Process Analysis 129
Process Equipment 31
Process Instructions 33
Process Integration 121
Processing Time 180, 241
Processor 42
Product Mix 171
Production Control 213
Production Planning 32
Production Scheduling 27
Production-Driven Activity 288
Properties 78
Property 72
Protocol 144
Protocols 7

Q

QC 289
Quality Control 22, 27
  TQC 15
Quality control..i.QC 2
Queue 122
Queue Time 180
Queueing Network 323
  Arrival Instant Theorem 326

R

RFP 18
Raw Materials 117
Record
  Occurrence 78
  Type 78
Record Keeping 70
Recursion 93
Relation 69
  Arity 76, 79
  As Table 75
  Attributes 78
  Between Objects 78
  Binary 76
  Functional Dependency 76
  Many-to-Many 76
  many-to-one 76
  Mapping 76
  Relational Algebra 77, 81
    Division 82
    Join 82
    Projection 82
    Selection 81
Relational Algebra
  Null Values 86
Relations 74
Reorder Point 140, 211
Repetitiveness 285
Reports 31, 44
Request for Proposals 18
Resource
  Indivisible 196, 294
  Infinitely Divisible 195, 294
Resource Model 261, 276
Retained Data Model 69
Retrieval 136
Reverse Salami Tactic 286
Rework 127
River Rouge 144
Routings 275

S

SPT 233, 324
Sales Rep. 18
Sampling 299
Schedule
  Non-Delay Schedule 276
Scheduling
  Maintenance 2
  Production 2
Scrap 125, 127
Second Sourcing 304
Secretiveness 22
Sectionalism 6
Security 81
  Company Secrets 81
  Excess of 81
Selection 101
Selection Committee 18

Semantics 25
Sequence 101
Set
Aggregate 69
Element 71
Entity 69
Infinite Set 71, 73
Object 69
Pragmatic Neutrality 71
Property 71
Subset 71, 73
Subset Selection 72
Theory 69, 70, 84
Type 69
Union 71
Setup
External 295
Internal 295
Matrix 197
Time 197
Setup Time 9, 198, 294
Setup Time Reduction 303
Setups 276
Shortest Job Next 10
Sink 42
Slack 234
Software 135
Software Design 95
Software Engineering 28
Software Marketing 290
Software Quality 15
Software Requirements 18
Software Test 104
Source 42
Spare Parts 138
Specification 42
External Context 44
Standard Throughput Time 266
Start Schedule 212
Start Signal 48
State
Sink 61
Source 61
State Transition
Markov Property 59
Model Review 61
Models 54
State Transition Diagrams 36
State Variables 78
Structured Analysis 30
Structured Natural Language 36
Subway Tokens 135
Switch 48
Syntax 25
Systems Analysis 18
Analysis Paralysis 21
Review 22
Systems Engineers 20
Systems Nightmare 4
Systems Policy 21

T

Terminals 31
Thoughtware 272
Thread of Execution 47
Throughput 180, 272
Parameter 323
Throughput Time 178, 240
Time Bucket 243
Timeline 123
Timetable Planning 304, 316
Top-Down 42
Tracking
Equipment 2
Materials 2
Transaction 45
Transport 123, 128
Transport System 121
Turnaround Time 179
256KDRAM 242

U

Unit Cost 157
Unscrap 126
Update 136
Utilization Law 236

V

Variability
Internal vs. External 253
Variant Structure 92, 276

Vehicle 63

W

Waiting State 123
Waiting Time 180
Walkthrough 137
Withdrawal 136
Work Cell 189
  Controller 109
  Essential Model of 108
  Modeling 111
Work Center 12, 189
Work in Process 117, 260

Y

Yield 89
Yield Modeling 242

Z

Zero Inventories 285